CRYSTAL GROWTH OF Si INGOTS FOR SOLAR CELLS USING CAST FURNACES

CRYSTAL GROWTH OF Si INGOTS FOR SOLAR CELLS USING CAST FURNACES

KAZUO NAKAJIMA

Emeritus Professor, Tohoku University, Aoba-ku, Sendai, Japan; PAO YU-KONG Chair Professor, Zhejiang University, Hangzhou, People's Republic of China

ELSEVIER

Elsevier
Radarweg 29, PO Box 211, 1000 AE Amsterdam, Netherlands
The Boulevard, Langford Lane, Kidlington, Oxford OX5 1GB, United Kingdom
50 Hampshire Street, 5th Floor, Cambridge, MA 02139, United States

Copyright © 2020 Elsevier Inc. All rights reserved.

No part of this publication may be reproduced or transmitted in any form or by any means, electronic or mechanical, including photocopying, recording, or any information storage and retrieval system, without permission in writing from the publisher. Details on how to seek permission, further information about the Publisher's permissions policies and our arrangements with organizations such as the Copyright Clearance Center and the Copyright Licensing Agency, can be found at our website: www.elsevier.com/permissions.

This book and the individual contributions contained in it are protected under copyright by the Publisher (other than as may be noted herein).

Notices

Knowledge and best practice in this field are constantly changing. As new research and experience broaden our understanding, changes in research methods, professional practices, or medical treatment may become necessary.

Practitioners and researchers must always rely on their own experience and knowledge in evaluating and using any information, methods, compounds, or experiments described herein. In using such information or methods they should be mindful of their own safety and the safety of others, including parties for whom they have a professional responsibility.

To the fullest extent of the law, neither the Publisher nor the authors, contributors, or editors, assume any liability for any injury and/or damage to persons or property as a matter of products liability, negligence or otherwise, or from any use or operation of any methods, products, instructions, or ideas contained in the material herein.

Library of Congress Cataloging-in-Publication Data
A catalog record for this book is available from the Library of Congress

British Library Cataloguing-in-Publication Data
A catalogue record for this book is available from the British Library

ISBN: 978-0-12-819748-6

For information on all Elsevier publications visit our website at https://www.elsevier.com/books-and-journals

Publisher: Matthew Deans
Acquisition Editor: Christina Gifford
Editorial Project Manager: Thomas Van Der Ploeg
Production Project Manager: R.Vijay Bharath
Cover Designer: Isabella C. Silva

Typeset by TNQ Technologies

Contents

Biography		*ix*
Foreword		*xi*
Preface		*xv*
Acknowledgments		*xxv*
Nomenclature		*xxvii*

1. Basic growth and crystallographic quality of Si crystals for solar cells
1

1.1	Si single and multi-crystals	1
1.2	Basic growth of Si crystals	2
1.3	Crystallographic structure and defects of Si crystals	14
1.4	Impurities and their activities in defects	39
1.5	Strain and stress	52
1.6	Appendixes	53
	References	55

2. Basic characterization and electrical properties of Si crystals
63

2.1	Methods to measure electrical properties	63
2.2	Electrical properties	65
2.3	Optical measurement method and optical properties of Si crystals	77
2.4	Processes to control electrical and optical properties of Si crystals	78
2.5	Si crystal solar cells	88
2.6	Theoretical estimation for the characterization of crystals	94
	References	95

3. Growth of Si multicrystalline ingots using the conventional cast method
101

3.1	Unidirectional growth of Si multicrystalline ingots	101
3.2	Growth behavior of Si multicrystalline ingots	110
3.3	Crystal defects and impurities in Si multicrystalline ingots	118
3.4	Si_3N_4 coating materials	133
3.5	Electrical properties and solar cells of Si multicrystalline ingots	143
3.6	Growth of large-scale ingots in industry	145
3.7	Key points for improvement	147
	References	148

v

vi Contents

4. Dendritic cast method ... 155

4.1 Motivation to develop the dendritic cast method ... 155
4.2 Growth and behavior of dendrite crystals using the in-situ observation system ... 156
4.3 Ingot growth controlled by dendrite crystals grown along the bottom of a crucible ... 165
4.4 Arrangement of dendrite crystals ... 169
4.5 Generation of dislocations ... 175
4.6 Quality and solar-cell performance of Si ingots using the dendritic cast method ... 179
4.7 Pilot furnace for manufacturing industrial scale ingots ... 184
4.8 Key points for improvement and impact ... 187
References ... 190

5. High performance (HP) cast method ... 195

5.1 Concept of the HP cast method ... 195
5.2 Control of grain size, grain orientation and grain boundaries using assisted seeds ... 196
5.3 Behavior and control of dislocations and dislocation clusters in ingots ... 204
5.4 Structure and defects in Si ingots using the HP method ... 207
5.5 Electrical properties and solar cells ... 216
5.6 Key points for improvement ... 219
References ... 221

6. Mono-like cast method ... 225

6.1 Concept and feature of the mono-like cast method ... 225
6.2 How to control to obtain a large single grain ... 226
6.3 Growth and control of small grains appeared from crucible wall ... 230
6.4 Behavior of dislocations and precipitates in mono-like ingots ... 236
6.5 Quality of Si ingots using the mono-like cast method ... 249
6.6 Key points for improvement ... 254
References ... 255

7. Growth of Si ingots using cast furnaces by the NOC method ... 259

7.1 Development of the NOC method ... 259
7.2 Establishment of the low-temperature region in a Si melt ... 266
7.3 Growth of Si ingots using Si_3N_4 coated crucibles by the NOC method ... 278
7.4 Growth of Si single ingots using the NOC method ... 281
7.5 Electrical properties and solar cells of Si single ingots ... 299
7.6 Key points for improvement of the NOC method ... 308
References ... 312

Contents vii

8. Future technologies of Si ingots for solar cells — 317

8.1 Proper grain size and stress in Si multicrystalline ingots — 317

8.2 Novel technologies for dislocation-free single ingots with large diameter and volume — 318

8.3 Growth of square-shaped ingots using the NOC method — 326

8.4 Ga-doped Si multicrystalline ingots — 334

8.5 **Si-Ge** multicrystalline ingots — 337

References — 345

Index — *349*

Biography

Professor Kazuo Nakajima is the professor emeritus of IMR Tohoku University, Japan as well as the Pao Yu-Kong Chair with Zhejiang University's State Key Laboratory of Silicon Materials, China. He is a member of the Japan Society of Applied Physics and the Japanese Association for Crystal Growth and series on the Editorial Board of the Journal of Crystal Growth. In 2006, Dr. Nakajima founded and chaired the 1st International Workshop on Science and Technology of Crystalline Si Solar Cells, a workshop that continues to be held annually. In 2007−08, he chaired the 4th Asian Conference on Crystal Growth and Crystal Technology (CGCT-4). For his achievements of the crystal growth, the Japanese Association for Crystal Growth awarded him "The 12th Achievement Award & Isamu Akasaki Award" at 2017, and the International Organization for Crystal Growth (IOCG) awarded him "Laudise Prize" at 2019. He has written over 16 books and handbooks and published over 350 papers, as well as being the inventor or co-inventor of 64 registered patents in the areas of III-V liquid phase epitaxial growth, crystal growth for semiconductors, optical devices, solar cells and plastic deformation.

Foreword

Silicon (Si) was discovered in 1810 by Swedish scientist, Jöns Jakob Berzelius, and has widely applied as the basic material for electronic industry, special for integrated circuits (ICs), since the 1950's. In the last decades, ICs have played the extreme important role in the development of high technology in the world, and have enhanced our era to get into the age of information technology. It is obvious that Si is one of the most important materials in modern industry.

Si is also the main material for photovoltaic (PV) industry. As we know, the resource shortage, environment pollution et al. on the earth have become the great challenges of our humanity. To protect the resource and environment on the earth, man has to change the production mode of modern industrialized energies, and to devote to the clean and renewable energies. Among lots of efforts, solar photovoltaic which can directly transform solar energy to electrical energy has attracted more and more attention, and developed into one of the most important renewable energy.

Date back to 1954, the first silicon solar cell was invented by D. M. Chapin, C. S. Fuller and G. L. Pearson in Bell Lab in USA, which triggers off the research and application of modern photovoltaic industry. Since then, solar cells have been widely used in different fields all over the world, e.g. in satellites and space stations, remote prairies, mountains and islands, to offer off-grid electricity. They have also been installed on the roofs of houses, apartments and public building to generate in-grid electricity. Specially, they have been installed together to form large power stations with output of mega-watt (MW) or even giga-watt (GW) electricity. In 2018, 94.3 GW solar cells are emerged, and the total installation of solar cells is more than 480 GW in the world. In current, photovoltaic on the bases of Si crystal solar cells, which occupies more than 90% market, has become one of the fleetly developed industry. It is clearly that Si crystal determining the quality, cost and even efficiency of Si solar cells will play a major role in the future of PV industry.

Si materials including crystalline Si, poly-crystalline thin film Si, amorphous thin film Si, nano-Si et al. are largely used to fabricate solar cells. For crystalline Si, there are four types, i.e. float zone Si (Fz-Si), Czochralski Si (Cz-Si), multicrystalline Si (mc-Si) and ribbon Si. Fz-Si has relative perfect crystal lattice and higher quality, which could be applied to fabricate high efficiency solar cells. But, it is too expensive due to the complicative and exorbitant cost of crystal growth process. Moreover, ribbon Si contains higher density defects and impurities induced during crystal growth, which seriously deteriorates the efficiency of solar cells. Therefore, only Cz-Si and mc-Si (including quasi-single crystal Si, also called as mono-like Si) have been widely used in PV industry to fabricate Si solar cells. Since 1990's, mc-Si occupies more and more PV markets due to the advantages of low cost, low energy consumption, higher impurity tolerance for raw Si materials and so on. In 2018, mc-Si still holds more than 60% market shares in PV industry even if the market share of Cz-Si increases in recent years.

Prof. Kazuo Nakajima has engaged in the growth of high-quality crystalline Si for high efficiency solar cells for about 20 years. He has got lots of achievements in this field, e.g., developed the dendritic cast method by controlling the crystallographic structure to grow up mc-Si ingots, and also developed the NOC method to grow up high quality Si single ingots inside Si melt without contacting the quartz crucible wall. As a colleague in Si materials, I have known him for a the long time, and considered that he is a well-known and distinguished scientist in the world.

This book illustrates the principle and technology of mc-Si growth with cast methods for solar cells. It firstly describes the basic concept about supercooling, defects and impurities during Si crystal growth. And then electrical properties of Si crystals related doping, thermal donors and B-O complexes and basic characterization techniques are introduced. The book detailly discuss the mc-Si crystal growth of different cast technologies, including the conventional cast method, dendritic cast method, and seed-assisted cast method (so-called high performance mc-Si). Moreover, mono-like (quasi-single crystal) Si produced by cast methods and single crystal Si using cast furnaces by the NOC method also are depicted in the book. Finally, future technologies of Si ingots for solar cells are considered.

The book is well organized, and designed for the wide readers, including students, Master/Ph. D students, researchers in universities and research institutes, and engineers for silicon crystal growth and silicon solar cell process in PV industry. The extensive reference list after each chapter is giving indispensable references for understanding the state of the art in cast Si crystal used for solar cells and for discussing deeply Si crystal growth.

Finally, I truly believe that this book contains massive and valuable knowledges about the crystal growth of cast Si materials, and enthusiastically recommend to readers.

Prof. Dr. Deren Yang
Academician of Chinese Academy of Science
Cheung Kong Professor
Director, State Key Lab of Silicon Materials
Zhejiang University
Hangzhou, China
April 2019

Preface

The estimated amounts of the reserved oil-, natural gas- and coal-energy resources on the earth are 2.8×106, 2.5×106 and 8.4×106 TWh at 2017, respectively [1]. The total amount is 13.7×106 TWh and it will be exhausted within the next 100 years. So, we must rapidly develop the new environment-friendly clean energy conversion system. Solar energy is the only ultimate natural energy source and the total solar energy reached to the earth is about 1×105 TW (solar energy density on the surface = 1 kW m^{-2}). 30% of total solar energy is reflected from the surface, but 70% of total solar energy is available for us as shown in Fig. P.1. The world's energy consumption is 18 TW at 2016 (which corresponds to 1.58×105 TWh/year) and it increases by1 TW within two years from 2014 to 2016 [1]. This world's energy consumption corresponds to only 0.026% of the available solar energy. When solar cells with the efficiency of 20.0 % are used, the total area is estimated as 640,000 km^2 and it is equal to a square of 800 km. Solar energy used as solar cells does not supply additional energy on the earth. Therefore, solar cells have been expected to be one of the solutions to solve the problems of the global environment

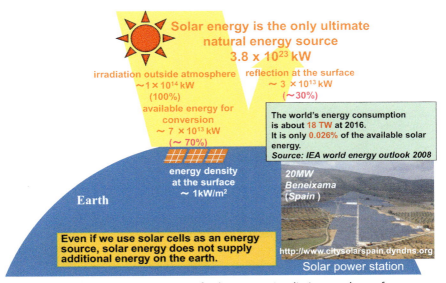

Fig. P.1 Schematic illustration of solar energy irradiating on the surface.

and energy and to realize sustaining human development and providing high living standards. So, solar cells must be effectively developed as an environment-friendly clean energy conversion system.

The present market of solar cells expands to 100 GW/year and the silicon (Si) crystal solar cells occupy more than 90% in it. In the present market of Si solar cells, p-type multicrystalline wafers are used more than 60% of total Si wafers and p-type single (mono)-crystalline wafers are used about 30% of them as shown in Fig. P.2 [2]. Now, Si p-type multi-crystalline wafers grown by the conventional and high-performance (HP) cast methods occupy 20 and 40% of the total market of Si wafers, respectively. n-Type wafers will increase their share to near 10% at 2019. The price of Si multicrystalline wafers is slightly lower than that of Si single wafers. The Si multicrystalline wafers have a larger share than single wafers because of their high productivity. But they lose 1 % efficiency because of their inefficient texture structure. The next target of the cast method is in the direction to develop a uniform large and wide Si single ingot using a cast furnace. The Czochralski (CZ) method is also in the direction to develop a uniform large and long Si single ingot using a CZ furnace.

To largely expand solar cells as an energy source on the earth, the energy cost of solar cells should be much lower comparing to the present

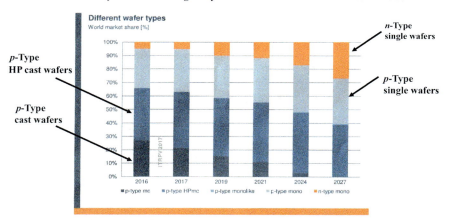

Fig. P.2 Share of Si multicrystalline and single (mono)-crystalline wafers in the present market of Si solar cells. The *p*-type multicrystalline and single-crystalline wafers occupy about 60% and 30% of total Si wafers at present, respectively. *Referred from Fig. 36 in International Technology Roadmap for Photovoltaic Results. ITRPV Eighth Edition 2017 (2016 Results).*

energy cost. To reduce the cost of solar cells, their productivity, yield and conversion efficiency should be largely enhanced. If we will use solar energy for 10% of the world's energy consumption (more than 1700 GW) at 2050, we must manufacture 60–80 GW of solar cells every year for 30 years. The required Si feedstock will be more than 800,000 ton per year. In this mass production, Si is the most proper candidate to manufacture more than 80 GW of solar cells per year because Si is the second rich element in natural resources on the earth. The optimum band gap energy to obtain the maximum conversion efficiency of solar cells is approximately 1.4 eV under air mass (AM) 1.5 spectral irradiation [3]. Even through the band gap of Si is 1.1 eV, Si can cover this band gap energy. The total amount of installed solar cells has already come up to 75 and 100 GW all over the world in 2016 and 2017, respectively [4]. The share of solar cells prepared by Si crystals will have reached to 95 % in 2017. Most of Si crystal solar cells are made in China and Taiwan, and the share of Chinese companies is more than 70 %. In this situation, one of main technologies to be developed is the low-cost production technology to grow large and high-quality Si ingots with sufficient quality to obtain solar cells with a high yield and a high-conversion efficiency. Almost half the cost of a crystalline Si based solar module comes from producing Si wafers which includes purifying Si source materials, preparing an ingot and sawing into wafers [5]. The development of these low-cost technologies is one of the most important subjects for the solar cell industries.

The history of several Si cast methods for Si multicrystalline ingots for solar cells is briefly introduced. For a long time, Si multicrystalline ingots were manufactured by the unidirectional solidification cast method (call as a conventional cast method). Si multicrystals have many grains with different orientations and different sizes, and they have many grain boundaries with different characteristics as shown in the upper left figure in Fig. P.3. A Si multicrystalline ingot was unidirectionally grown using the conventional cast method and the grown ingot had columnar structure which was effective to reduce parallel grain boundaries in the growth direction. An ingot with columnar structure is shown in the lower left figure in Fig. P.3. Generally, grain boundaries deteriorate the electrical properties of a Si multicrystalline ingot. Therefore, realization of a Si ingot with large grains and few grain boundaries is one of solutions to obtain the higher-quality ingots and higher-efficiency solar cells. If we can obtain a Si multicrystalline ingot which has grains with same orientation and proper sizes, and grain boundaries with electrically inactive characteristics as shown in the

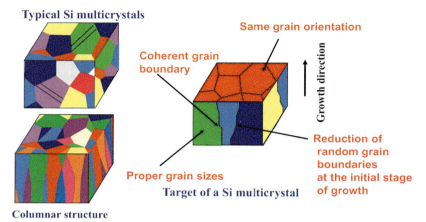

Fig. P.3 Target points of the structure control for the growth of a Si multicrystalline ingot for solar cells. The ideal structure is an ingot which has grains with same orientation and proper sizes, and grain boundaries with electrically inactive characteristics.

right figure in Fig. P.3, the quality of the ingots should be largely improved as that of a CZ single ingot. Such a concept of the structure control has not been tried yet for the growth of Si multicrystalline ingots for solar cells.

To obtain a Si high-quality multicrystalline ingot for solar cells, a new concept was proposal by Nakajima et al. [6,7] that the Si multicrystalline ingot should be prepared on the view point of controlling crystallographic structure of the ingot by controlling nucleation sites on the bottom of crucible. To obtain novel information for such structure control, Fujiwara et al. [8,9] have succeeded in developing the in-situ observation equipment for the first time in the world, by which the growth mechanism of a Si crystal was able to be directly observed. Using this equipment, a novel growth mode was found in which at first thin dendrite crystals grew along the bottom of crucible and then a Si multicrystal simultaneously grew on the upper surface of the dendrite crystals [6,7]. To apply this growth mode for the structure control of Si ingot growth, the dendritic cast method was invented as an advanced cast method for the structure control of Si multicrystalline ingots and the structure control was realized for the first time using the dendritic cast method [6,7,10−13]. In this method, dendrite crystals are used as as-grown seed crystals to control initial nucleation sites

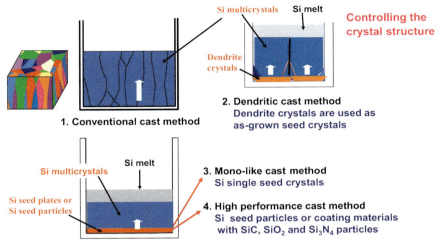

Fig. P.4 History of Si cast methods to prepare higher-quality Si ingots for solar cells by controlling the crystal structure.

on the bottom of the crucible as shown in Fig. P.4. The dendritic cast method proved at the first time that the structure control was very effective to improve the quality and yield of the Si multicrystalline ingots [6,7,10−13] and the conversion efficiency of solar cells prepared using the ingots was kept high over almost all parts of the ingots and the uniformity of the ingots were largely improved [12]. These data showed for the first time that high-quality and uniform Si multicrystalline ingots could be obtained even using the cast method as long as the structure control was executed. These data gave a strong impact to promote the structure control in the research field of the ingot growth for multicrystalline solar cells and gave a large motivation for the worldwide researchers to start the ingot growth on the basis of the structure control. The effectiveness was proved in many institutes and companies all over the world. The initial nucleation control was started by studying several kinds of seed including dendrite crystals for the growth of high-quality Si multicrystalline ingots for solar cells. The possibility of the structure control shown by dendritic cast method became an abetter for the following development of several methods using various kinds of seed on the bottom of crucible.

Then, several significant technologies for the growth of Si multicrystallie or mono-like ingots have been developed to control the crystallographic structure on the basis of the initial nucleation control using

xx   Preface

seeds. To more easily control the crystal structure, the mono–like or MONO2 cast method [14,15] was proposed using Si single seed plates set on the bottom of a crucible as shown in Figs. P.4 and 6.1. The mono–like cast method is widely investigating to obtain a high–quality Si large single ingot using a cast furnace with high productivity. To obtain more uniform crystal structure and more easily reduce dislocations, the HP cast method or the S2 method was proposed in Taiwan and China using Si seed particles or coating materials with SiC, SiO_2 and Si_3N_4 particles mainly on the bottom of crucible as shown in Fig. P.4 [16,17]. Realization of a Si ingot with many small grains and many random grain boundaries is another solution to obtain higher–quality ingots and higher–efficiency solar cells. Many random grain boundaries are effective to reduce stress and prevent movement of dislocations. The concept of the HP cast method is in the opposite side of that of the dendritic cast method, but it was incidentally found during studying the dendritic cast method [18]. Now, the HP cast method is practically used in China and Taiwan because the controllability to manufacture uniform large ingots is comparably easy and the conventional cast furnace can be basically available to keep high productivity. Although the dislocation density in HP ingots became relatively lower than that in ingots grown by the conventional cast method, the HP ingots were still in contact with crucible wall and stress remained.

The conventional cast method and the dendritic cast method can be regarded as the fully–melted cast method, and the HP cast method and the mono–like cast method can be regarded as the half–melted cast method (see Section 5.2.2). An ingot with small grains has small stress and many random grain boundaries. This concept is used for the HP cast method. An ingot with a large grain has large stress and few random grain boundaries. This concept is used for the mono–like cast method (see Section 4.8.1). From the viewpoint of the uniformity, crystal defects and cost–performance of ingots grown by the mono–like and HP cast methods, it is still not clear which method is better for solar cells with high performance and high yield.

The target of the mono–like cast method is to obtain an ingot with higher minority carrier lifetime than the conventional cast method by avoiding formation of grain boundaries. A Si ingot grown by the mono–like cast method has a structure with a quite large single–crystal–like grain in its center surrounded by small grains grown from crucible wall. The ingot is effective to make efficient texture structure, but its uniformity is still not good. The alkaline texturing can be applied for solar cell fabrication to reduce the sunlight reflectance and it can increase the conversion efficiency near 1%. The light induced degradation in solar cells is low due to its low oxygen (O) concentration. Comparing to the CZ wafers, the square shape of single crystalline wafers increases the effective area ratio in the cell modules, which are beneficial for improvement of the electricity output. However, dislocations are generated between adjacent seed plates and nucleated from precipitates or particles, and they spread in cascades from 30% or 50% of the ingot height toward the top of the ingot. The containing of many dislocations is still a serious problem. Moreover, the realization of dislocation-free ingots is a future problem for the mono–like cast method to overcome the CZ method.

For the HP cast method, random grain boundaries can prevent movement of dislocations and become almost harmless by hydrogenation. Such effects are very lucky for the HP cast method. In the bottom part of the HP ingot, grains are smaller and more uniform and the etch–pit density (EPD) of dislocations is lower comparing with those of the ingot grown by the conventional cast method. However, the grain size becomes larger and dislocations increase with the ingot height. In the top part of the ingot, the average grain size is very similar to that of the conventional cast ingot. The ingot still has many random grain boundaries.

When using these methods to grow Si ingots, the dislocation density in both ingots is still locally high because the ingots are always in contact with the crucible wall during growth and contain large residual stress. Stress generally occurs due to the large temperature gradients during crystallization and cooling. The expansion of Si also causes large stress due to solidification of a Si melt in an inflexible crucible. The stress results in plastic deformation and generation of dislocations in the ingots. The relaxation of stress is still an important problem to be solved in the field of the ingot growth for solar cells using the cast furnace. The noncontact crucible (NOC) method was proposed to obtain a uniform and large Si single ingot without any contact to the crucible wall and to reduce the stress in the ingot. The NOC method can be comparable to the CZ method, in which an ingot is grown above the surface of a Si melt without contact with the crucible wall and has few stress due to the expansion of Si. Essentially, the ingot growth is performed in equilibrium inside the Si melt for the NOC method and it is performed in non–equilibrium outside of the Si melt or above the Si melt for the CZ method. From this point, defect formation mechanisms will be very different for both methods.

The target market of solar cells prepared mainly by the Si multi-crystalline and mono–like ingots grown using the HP and mono–like cast methods is the largest one such as the mega-PV park. The high productivity and low-cost are required in this market. The conventional and HP cast method can prepare Si multicrystalline ingots with the lowest cost using a cast furnace which can be easily enlarged to G8. On the other hand, the market of solar cells prepared by CZ single ingots is the relatively smaller one such as the roof-top market, but the higher-efficiency solar cells are required in this market. The low cost is also required in this market.

The technologies of Si ingot growth using cast furnaces have been widely developed and used to manufacture ingots for solar cells for a long time. However, there are few books which basically and wholly describe the concepts and technologies on the view point of comparison between strong and week points for them. In this book, the basic growth and basic characterization of Si crystals are simply described to understand the ingot growth technologies. The history of dominant Si cast methods for Si ingots for solar cells is briefly introduced. The concepts, growth mechanisms, growth technologies and problems to be solved of these methods are clearly explained in an easy-to-understand manner. Moreover, the concept and main technologies to create distinct low-temperature region in a Si melt for the NOC method is precisely described in the text for the first time.

This book is written for industrial engineers, beginners and students who are leaning and working for the research & development of the Si ingot growth, crystal defects and characterizations of wafers for solar cells, to develop the higher-quality multi- or single Si ingots using low-cost cast furnaces. It covers the concept of crystal growth and contains main growth technologies such as the conventional cast method, the dendritic cast method, the HP cast method, the mono-like cast method and the NOC method. These methods are still in progress development. The quality of Si ingot is discussed mainly on the view point of crystal defects and controllability of them. To help for understanding the frame work and key points of this field and conceiving new theme, many papers written by other researchers are referred as much as possible in this book. I would like to express respectful acknowledgments for all of them.

References

[1] BP Statistics 2018, and Energy White Paper from Agency for Natural Resources and Energy under the Ministry of Economy, Trade and Industry (METI), Japan, 2018.
[2] International Technology Roadmap for Photovoltaic Results 2016, ITRPV Eighth Edition, 2017.

[3] N. Usami, W. Pan, K. Fujiwara, M. Tayanagi, K. Ohdaira, K. Nakajima, Sol. Energy Mater. Sol. Cells 91 (2007) 123.

[4] Report of RTS Corporation, vol. 26, p. 21 (2017).

[5] A. Appapillai, E. Sachs, J. Cryst. Growth 312 (2010) 1297.

[6] K. Nakajima, K. Fujiwara, W. Pan, M. Tokairin, Y. Nose, N. Usami, in: Conference Record of the 2006 IEEE 4th World Conference on Photovoltaic Energy Conversion, vols. 1 and 2, Hilton Waikoloa Village, Waikoloa, Hawaii, USA, May 7-12, 2006, pp. 964—967.

[7] K. Fujiwara, W. Pan, N. Usami, K. Sawada, M. Tokairin, Y. Nose, A. Nomura, T. Shishido, K. Nakajima, Acta Mater. 54 (2006) 3191.

[8] K. Fujiwara, Y. Obinata, T. Ujihara, N. Usami, G. Sazaki, K. Nakajima, J. Cryst. Growth 266 (2004) 441.

[9] K. Fujiwara, Y. Obinata, T. Ujihara, N. Usami, G. Sazaki, K. Nakajima, J. Cryst. Growth 262 (2004) 124.

[10] K. Fujiwara, W. Pan, K. Sawada, M. Tokairin, N. Usami, Y. Nose, A. Nomura, T. Shishido, K. Nakajima, J. Cryst. Growth 292 (2006) 282.

[11] K. Nakajima, K. Kutsukake, K. Fujiwara, N. Usami, S. Ono, I. Yamasaki, in: Proceedings of the 35th IEEE Photovoltaic Specialists Conference, Hawaii Convention Center, Honolulu, Hawaii, USA, June 20-25, 2010, pp. 817—819.

[12] K. Nakajima, N. Usami, K. Fujiwara, K. Kutsukake, S. Okamoto, in: Proceedings of the 24th European Photovoltaic Solar Energy Conference and Exhibition, CCH–Congress Center and International Fair Hamburg, Germany, September 21—24, 2009, 2009, pp. 1219—1221.

[13] K. Nakajima, K. Kutsukake, K. Fujiwara, K. Morishita, S. Ono, J. Cryst. Growth 319 (2011) 13.

[14] N. Stoddard, B. Wu, I. Witting, M. Wagener, Y. Park, G. Rozgonyi, R. Clark, Solid State Phenom. 131—133 (2008) 1.

[15] B. Marie, S. Bailly, A. Jouini, D. Ponthenier, N. Plassat, L. Dubost, E. Pihan, N. Enjalbert, J.P. Garandet, D. Camel, in: The 26th European Photovoltaic Solar Energy Conference (26th EU PVSEC), Hamburg, September 5—9, 2011, 2011.

[16] X. Tang, L. Francis, L. Gong, F. Wang, J.-P. Raskin, D. Flandre, S. Zhang, D. You, L. Wu, B. Dai, J. Cryst. Growth 117 (2013) 225.

[17] D. Zhu, L. Ming, M. Huang, Z. Zhang, X. Huang, J. Cryst. Growth 386 (2014) 52.

[18] C.W. Lan, C.F. Yang, A. Lan, M. Yang, A. Yu, H.P. Hsu, B. Hsu, C. Hsu, CrystEngComm 18 (2016) 1474.

Acknowledgments

The work was performed in Institute for Materials Research (IMR) Tohoku University, Graduate School of Energy Science Kyoto University and FUTURE-PV Innovation Japan Science and Technology Agency (JST). The work was supported by the Grant-in-Aid for Scientific Research from the Ministry of Education, Culture, Sports, Science and Technology of Japan (MEXT), the New Energy and Industrial Technology Development Organization (NEDO) under the Ministry of Economy, Trade and Industry (METI), and the Japan Science and Technology Agency (JST) under MEXT.

I thank for many peoples for their helpful and significant supports for this work. In IMR Tohoku University, Prof. T. Fujimori, Prof. A. Inoue, Prof. T. Fukuda, Prof. T. Sakurai, Prof. I. Yonenaga, Prof. M. Suezawa and Prof. T. Shishido for their managing leadership and scientific supports, Dr. G. Sazaki, Dr. T. Ujihara, Dr. N. Usami, Dr. K. Fujiwara, Dr. Y. Nose, Dr. W. Pan, Dr. K. Odaira, Dr. I. Takahashi, Dr. Y. Azuma, Dr. A. Alguno, Dr. M. Kitamura, Dr. M. Tayanagi, Dr. M. Tokairin, R. Nihei, and Dr. K. Maeda as co-workers. In Graduate School of Energy Science Kyoto University, Prof. H. Shingu and Prof. K. Ishihara for their managing leadership and scientific supports, Dr. K. Morishita, Dr. K. Kutsukake and Dr. R. Murai as co-workers. In FUTURE-PV JST and FREA, AIST, Prof. M. Konagai for his managing leadership, Mr. S. Ono, Mr. Y. Kaneko, Dr. H. Takato and Dr. T. Fukuda as co-workers.

I collaborated with many persons in Japanese companies, Dr. H. Ito and of Dai-Ichi Kiden Corporation, Dr. S. Okamoto and Dr. I. Yamasaki of Sharp Corporation, Dr. S. Matsuno of Mitsubishi Electric Ltd., Dr. K. Shirasawa of Kyosera Ltd. and Dr. E. Maruyama of Sanyo Electric Ltd. I collaborated with many persons in foreign universities and institutes. I thank for Prof. T. Buonassisi, Dr. D. M. Powell, Dr. S. Castellanos, Dr. M. Kivambe, Dr. M. A. Jensen, Dr. A. Youssef, Dr. J. Schön and his other group members of Massachusetts Institute of Technology, Dr. A. Jouini, Dr. F. Jay, Dr. Y. Veschetti, Dr. B. Drevet and his other group members of CEATECH, LITEN, INES, Prof. C. W. Lan of National Taiwan University, and Prof. D. Yang of Zhejiang University. Prof. L. Arnberg, Prof. M. Di Sabatino, Dr. G. Stokkan and Prof. T. Gabriella in NTNU.

I was given several chances to contribute books by Prof. T. Nishinaga and Prof. P. Rudolph as editors. I thank for their significant supports.

I have referred many papers written by the researchers in this field. Especially, I thank Dr. K. Adamczyk, Prof. K. Arafune, Dr. A. Autruffe, Dr. G. W. Alam, Prof. D. G. Brandon, Dr. I. Brynjulfsen, Dr. J. Chen, Dr. B. Chen, Dr. G. Coletti, Dr. D. Camel, Prof. G. Chichignoud, Dr. J. Ding, Dr. M. Dhamrin, Prof. T. Duffar, Dr. B. Dai, Dr. T. Ervik, Dr. K. E. Ekstøm, Dr. J. Friedrich, Dr. D. P. Fenning, Dr. B. Freudenberg, Dr. L. J. Geerligs, Dr. S. Gindner, Prof. B. Gao, Dr. L. Gong, Prof. P. Geiger, Dr. C. C. Hsieh, Prof. X. Huang, Prof. T. Hoshikawa, Dr. C. Hsu, Prof. A. A. Istratov, Dr. T. Iwata, Prof. M. Kohyama, Dr. H. Klapper, Prof. K. Kakimoto, Dr. I. Kupka, Dr. F.-M. Kiessling, Dr. E. Kuroda, Dr. T. Lehmann, Prof. N. Mangelinck-Nöel, Prof. G. Müller, Dr. H. J. Möller, Dr. T. Muramatsu, Dr. D. H. Macdonald, Dr. J. D. Murphy, Dr. W. Miller, Dr. D. B. Needleman, Prof. K. Nagashio, Prof. S. Nakano, Dr. A. Nouri, Prof. W. Obretenov, Dr. V. A. Oliveira, Prof. A. Otsuki, Dr. E. J. Øvrelid, Prof. A. Ogura, Dr. S. P. Phang, Dr. E. Pihan, Dr. R. R. Prakash, Prof. M. Rinio, Dr. V. Randle, Dr. C. Reimann, Dr. H. C. Sio, Dr. C. Sun, Dr. M. C. Schubert, Dr. J. Schmidt, Dr. T. Sekiguchi, Dr. V. Stamelou, Dr. R. Sachdeva, Dr. N. Stoddard, Dr. A. K. Søilanda, Prof. T. Saitoh, Dr. M. Trempa, Dr. M. G. Tsoutsouva, Prof. M. Tajima, Dr. H. W. Tsai, Dr. T. Tachibana, Prof. S. Uda, Dr. A. Voigt, Prof. E. R. Weber, Dr. T. Y. Wang, Dr. Y. T. Wong, Dr. Y. C. Wu, Dr. F. Wang, Dr. P. Wang, Prof. X. Yu, Dr. S. Yuan, Dr. Y. M. Yang, Dr. K. M. Yeh, Dr. H. Zhang, Prof. K. Zaidat for their significant results and discussions.

Finally, I thank my wife, Reiko for her enduring supports for my working.

Nomenclature

Preface

Si	Silicon
HP	The high-performance cast method
CZ	The Czochralski method
AM	Air mass
O	Oxygen
EPD	Etch-pit density
NOC	The noncontact crucible method

Chapter 1

Ar	Argon
2-D	Two-dimensional
SiO$_2$	Silicon dioxide
In	Indium
B	Boron
Ge	Germanium
SSL-TJ	Solid-solid-liquid tri-junction
P	Phosphorous
Ga	Gallium
As	Arsenic
Fe	Iron
EBSD	Electron backscattering diffraction measurements
CSL	Coincidence site lattice
XRC	X-ray rocking curve
TEM	Transmission electron microscopy
EBIC	Electron beam induced current
FTIR	Fourier transform infrared spectrometry analysis
Mp	The melting point of Si
O$_i$	Interstitial oxygen
C	Carbon
C$_s$	Substitutional carbon
O$_{2i}$	Oxygen dimer
V	Vacancy
C$_i$	Interstitial carbon
CO	Carbon monoxide
N	Nitrogen
PL	Photoluminescence
SiC	Silicon carbide
Si$_3$N$_4$	Silicon nitride
Ni	Nickel
Cu	Copper

Co	Cobalt
Cr	Chromium
Mo	Molybdenum
Al	Aluminum
Mn	Manganese
Ta	Tantal
Zr	Circonium
Nd	Neodymium
Ag	Silver
Sc	Scandium
Th	Thorium
Ce	Cerium
Sb	Antimony
Ca	Calcium
Sn	Tin
Au	Gold
Zn	Zinc
Cd	Cadmium
Ti	Titanium
Li	Lithium
ICP-MS	Inductively coupled remote plasma mass spectroscopy
$POCl_3$	Phosphoryl chloride
Cu_i	Interstitial copper
Ni_i	Interstitial nickel
Ni_s	Substitutional nickel.
H	Hydrogen
RT	Room temperature
Fe_s	Substitutional iron
Fe_i	Interstitial iron
Fe_i^0	Neutral interstitial iron
Fe_i^+	Interstitial ionized iron
DLTS	Deep-level transient spectroscopy
B_s	Substitutional boron
SIRP	Scanning infrared polar-iscope
EFG	Edge-defined film-fed growth

Chapter 2

μ-PCD	Microwave photoconductive decay
QSSPC	Quasi steady state photoconductivity
PDG	Diffusion gettering
Al_2O_3	Aluminum oxide
ALD	Atomic layer deposition
SEM	Scanning electron microscopy
EL	Electroluminescence
XRF	X-ray fluorescence
SPV	Surface photovoltage

Nomenclature **xxix**

FZ	The float-zone method
IQE	Internal quantum efficiency
μ-XRF	Micro-X-ray fluorescence
V	Vanadium
Pt	Platinum
PECVD	Plasma-enhanced chemical vapor deposition
Fe_p	Precipitated iron
TR	Tabula Rasa
EXT	Extended
Al-BSF	Al back-surface-field
PERC	Passivated emitter and rear contact
PERL	Passivated emitter rear locally-diffused
B_i	Interstitial boron
B_iO_i	B_i-O_i pairs
B_iC_s	B_i–C_s pairs

Chapter 3

TMFs	Traveling magnetic fields
W	Tungsten
CH_4	Methane
3-D	Three-dimensional
PVA	Polymer binder
SiO_2	Silica
EDX	Energy-dispersive X-ray spectroscopy
Na	Sodium
Mg	Magnesium
Cr	Chromium
K	Potassium
Hf	Hafnium
$FeSi_2$	Iron di-silicide
Fe_2O_3	Di-iron trioxide
Cu_3Si	Copper silicide
XRD	X-ray diffraction
XBIC	X-ray beam-induced current
MgO	Magnesium oxide
BN	Boron nitride
Si_2N_2O	Sinoite
RIE	Reactive ion etching
RTP	Rapid thermal processing

Chapter 6

$Si_xC_yO_z$	Oxycarbide
DLTS	Deep Level Transient Spectroscopy
SMART	Seed manipulation for artificially controlled defect technique
PERT	Passivated emitter, rear totally diffused
HET	Heterojunction
INES	National Solar Energy Institute

Chapter 7

MCZ	Magnetic-field-applied Czochralski
HF	Hydrogen fluoride
HNO$_3$	Nitric acid
MIT	Massachusetts Institute of Technology
FREA	Fukushima Renewable Energy Institute
AIST	National Institute of Advanced Industrial Science and Technology
KOH	Potassium hydroxide

Chapter 8

RF	Radiofrequency
BaO	Barium oxide
CCZ	Continuous-feeding CZ
RP-PERC	Random pyramid passivated emitter and rear cell

CHAPTER 1

Basic growth and crystallographic quality of Si crystals for solar cells

1.1 Si single and multi-crystals

A Si single crystal has a long regularity of atomic arrangement and a Si multi-crystal has a short regularity of atomic arrangement because it has many grains and many grain boundaries. The Si single ingot is usually prepared by the CZ method [1,2] and the Si multicrystalline ingot is usually prepared inside a quartz crucible by the cast method [3]. The Si feedstock is normally produced as polysilicon by the Siemens process. The growth mechanisms of both methods are very different as shown by the O concentration in ingots as an example. For the CZ method, the O concentration in an ingot strongly depends on the rotation rate of a crucible, the crucible temperature and the argon (Ar) gas flow rate at the melt surface [2]. For the cast method, the O concentration can be largely reduced using a silicon nitride (Si_3N_4) coating on the crucible inner wall and strongly depends on the Ar gas flow rate at the melt surface.

Si crystal has the Diamond structure with closed packed faces of $\{111\}$. Physical properties of Si are more precisely listed in Appendixes. Each atom in a perfect diamond lattice has 4 nearest neighbors at a distance $d_1 = \sqrt{3}$ should be written as more complete style as others in the same line. $\sqrt{3}\, a_0/4$, 12 second neighbors at a distance $d_2 = \sqrt{2}\, a_0/2$ and third neighbors at a distance $d_3 = \sqrt{11}\, a_0/4$ (the lattice constant of Si, $a_0 = 5.4307$ Å) [4]. The surface energy of the closed packed $\{111\}$ face is the smallest and the $\{111\}$ face is the most stable. The estimated crystal-melt interfacial free energies for the $\{100\}$, $\{110\}$ and $\{111\}$ faces are about 0.42, 0.35 and 0.34 J m^{-2}, respectively [5]. The averaged value is 0.37 J m^{-2}. As the interfacial free energy for the $\{111\}$ face is smaller than that for the $\{100\}$ face, the $\{100\}$ interface develops many steps with $\{111\}$ facets and becomes rough.

Crystal Growth of Si Ingots for Solar Cells Using Cast Furnaces
ISBN 978-0-12-819748-6
https://doi.org/10.1016/B978-0-12-819748-6.00001-3

Copyright © 2020 Elsevier Inc.
All rights reserved.

Larger driving force is required for growth on a crystal with lower interfacial free energy. The {100} face is usually used for the growth direction of a CZ single ingot and the wafer surface of solar cells with efficient texture structure.

On the surface of Kossel crystal [6,7], an atom on the facet has ψ, where ψ is an energy required to break the bond between the first neighbor. An atom on the step and kink has 2ψ and 3ψ, respectively. An atom on the kink position is called as a half-crystal because it has $\varphi_{1/2}$ ($= 3\psi$), where φ ($= 6\psi$) is the total bond energy per atom in the crystal. By repetitive attachment and detachment of atoms to and from this position, the whole crystal can be built up or disintegrated into single atoms [7]. As shown in Fig. 1.1, the {111} face has the stable face plane and the two-dimensional (2-D) nucleation is required to make steps and kinks to grow crystal on the surface. The {110} face contains many steps, and kinks can be more easily formed on the steps. The driving force for crystal growth on the {110} surface is smaller than that on the {111} face. The {100} face contains many kinks and the driving force for crystal growth is smallest among the {111} and {110} faces.

1.2 Basic growth of Si crystals

1.2.1 Degree of supercooling

1.2.1.1 Experiments about the degree of supercooling of a Si melt

The supercooling or undercooling is an essential factor to determine the crystal structure and growth behavior such as growing interface, growth rate,

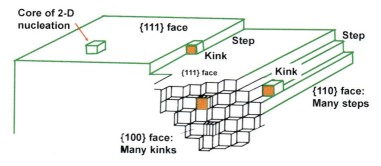

{111}:facet face, {110}:step face, {100}: kink face

Fig. 1.1 Stable {111} face required the 2-D nucleation to make steps and kinks to grow crystal on the surface. The {110} face with many steps, and kinks can be more easily formed on the steps. The {100} face with many kinks and the driving force for crystal growth is smallest among the {111} and {110} faces.

grain size, grain boundary character, dendrite growth, crystal defects and amounts of impurity content. The supercooling consists of the thermal, solute and curvature ones and the total supercooling is given by the sum of them [8]. A good wetting between solid and substrare leads to a lower energy barrier for nucleation, thus a lower supercooling for nucleation [9]. The supercooling limit of a Si melt is measured using the monochromatic and bi-chromatic pyrometers together for a Si droplet on zyarock and fused silica substrates [10]. The measured temperatures span over several magnitudes such as 2−237 K. The largest supercooling of 237 K is achieved by a Si droplet using a levitation technology [11]. The degree of supercooling, ΔT_s (K) is strongly affected by materials used as a substrate. ΔT_s is between 41 and 49 K on the zyarock substrate and between 45 and 89 K on the fused silica substrate [10]. Zyarock is the standard material used for industrial crucibles in the photovoltaic industry, which is the Vesuvius brand of all fused silica technical solutions. The supercooling limit of a Si melt is also measured using an infrared single-color pyrometer with a small spot size (0.35 mm) to reduce the processing time (several minutes) and the envi-ronmental heat effect [12]. It is about 50 K on a silicon dioxide (SiO_2) substrate and about 2.5−4.2 K on a SiO_2 substrate with Si_3N_4 coating [12]. By measurement of differential scanning calorimetry, the supercooling limit on a Si_3N_4 coating is confirmed to be lower than that on a dry SiO_2 coating [13]. The degree of supercooling of a Si melt on Si_3N_4 coated SiO_2 is reported to be between 12 and 37 K, and the supercooling does not depend on the O concentration in the coating [14]. When amorphous Si_3N_4 particles with 20−40 nm diameters are added in a Si melt and they are transformed into crystalline Si_3N_4 particles which act as nucleation sites, the degree of supercooling of the Si melt becomes lower to 2−6 K when that of a pure Si melt is 12−42 K [15]. The larger–supercooling samples have smaller grains comparing with the smaller-supercooling samples. The grain size on the surface of a droplet is larger on the Si_3N_4 coating than on the SiO_2 because of fewer nucleation sites by lower supercooling. The SiO_2 coating has a higher interfacial stability and does not cause heterogeneous nucleation even at a higher supercooling. On the other hand, the Si_3N_4 coating can easily cause nucleation at a lower supercooling. Thus, the oxides have a higher supercooling than Si_3N_4 because of their higher interfacial stability. While the SiO_2 substrates reach maximum supercooling near 100 K, the Si_3N_4 substrates appear to favorable supercooling below 40 K, suggesting that the Si_3N_4 substrates are more potent inoculants [16]. β-Si_3N_4 substrates lead to preferable nucleation of Si as potent inoculants because of its quite lower supercooling of 0.1–0.3 K comparing with that of α-Si$_3$N$_4$ substrates (6–11 K) [16] (see Section 3.4).

A higher cooling rate gives higher values of supercooling resulting in a higher growth rate [17]. The growth rate of Si rapidly increases with the increased amount of supercooling and then slowly increases after the supercooling reaches about 200 K because of the latent heat of instantaneous dendrite growth [11]. The relationship between the growth rate and the amount supercooling is classified into three regions such as the region I, II and III. The region I is between amounts of supercooling of 0 and 100 K, the region II is between 100 and 210 K, and the region III is above 210 K. Stepwise-lateral growth occurs at a faceted interface in a single plane in the region I, growth occurs at interfaces in multiple planes accompanied by few dendrite crystals in the region II and many dendrite crystals appear by in the region III.

1.2.1.2 Estimation of the degree of supercooling in a Si melt near a growing interface

The relationship between growth rate and supercooling for pure liquid can be known in an early paper in which the growth rate is basically expressed by a function of the supercooling, entropy of fusion, temperature and diffusion coefficient [18]. As shown in Section 3.2.1, the growth rate on the (111) side plane of a dendrite crystal is determined to be $V = 6.9 \times 10^{-4}$ (m s^{-1}) by the direct observation [19]. In this case, the initial melt has a positive temperature gradient before crystal growth, and the melt closed to the growing interface has a negative temperature gradient and a local supercooling because of the latent heat of the grown crystal. The growth on the Si {111} face which is atomically smooth needs a certain degree of supercooling because two-dimensional nucleation is necessary to start a new layer in case of low or zero dislocation density [20]. The degree of supercooling can be estimated using the two-dimensional nucleation model [21]. Obretenov et al. [22] derived an analytical expression between growth rate and degree of supercooling based on the two-dimensional nucleation of mononuclear growth and polynuclear growth. In the mononuclear growth, each nucleus formed on the crystal face gives rise to a monoatomic layer. In the polynuclear growth, other nuclei are formed next to the spreading nucleus layer before the monolayer is complete [22]. The vertical growth rate at the steady state, V (ms^{-1}) is expressed as

$$V = hJS/N, \tag{1.1}$$

where N is the average number of nuclei taking part in the formation of one monolayer, h (m) is the monolayer step height of 3.13×10^{-10} m, J $(m^{-2}s^{-1})$ is the two-dimensional nucleation rate and S (m^2) is the facet area of the crystal. The product JS determines the total number of nuclei formed per unit time. JS/N means the number of monolayers grown per unit time. N and x are expressed as [23].

$$N = (3/b)^{1/3}\Gamma(4/3)x, \tag{1.2}$$

$$x = (J/v)^{2/3}S, \tag{1.3}$$

where Γ is the complete gamma function, v (ms^{-1}) is the spreading velocity of monoatomic steps and b is a geometrical factor of the nucleus shape ($b = \pi$ for circle, $b = 4$ for square). x plays the role of total number of nuclei. N is expressed as the number of nuclei which takes part in the formation of the upper monolayers as

$$N = a_0 + a_1 x, \tag{1.4}$$

where a_0 and a_1 are the coefficients which correspond to the mononuclear ($x \rightarrow 0$) and polynuclear growth ($x \rightarrow \infty$) under extreme conditions, respectively.

$$a_0 = 1, \tag{1.5}$$

$$a_1 = 1/\beta b^{1/3}, \tag{1.6}$$

where β is a numerical coefficient of $\beta = 0.94 \pm 0.02$ [22,24,25]. From the least squares best fit of simulation data, a_0 and a_1 are estimated as 0.71 ± 0.46 and 0.728 ± 0.012, respectively [22]. From Eqs. (1.3−1.6), N can be expressed as

$$N = 1 + \left(S/\beta b^{1/3}\right)(J/v)^{2/3}, \tag{1.7}$$

Substituting Eq. (1.7) into Eq. (1.1), V can be obtained as

$$V = hJS\Big/\left\{1 + \left(S/\beta b^{1/3}\right)(J/v)^{2/3}\right\}, \tag{1.8}$$

where J and v are functions of supercooling and the temperature is close to maximum supercooling. Eq. (1.8) corresponds to both mononuclear growth (small values of x) and polynuclear growth (large values of x) as a general equation for the nucleation mechanism [22]. Beatty and Jackson [26] simulated the growth on the Si {100} and {111} face under different

6   Crystal Growth of Si Ingots for Solar Cells Using Cast Furnaces

degrees of supercooling, ΔT_s using the Monte Carlo method, and showed v and J as

$$v = 0.12\Delta T \quad \text{for } \{100\}, \tag{1.9}$$

$$v = 0.3\Delta T \quad \text{for } \{111\}, \tag{1.10}$$

$$J = 1.15 \times 10^{24} \, \exp(-140/\Delta T_s) \quad \text{for } \{111\} \tag{1.11}$$

From Eqs. (1.9) and (1.10), v for $\{111\}$ is significant faster than v normal to $\{100\}$. By substituting Eqs. (1.10) and (1.11) into Eq. (1.8), it is known that the growth rate or growth velocity varies with the degree of super-cooling. On the other hand, by introducing the experimentally determined $V = 6.9 \times 10^{-4}$ (m s^{-1}) into Eq. (1.8), the degree of the supercooling can be estimated as $\Delta T_s = 10$ K. Miller estimated the degree of the super-cooling as 4.7−5.9 K for $V = 2.5 \times 10^{-5}$ (m s^{-1}) using Eq. (1.8) [20].

For growth of facet dendrite crystals, the relationship between the supercooling and the growth rate is experimentally determined [27]. The growth rate linearly increases with the degree of supercooling, which is expressed as

$$V(\text{ms}^{-1}) = 7.5 \times 10^{-6}\Delta T(\text{K}) \tag{1.12}$$

Using Eq. (1.12), the critical supercooling is estimated to be between 10 K at $V = 7.4 \times 10^{-5}$ (m s^{-1}) and 13 K at $V = 9.7 \times 10^{-5}$ (m s^{-1}) [27] (see Sections 3.2.1and 3.2.2).

1.2.1.3 Experiments to determine the degree of supercooling in a Si melt near a growing interface

The supercooling near the growing interface depends in the temperature gradient in the melt. When the temperature gradient is positive in the growth direction, a large cooling rate is required to start crystal growth by locally generating a small supercooling near the growing interface as shown in Fig. 1.2. When the temperature gradient is negative, a small cooling rate is enough to start crystal growth by generating a supercooling. When the temperature gradient is positive, the supercooling in a melt and the thermal gradient in a crystal to the growth direction can be increased with the cooling rate or the growth rate [28]. When the cooling rate is as fast as 30 K min^{-1}, the growing interface becomes irregular and some protrusions or wave-peaks appear because of the high supercooling in the Si melt near the interface. The amount of supercooling near the growing

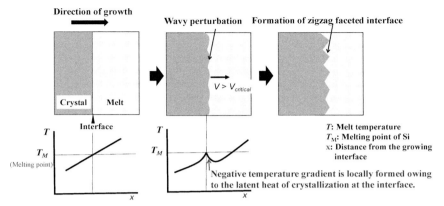

Fig. 1.2 Growing interface during growth. At a low growth rate, it keeps planar under a positive temperature gradient along the growth direction which is common for the cast growth. A wavy perturbation is introduced by amplification of instability at a high growth rate, resulting in the zigzag interface. A negative temperature gradient promotes the amplification because of the large supercooling at the top of the protruding portion on the interface. The negative temperature gradient is locally formed at the growing interface owing to the latent heat of a high growth rate. *(Referred from Fig. 9 in K. Fujiwara et al. Acta Meter. 59 (2011) 4700.)*

interface can be experimentally estimated as 7 K on {111} plane when the cooling rate is 40 K min^{-1} [23]. Under the temperature gradient in the melt of 3.1 K mm^{-1}, the amount of supercooling near the interface is estimated between 1 and 4 K on {100} plane in the range of the cooling rate from 10 to 27 K min^{-1} using an infrared pyrometer [28]. Generally, the cooling rate becomes larger as the positive temperature gradient becomes larger to generate a small local supercooling to start growing near the interface. The amount of supercooling is essentially larger on the stable {111} facet plane than on the {100} kink plane. For both facets on both sides at a wave-peak on the growing interface, the average supercooling of a smaller facet is higher than that of a larger facet, so the growth rate on the smaller facet is higher than that of the larger facet [29]. Finally, the initial smaller facet becomes larger than the initial larger facet, and the shape of the wave with such facets changes to a turnover shape (see Section 3.2.2).

1.2.1.4 Constitutional supercooling

The constitutional supercooling occurs near the growing interface for ingot growth from a highly doped Si melt under the condition of the quite small segregation coefficient of the dopant. The equation for steady-state diffusion in the coordinate system moving with the interface at a velocity or a growth rate, v is shown as follows,

$$D_d d^2 C/dz^2 + v dC/dz = 0, \qquad (1.13)$$

where C is the composition of the dopant in the Si melt, D_d is the diffusion coefficient of the dopant in the melt and z is the distance from the interface. C can be derived as [30,31],

$$C = C_\infty + C_\infty (1/k_e - 1) \, \exp(-vz/D_d), \qquad (1.14)$$

where C_∞ is the initial concentration of the dopant in the melt and k_e is the equilibrium segregation coefficient of the dopant.

$$k_e = C_S/C_L, \qquad (1.15)$$

where C_S is the composition of the dopant in the solid and C_L is the composition of the dopant in the melt at the interface. Using Eq. (1.14), the liquid temperature, T_L can be known from the C as a function of z by the T_L-C solubility curve or liquidus curve of the dopant in the melt. The constitutional supercooling occurs when the temperature gradient, G, should be less than the value given by

$$G < mC_\infty (1/k_e - 1) \, v/D_d, \qquad (1.16)$$

where m is the temperature gradient of the liquidus curve at $C \rightarrow 0$ of the dopant-Si phase diagram. The right side of Eq. (1.16) can be obtained by differentiating Eq. (1.14) by z at $z \rightarrow 0$. The constitutional supercooling can be avoided by largely increasing the temperature gradient near the interface using some hot zone in the furnace and by decreasing the pile up of dopants using a low growth rate and a large convection [31]. However, for a given hot zone, the temperature gradient generally decreases with increasing the diameter of an ingot and it becomes lower in the center of the ingot because of the increasing heat content due to the lager volume of the ingot [31]. Various interface morphologies from planar, cellular and dendritic can be described by the temperature gradient, G and the growth rate, v because they determine the area of large supercooling [32].

For growth of a p-type Si ingot, the heavily boron (B) doping with a higher initial B concentration than 1×10^{18}–1×10^{21} atoms cm^{-3} (normally 3×10^{20} atoms cm^{-3}) generates the constitutional supercooling which causes cellular growth or polycrystallization in the ingot when the growth rate is approximately 1 mm min^{-1} [33]. The polycrystallization follows the cellular growth. The cellular structure consists of a honey–comb–like structure, elongated in the growth direction [30]. The segregation coefficient of B decreases from 0.8 to 0.3 with the heavy B doping [2]. The decrease of the segregation coefficient of B affects the occurrence of the constitutional supercooling. The highly indium (In) doping also causes the constitutional supercooling which results in cellular growth in the Si ingot because the small In segregation coefficient of 0.0004 largely increases the In concentration in the Si melt near the growing interface even though the low solubility of In in the Si melt [34].

1.2.2 Growing interface and grooves

Basically, the degree of surface roughness on the atomic scale affects the shape of the growing interface [35]. The Si {100} growing interface basically has the kink face. It is atomically rough and has zigzag facets surrounded by stable {111} faces as shown in Fig. 3.5 [23]. The specific interfacial free energy of a step edge on the {111} face in contact with a Si melt is about one-10th of the specific interface free energy of the {111} facet [26]. Even for the {100} face, at a low growth rate, the growing interface keeps planar under a positive temperature gradient along the growth direction which is common for the cast growth. However, a wavy perturbation is introduced to the planar interface during growth and is amplified by instability at a high growth rate as shown in Fig. 3.6 [36,37]. Thus, at first the wavy perturbation appears on the planar interface and it is amplified to form the zigzag interface (see Section 3.2.2). A negative temperature gradient promotes the amplification because of the large supercooling at the top of the protruding portion of the growing interface as shown in Fig. 1.2 [38,39]. The negative temperature gradient is locally formed at the growing interface owing to the latent heat of a high growth rate [36]. The periodicity of the zigzag facets or the initial wavelength of the wavy perturbation becomes shorter as the growth rate increases [36,37]. After merging of these facets, however, the stabilized wavelength is proportional to the growth rate, which is determined by the interface kinetics [40]. The {110}, {112} and {100} growing interfaces can be zigzag at a high growth rate, but the {111} interface maintains planar because of its closed packed stable facet plane and its stable step-flow mode to grow faster in the lateral direction. The critical growth rate for the transformation to the zigzag shape is between 102 and 147 $\mu m\ s^{-1}$ for these three faces [39].

The critical growth rate decreases as the temperature gradient decreases because the local negative temperature gradient can be easily generated near the growing interface and the wavy perturbation can be easily occurred for such low temperature gradient [36,37]. The initial wavelength of the wavy perturbation or the size of the initial facets becomes larger as the critical growth rate becomes smaller or the temperature gradient becomes smaller [36,37]. The amplification of the wavy perturbation is promoted at an interface with twin boundaries [41]. In this case, the instability occurs at the lower growth rate than the critical growth rate, and the wavelength of the perturbation is determined by the spacing between twin boundaries and is independent of the growth rate [41].

For the <100> growth of Si-germanium (Ge) from a Si-rich Si-Ge melt with a large Ge content (8.4 at. % of Ge in crystal grown from 15 at. % of Ge in melt), the critical growth rate is much smaller that of Si and is between 1.0 and 5.8 μm s^{-1} because the interface instability at the planar interface easily occurs in the binary system owing to the constitutional supercooling [42]. The equilibrium segregation coefficient of Ge in Si (Ge in Si crystal/Ge in Si melt) is 0.42–0.56 [42,43] (see Section 1.4.3). For the <110> growth of SiGe with a small Ge content (1.0 at. % of Ge in crystal grown from 1.8 at. % of Ge in melt), the critical growth rate is much larger of 55 μm s^{-1} because the planar interface is stably maintained at the smaller Ge content [44]. When the growth rate reaches the critical growth rate, the constitutional supercooling occurs near the interface. When the growth rate is higher than the critical growth rate, the growing interface changes from planar to zigzag facets due to the amplification of interface perturbation by the constitutional supercooling. Finally, a faceted cellular interface appears after the zigzag facet formation at a high growth rate because of the Ge segregation at the valleys of zigzag facets and the local constitutional supercooling at them [42]. At the onset of instability, the periodicity of the zigzag facets becomes shorter or the grooves grow wider as the growth rate and Ge concentration increase [45,46]. The critical growth rate decreases with Ge concentration for both <100> and <110> growth and is almost independent of the growth orientation [44] (see Section 1.3.2).

When a growing interface with {111} facet planes encounters another {111} growing interface, an {111} $\Sigma 3$ grain boundary which means a $\Sigma 3$ grain boundary with a {111} boundary plane is formed between the impinged growing interfaces as a twin grain boundary [47]. When a growing interface with the zigzag shape encounters another zigzag growing interface, a random grain boundary is formed between them [47]. For both cases, the impinged interface becomes linear because the temperature of a thin melt enclosed between two interfaces increases just before the impingement due to the trapped latent heat.

At the growing interface between crystal and melt, grooves are generally formed at tips of high-Σ grain boundaries such as $\Sigma 27$ and random grain boundaries during growth as shown in Fig. 1.3A [38,45]. These grain boundaries have rough boundary planes with dangling bonds. Stable {111} facets or rough planes appear as the inner surface of grooves to form these grooves by such instability [38,48]. The depth of the grooves increases during growth because the growth rate on the inner surface is lower than that on the growing interface because of smaller supply of solute elements. Such grooves are called as grain boundary grooves. Finally, crystallization occurs inside the deep grooves due to increase of the supercooling [49]. A new groove can be formed inside the original grove due to the crystallization of a new grain on the inner surface as shown in Fig. 1.3B. One of the new grain boundaries is the {111} $\Sigma 3$ grain boundary or twin grain boundary formed on an {111} facet of the original groove. When the new grain has a random grain boundary, the new groove changes from a facet-facet groove to a rough-facet groove. The original random grain boundary

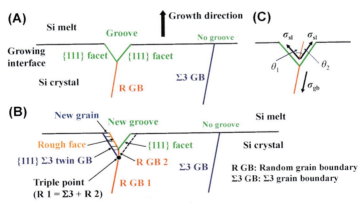

Fig. 1.3 Growing interface between crystal and melt. A groove is difficult to be formed at the tip of the {111} $\Sigma 3$ boundary. (A) Groove formed at the tip of a random grain boundary during growth. Stable {111} facets appear as the inner surface of grooves. (B) New groove formed due to the crystallization of a new grain on the inner surface. One of the new grain boundary is the {111} $\Sigma 3$ grain boundary. When the new grain has a random grain boundary, the new groove changes from a facet-facet groove to a rough-facet groove. The original random grain boundary splits to a random grain boundary and an {111} $\Sigma 3$ grain boundary (R = R + $\Sigma 3$) at the triple point. (C) Force balance between σ_{sl} and σ_{gb} at the SSL-TJ in the groove formed from a grain boundary.

(R 1) splits to a random grain boundary (R 2) and an $\{111\}$ $\Sigma 3$ grain boundary (R $1 = $ R $2 + \Sigma 3$) at the triple point or solid–solid–liquid tri-junction (SSL-TJ). Even at the growing interface, however, stable grain boundary grooves are difficult to be formed at tips of the $\{111\}$ $\Sigma 3$ boundaries which have originally stable $\{111\}$ planes as shown in Fig. 1.3. Impurities easily accumulate at such grain boundary grooves and the grooves grow wider as the impurity concentration at the grooves increases because the step growth on the inner surface of the grooves becomes faster due to the constitutional supercooling of the impurity accumulation [45,50]. Thus, the fact that impurity concentrations become high at the high-Σ grain boundaries is consistent with the high recombination activity of these grain boundaries (see Section 2.2.3).

At the equilibrium growth condition or the zero-growth rate, the groove formed from a grain boundary can be used to measure the crystal/melt interfacial energy, σ_{sl}, and the grain boundary energy, σ_{gb}, because the shape of the groove is determined by the force balance between σ_{sl} and σ_{gb} [51]. The relation between σ_{sl} and σ_{gb} is shown in Fig. 1.3C. The force balance can be written as follows [51],

$$\sigma_{gb} = (\cos\theta_1 + \cos\theta_2)\sigma_{sl}, \tag{1.17}$$

where θ_1 and θ_2 are each angle between these three grain boundaries at the SSL-TJ, respectively, as shown in Fig. 1.3C. Variation of the groove shape can be directly observed using the in-situ observation system during growth (see Section 3.2.2).

1.2.3 Dopant distribution in Si multicrystalline ingots

The distribution of the resistivity or doping elements such as B and phosphorous (P) in an ingot is essential for the ingot growth. B and gallium (Ga) are p-type dopants and P and arsenic (As) are n-type dopants. The segregation coefficients of B, Ga, P and As in a Si crystal solidified from a Si melt are 0.7–0.8, 0.008, 0.35 and 0.8, respectively [43,52–54]. As a dominant impurity in Si, the segregation coefficient of iron (Fe) in Si is 8×10^{-6}, which is quite small comparing with those of the dopants [43]. The covalent radius of B is 0.88Å, that of Si is 1.17Å and that of Ga is 1.26Å [53]. The B and Ga concentrations monotonically increase with ingot height comparing with slow increase of the Fe concentration in the ingot because of their relative larger segregation coefficients [55] (see Section 8.4). The distribution of dopant is affected by the different temperature gradient in a Si melt for the different growth method such as the dendritic casting and NOC methods

[56]. For example, in the NOC method which has a large low-temperature region in the Si melt, the temperature gradient is affected by the slow rotation of an ingot (1.0–2.0 rpm) and the low pulling rate (0.2 mm min^{-1}) due to keeping the growth in the low-temperature region. To know the B distribution or resistivity in several ingots grown by these methods, many p-type ingots were prepared by the dendritic casting method [9,57,58] and the NOC method [56,59–62]. These ingots were cut into small blocks and sliced into wafers (10 cm × 10 cm or 15.6 cm × 15.6 cm). The resistivity of the wafers cut from each ingot was measured using the conventional four-point probe method. Fig. 1.4 shows the resistivity in these p-type ingots as a function of the solidification ratio [56]. Three ingots were grown by the NOC method from the upper part of Si melt, and three were grown by the dendritic cast method from the lower part of Si melt. The solid and broken lines are results estimated using Scheil's equation as shown in Eq. (1.20) [63], in which local equilibrium of the advancing solidification front at the solid–liquid interface and complete mixing of the solute in the whole melt are assumed.

$$(C_L - C_S)\mathrm{d}f_S = f_L\ \mathrm{d}C_L, \tag{1.18}$$

$$f_S + f_L = 1, \tag{1.19}$$

$$C_S = k\ C_0(1-f_S)^{(k-1)}, \tag{1.20}$$

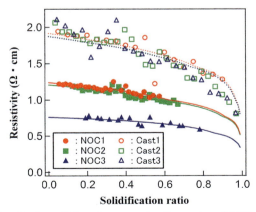

Fig. 1.4 Resistivity of p-type Si ingots grown by the cast and NOC methods as functions of the solidification fraction. The solid and broken lines show the estimated results using Scheil's equation. The distribution of the resistivity of these ingots can be well expressed by the segregation coefficient of the dopant. *(Referred from Fig. 1 in K. Nakajima et al. J. Cryst. Growth **372** (2013) 121.)*

where C_S is the dopant concentration in the solid, C_L is the dopant concentration in the liquid and C_0 is the initial dopant concentration in the liquid ($C_L = C_0$ at $f_S = 0$). C_S and C_L are equilibrium with each other at the growing interface, k is the segregation coefficient of the dopant ($= C_S/C_L$), and f_S and f_L are the solid and liquid fractions ($0 \leqq f_S \leqq 1$), respectively. The distribution of the resistivity of these ingots grown by the cast and NOC methods can be well expressed by Scheil's equation even though the staring melt compositions are different from each other. This means that the distribution of dopant such as B in the Si ingots is essentially determined by the segregation coefficient of the dopant even though the low-temperature region exists in the Si melts. In this case, the clear gradient of the B distribution is observed because the Si melt thickness is only 3−4 cm and it is not sufficiently thick to obtain more flat distribution.

1.3 Crystallographic structure and defects of Si crystals

1.3.1 Determination of crystallographic structure using seed plates and particles

The crystallographic structure such as the grain size and orientation is largely affected by that of the crystal plates or particles used as seed. To simply know this relation, Azuma et al. [64] studied Ge or SiGe ingots grown using the Ge single crystal plates or particles (<0.2 cm) used as seed. The growth rate is 0.0016 mm min^{-1} and it is quite small like a near equilibrium condition. This combination of SiGe and Ge clearly shows this relation because of their habit planes. A SiGe ingot grown on Ge seed particles has a preferential orientation of <110> and the largest area fraction of {110} faces. The area fraction of the {110} face rapidly increases and that of the {100} or {111} face decreases as the ingot height. For a SiGe ingot grown on a Ge {110} single seed plate, no additional grains appear from the crucible wall during growth. On the other hand, for a SiGe ingot grown on a Ge {111} single seed plate, many grains additionally appear from the crucible wall and the area fraction of the {111} face drastically decreases. These results mean that polycrystallization can be effectively prevented using the preferential oriented seed plates. The preferential orientation has the largest increasing rate of the area fraction. Fig. 1.5 shows a cross section of a directionally grown Ge ingot from Ge seed particles with random orientation [64]. The diameter and length of the ingot are 1.5 and 5.0 cm, respectively. From the two points shown by the yellow arrows (1.4 and

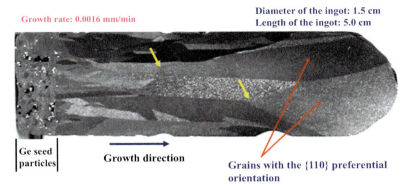

Fig. 1.5 Cross section of a directionally grown Ge ingot from Ge seed particles with random orientation. The diameter and length of the ingot are 1.5 and 5.0 cm, respectively. From two points shown by the yellow arrows (1.4 and 2.6 cm from the seed particles), two large grains with the {110} preferential orientation appears. No additional grains appear from the crucible wall during growth. *(Referred from Fig. 1 in Y. Azuma et al. J. Cryst. Growth, **276** (2005) 393.)*

2.6 cm from the seed particles), two large grains with the {110} preferential orientation appear. The area fraction of the {110} face drastically increases with the ingot height. No additional grains appear from the crucible wall during growth. The Ge ingot grown on Ge seed particles finally has a preferential orientation of <110> and the largest area fraction of {110} faces during growth as shown in Fig. 1.5.

From these results for the growth of the SiGe and Ge ingots using seed particles, we can see that the grain size always increases with the ingot height and the fractions of grains with preferential orientations or stable surface energies increases with the ingot height. This information has given an important implication for the development of the HP cast method as one of pioneer works (see Section 5.2.1). For the growth of the SiGe and Ge ingots using Ge single crystal seed plates, we can see that a large single grain can be obtained by selecting a preferential orientation as the face of the seed plates. This information is also very effective for the development of the mono-like cast method as one of pioneer works (see Section 6.1).

1.3.2 Grains in Si crystals

A Si multicrystalline ingot has many small grains with the different sizes and orientations and many grain boundaries with different characteristics. Such a structure is basically an obstacle for high-efficiency solar cells. Fig. 1.6 shows distributions of grain orientation and grain boundary on cross sections of

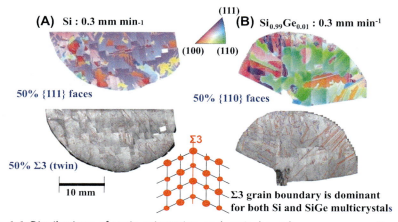

Fig. 1.6 Distributions of grain orientation and grain boundary on cross sections of (A) a typical Si multicrystalline wafer and (B) a typical SiGe multicrystalline wafer with the average Ge composition of 1%.

typical Si and SiGe multicrystalline ingots grown at the cooling rate of 0.3 mm min^{-1} by the conventional cast method. Reference directions of the sample are colored using the inverse pole figure triangle. The orientation of grains can be measured by the resolution of electron backscattering diffraction (EBSD) measurements. The detected electrons are back scattered from a sample surface irradiated by a focused electron beam. It is used for efficient mapping of crystal orientation, and its angle resolution is high and generally in the order of 1°. The Laue scanner based on X-ray diffraction together with the grain detector can be also used to determine the crystal orientation, misorientation, grain orientation and grain boundary type for the entire area of an uneven, unpolished and up to 50 cm thick Si wafer with an area of 38 cm × 40 cm [65]. The {111} face which includes {112} face colored by violet is dominant and more than 50 % of total grains have {111} face as shown in Fig. 1.6A. This means that the most stable {111} face forms under near equilibrium condition or low growth rate. An ingot grown by the cast method has an inhomogeneous distribution of grain size and grain orientation at the bottom of the ingot. In other words, grains with different sizes and random orientations coexist by inhomogeneous nucleation at random points on the bottom of the crucible. The grain size is typically between 0.5 and 3 cm. For the SiGe wafer, the same EBSD and Laue scanner measurements are used. The {110} face colored by green is dominant and more than 50 % of total grains have {110} face even for the average Ge composition of only 1% as shown in Fig. 1.6B (see Section 8.5). The mechanism why the dominant face of the SiGe grains changes to {110} depends on the behavior of the growing interface, but the detail is not still clear (see Section 1.2.2). As shown in Fig. 1.12, an {100} oriented grain has four {111} facets and an {110} oriented grain has two facets which make an angle of 35.27° with the {110} grain surface. An {111} oriented grain has four facets of which one face is parallel to the {111} grain surface and three facets make an angle of 70.53° with the {111} grain surface.

Recently, to investigate the nucleation and growth mechanisms of multicrystalline Si ingots, application of the weighted Voronoi diagram is tried to reproduce actual grain distribution by considering the difference of nucleation timing and anisotropy of the growth rate as weight [66]. The Voronoi diagram divides a plane into regions based on the distance to generating points which are arbitrarily arranged.

1.3.3 Grain boundaries in Si crystals
1.3.3.1 Small and large angle grain boundaries (Σ grain boundaries)

A Si multicrystalline ingot has many grain boundaries with different characteristics. The properties of grain boundaries depend on 5 crystallographic parameters. In other words, 5° of freedom are required to describe the geometry of a grain boundary [67]. Among them, 3° of freedom are assigned to the misorientation and 2 to the crystallographic orientation of the grain boundary plane [67]. Fig. 1.7 shows the category of grain boundaries by

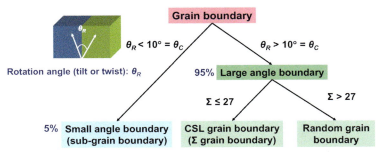

Fig. 1.7 Category of grain boundaries by rotation angle and Σ value. Grain boundaries are categorized as small and large angle grain boundaries. Most of grain boundaries are categorized in the large angle grain boundaries which contain the CSL grain boundaries with smaller Σ values than 27 and the random grain boundaries with larger Σ values than 27. *(Referred from Fig. 5 in K. Kutsukake, in: D. Yang (Ed.), Chapter 11, Section Two, Crystalline silicon growth, Hand Book of "Photovoltaic Silicon Material", Springer, Berlin Heidelberg, 2018. On line.)*

rotation angle and Σ value [68]. The Σ value reflects the periodicity of atomic arrangements at grain boundaries and gives information about the misorientation between grains. The Σ value increases as the coherency of grain boundaries decreases as explained in the below text. A perfect $\Sigma 1$ grain boundary has a rotation or misorientation angle (tilt and/or twist), θ_R of $0°$. Grain boundaries are categorized as small and large angle grain boundaries.

The small angle grain boundary has θ_R of less than $10°$ [69] or $15°$ as critical angle, θ_C [68,70,71] and called as sub-grain boundary. θ_C corresponds the angle at which the microstructure of grain boundaries changes from aligned dislocations to random structure [68]. The rotation or misorientation is caused by the tilt or twist rotation between two crystals. The small angle grain boundary with a tilt angle less than $1°$ fundamentally consists of an array of parallel edge dislocations with a certain distance (70−80 nm) and a certain density [70], and it is normally composed of tilt or twist dislocations [72]. The small angle grain boundary with $[1-10]$ axis and (110) boundary plane can be described as an array of dislocations without dangling bonds [73]. The average distance of a dislocation array formed in the small angle grain boundary is about 11.35 nm, causing a misorientation of about $2°$ [74]. The interval between edge dislocations, d_{ed} varies with respect to the misorientation angle, and it is inversely proportional to θ_R as follows,

$$d_{ed} = b/2\sin(\theta_R/2/) = b/\theta_R, \tag{1.21}$$

where b is a length of Burgers vector ($= 0.37$ nm) (see Section 1.3.4.1). About 5 % of total grain boundaries are small angle grain boundaries.

Low- and high Σ grain boundaries ($\Sigma = 3, 9, 27$ and other) and random grain boundaries are categorized as large angle grain boundaries. The large angle grain boundary has θ_R of larger than $10°$ as θ_C and smaller than $159.95°$ ($10° \leq \theta_R \leq 159.95°$). In the large angle grain boundary with misorientation angle above $10°-15°$, single dislocations overlap in such a coincidence site lattice (CSL) grain boundary or a random grain boundary as shown in Fig. 1.7 [68,75]. About 95% of total grain boundaries are large angle grain boundaries. Most of grain boundaries are categorized in large angle grain boundaries which contain the CSL grain boundaries with smaller Σ values than 27 ($\Sigma 3^n$) and the random grain boundaries with larger Σ values than 27 [71].

From Brandon [70], the coincidence relationship can be known in detail. Essentially, CSL grain boundaries exist for all odd numbers of Σ values, but coincidence has little significance if Σ values are large. Any coincidence relationship can be expressed by an axis–angle pair, and in the cubic system each such relationship can be described by 24 different axis–angle pairs, corresponding to the 24 symmetry elements of the cubic system. Thus, there are 24 different rotations with θ_R not exceeding 180° in the diamond lattice [73].

The Σ value is determined by relative orientation of grains which form a grain boundary as shown in Tables 1.1 and 1.2. Table 1.1 shows six relations between rotation axis, rotation angle, grain boundary plane and grain boundary character for tilt rotation. They are all twin grain boundaries which have different rotation axes and different grain boundary planes. Table 1.2 shows five relations between rotation axis, rotation angle, grain boundary plane and grain boundary character for twist rotation. Even for the $\Sigma 3$ grain boundaries, they are non-twin grain boundaries which have the rotation axes perpendicular to the grain boundary planes. The Σ value represents a relative size of the unit cell of the coincidence site lattice to that of the crystal lattice as shown Fig. 1.8. In Fig. 1.8 [68], the $\{111\}$ $\Sigma 3$ grain boundary with a rotation angle of 70.53° is shown as an example. The

Table 1.1 Relation between rotation axis, grain boundary plane and grain boundary character for tilt rotation (Twin boundaries).

Rotation axis	Rotation angle θR	Grain boundary plane	Grain boundary character
$\langle 110 \rangle$	70.53°	$\{111\}$	$\Sigma 3$
$\langle 210 \rangle$	131.81°	$\{121\}$	$\Sigma 3$
$\langle 111 \rangle$	60.00°	$\{211\}$	$\Sigma 3$
$\langle 100 \rangle$	53.13°	$\{012\}$	$\Sigma 5$
$\langle 100 \rangle$	36.87°	$\{013\}$	$\Sigma 5$
$\langle 110 \rangle$	38.94°	$\{221\}$	$\Sigma 9$

Table 1.2 Relation between rotation axis, grain boundary plane and grain boundary character for twist rotation (Non-twin boundaries).

Rotation axis	Rotation angle θR	Grain boundary plane	Grain boundary character
$\langle 110 \rangle$	70.53°	$\{110\}$	$\Sigma 3$
$\langle 210 \rangle$	131.81°	$\{210\}$	$\Sigma 3$
$\langle 100 \rangle$	36.87°	$\{100\}$	$\Sigma 5$
$\langle 100 \rangle$	53.13°	$\{100\}$	$\Sigma 5$
$\langle 110 \rangle$	38.94°	$\{110\}$	$\Sigma 9$

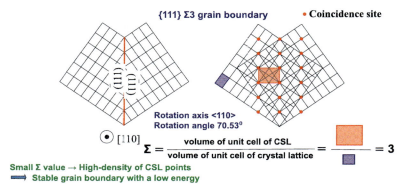

Fig. 1.8 CSL created by the coincidence site. The Σ value is determined by relative orientation of grains which form a grain boundary. It represents a relative size of the unit cell of CSL to that of the crystal lattice as shown by the {110}Σ 3 grain boundary. A small Σ value leads to a high density of CSL points. Thus, the Σ grain boundary is formed as a stable grain boundary with a low energy. *(Referred from Fig. 7 in K. Kutsukake, in: D. Yang (Ed.), Chapter 11, Section Two, Crystalline silicon growth, Hand Book of "Photovoltaic Silicon Material", Springer, Berlin Heidelberg, 2018. On line.)*

{111} Σ 3 grain boundary has the lowest energy minimum among Σ3 grain boundaries and other Σ grain boundaries. The grain boundary energy increases with the amount of distortion of bond lengths and angles and with the number of coordination defects [76]. The lowest-energy boundaries with quite small bond-length and angle distortions have no coordination defects. A small Σ value leads to a high density of CSL points on the grain boundary plane and to a stable grain boundary. Thus, the Σ grain boundary is formed as a stable grain boundary with a low energy. The rotation angle θ_R is also written on the table. Outside the table, the {211} Σ3 ($\theta_R = 109.47°$), {411} Σ9 ($\theta_R = 141.06°$), {311} Σ 11 ($\theta_R = 129.52°$), {322} Σ 17 ($\theta_R = 93.37°$), {511} Σ 27 ($\theta_R = 148.41°$), and {811} Σ 33 ($\theta_R = 159.95°$) are <110> tilt grain boundaries [77]. The boundary energies of the {211} Σ3 and {411} Σ9 grain boundaries are 2.81 and 3.38 J m^{-2}, respectively [77]. The grain energy of the {111} Σ3 tilt grain boundary is much lower as 0.03−0.06 J m^{-2} [78] and the grain energy of the {111} Σ3 twist grain boundary is 0.017 J m^{-2} [79]. The grain energies of the {221} Σ9 and {130} Σ5 grain boundaries are 0.29 or 0.32 J m^{-2} and 0.26 J m^{-2}, respectively [80,81]. The grain energies of non-Σ and Σ 27 are 0.48 and 0.38 J m^{-2}, respectively [79]. Among them, the {211} Σ3 grain boundaries are frequently observed together with the {111} Σ3 grain boundaries, and the former ones are more electrically active than the latter ones. The grain boundary energy of $<100>$ tilt grain boundaries is relatively higher than that of $<110>$ tilt grain boundaries, and the energy of the most stable $<100>$ tilt grain boundaries is also higher than the maximum energy of the $<110>$ tilt grain boundaries [79].

The annihilation interaction between a ΣA grain boundary and a ΣB grain boundary to form a new Σ grain boundary follows as

$$\Sigma A + \Sigma B = \Sigma(A \times B) \quad \text{or} \tag{1.22}$$

$$\Sigma A + \Sigma B = \Sigma(A/B) \quad (A/B: \text{integer and } A > B). \tag{1.23}$$

For example, when $A = 9$ and $B = 3$,

$$\Sigma 9 + \Sigma 3 = \Sigma 27 \quad \text{or} \Sigma 3 \tag{1.24}$$

The generation interaction of $\Sigma(A \times B)$ or $\Sigma(A/B)$ between a ΣA grain boundary and a ΣB grain boundary to form new Σ grain boundaries follows as

$$\Sigma(A \times B) = \Sigma A + \Sigma B, \tag{1.25}$$

$$\Sigma(A/B) = \Sigma A + \Sigma B \quad (A/B: \text{integer and } A > B) \tag{1.26}$$

For example, when $A = 3$ and $B = 3$ for Eq. (1.25) and $A = 27$ and $B = 3$ for Eq. (1.26),

$$\Sigma 9 = \Sigma 3 + \Sigma 3, \tag{1.27}$$

$$\Sigma 9 = \Sigma 3 + \Sigma 27. \tag{1.28}$$

(see Section 1.3.3.5) According to these reactions using Eq. (1.26) and Eq. (1.22), the most favorable $\Sigma 3 <111>$ grain boundary with the lowest boundary energy can be followed by a $\Sigma 9 <110>$ grain boundary and then followed by a $\Sigma 27 <110>$ grain boundary which is the most unfavorable of these three [82],

$$\Sigma 3 = \Sigma 3 + \Sigma 9 = \Sigma 27. \tag{1.29}$$

The $\Sigma 3 <111>$ grain boundary means a $\Sigma 3$ grain boundary with a rotation axis of $<111>$. However, an encounter between a $\Sigma 9$ and a $\Sigma 3$ gives a $\Sigma 3$ grain boundary rather than a $\Sigma 27$ in Eq. (1.24), and an encounter between a $\Sigma 9$ and a $\Sigma 27$ gives a $\Sigma 3$ grain boundary rather than a $\Sigma 243$ [67]. In general, the following reaction,

$$\Sigma 3^{n} + \Sigma 3^{n+1} = \Sigma 3 \tag{1.30}$$

is more preferred rather than the following reaction to produce higher $\Sigma 3^{n+2}$,

$$\Sigma 3^n + \Sigma 3^{n+1} = \Sigma 3^{2n+1}. \tag{1.31}$$

When a $\Sigma 3$ grain boundary meets a random grain boundary, the new gain boundary created at the new triple junction will have a Σ value reduced by a factor of three $(1/3)$ compared to the random grain boundary, according to Eq. (1.23). The twin has five degrees of freedom and the random grain boundary has zero degrees of freedom. The newly created grain boundary at the triple junction will have four out of five degrees of freedom [67]. Small segments of a $\Sigma 9$ grain boundary are required to act as bridging segments between mobile $\Sigma 3$ grain boundaries [67]. The grain boundary interactions in three dimensions for Si multicrystals can be tried to estimate using a phase-field model [79].

According to the Brandon criterion [70], the maximum allowed angle deviation, θ_{max} from the precise or essential rotation angle and rotation axis is given to be

$$\theta_{max} = 15°/\Sigma n^{1/2}. \tag{1.32}$$

θ_{max} decreases with the Σ value. For example, $\theta_{max} = 15°$ for the $\Sigma 1$ $(n = 1)$ grain boundary as the real lattice and $\theta_{max} = 8.7°$ for the $\Sigma 3$ $(n = 3)$ grain boundary. Increase in such angle deviation θ from the $\Sigma 3$ grain boundary deteriorates the solar cell performance due to increase in the bond–length and angle distortions resulting in carrier recombination velocity [83].

1.3.3.2 Effects of misorientation of grain boundaries on dislocation generation

From Fig. 1.9, the most stable symmetric grain boundary energies for various Σ values can be known as a function of the rotation angle (tilt) θ_R for the <110> axis [77]. (see Section 6.3) In the range smaller than $\theta_R = 10°$ in Fig. 1.9, grain boundaries can be regarded as small angle grain boundaries consisting of an array of edge dislocations. In the range of $\theta_R \geq 159.95°$, grain boundaries can be also regarded as small angle grain boundaries with an array of edge dislocations. A small angle grain boundary with small misorientation or a small rotation angle or a small angle deviation is close to the energy minimum as shown in these ranges of Fig. 1.9. Shallow cups can be found at the {211} $\Sigma 3$ $(\theta_R = 109.47°)$, {311} $\Sigma 11$ $(\theta_R = 129.52°)$ and {511} $\Sigma 27$ $(\theta_R = 148.41°)$ grain boundaries. Another energy minimum appears at the {110} $\Sigma 3$ $(\theta_R = 70.53°)$ grain boundary.

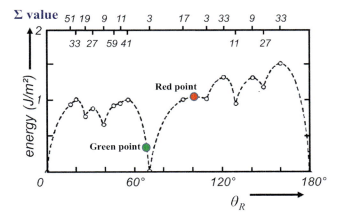

Fig. 1.9 Tilt grain boundary energy for several Σ values as a function of the rotation angle θ_R for the <110> axis. A non-perfect Σ3 grain boundary with a small angle deviation is close to the energy minimum like the Σ3 grain boundary as shown by a green bottom. A large angle grain boundary is too far away from the energy minima like a random grain boundary or any Σ grain boundary with large angle deviation as shown by a red bottom. (Referred from Fig. 5 in M. Kohyama et al. Phys. Status Solidi (b) 137 (1986) 387.)

A non-perfect {110} Σ3 ($\theta_R = 70.53°$) grain boundary with a small angle deviation is also close to the energy minimum as shown by the green point. It can generate dislocations or split a second grain boundary to minimize its energy as if it were a small angle grain boundary. A large angle grain boundary ($10° \leq \theta_R \leq 159.95°$) has a high grain boundary energy and it is too far away from the energy minima like a random grain boundary or any Σ grain boundary with large angle deviation as shown by the red point. The energy gains due to angle changing is not large enough to overcompensate the energy effort for dislocation formation or splitting a second grain boundary because the energy curve is almost flat near the red point. Therefore, the large angle grain boundary keeps the structure and causes no dislocation generation or grain boundary splitting during crystal growth. Thus, the microstructures of CSL grain boundary are sensitive for θ_R because the angle deviation from the perfect CSL grain boundary is tended to be relaxed along the grain boundary energy versus θ_R curve as shown by the case of the non-perfect Σ3 grain boundary in Fig. 1.9.

The precise ordered structure of twist Σ grain boundaries is simulated. The ordered configurations for {100} Σ25 ($\theta = 16°$), Σ13 ($\theta = 23°$), Σ17 ($\theta = 28°$) and Σ5 ($\theta = 37°$) twist grain boundaries with low-energy ordered structures consistent with experimentally observed cusps of the grain boundary energy as a function of the CSL misorientation or the rotation angle, θ [76,84]. In this case, the $\Sigma 13$ grain boundary has the deepest energy cusp among them and the $\Sigma 5$ grain boundary has also the quite deep energy cusp as shown in Fig. 6.3 [84]. Especially, the $\Sigma 5$ grain boundary has only five atoms per primitive cell in the grain boundary plane and makes the smallest primitive cell among the above four boundaries [76]. The $\Sigma 5$ grain boundaries are effectively used to prevent the propagation of dislocations in a mono-like cast ingot. The more precise minimum grain boundary energies are simulated as a function of θ_R for various Σ grain boundaries [85]. They show the very large number of metastable states (see Section 6.3).

From Brandon's coincidence model [70], the structure of a CSL grain boundary depends on five degrees of freedom. Three define the angle and axis of misorientation or rotation angle and two define the plane of the boundary. Deviations from the angular misorientation required for exact coincidence can be described by a sub-grain boundary network of dislocations. The sub-grain boundary dislocations have partial Burgers vectors in the coincidence lattice. Generally, the axis of misorientation of the sub-grain boundary will lie at some arbitrary angle to the axis of misorientation chosen to describe the coincidence lattice. The density of dislocations introduced into a CSL grain boundary is limited by the density of coincidence site lattice at the boundary as shown in Fig. 1.8. Since the density of coincident sites decreases with increasing Σ, the maximum permissible density of boundary dislocations must also decrease with Σ.

Iwata et al. [86] experimentally confirmed effects of several grain boundaries on dislocation generation using artificially designed Si seeds. Dislocations frequently generate from a non-perfect $\Sigma 3$ grain boundary with misorientation (3.36−3.44°) or small angle deviation, but dislocations are hardly observed around a perfect $\Sigma 3$ grain boundary. The misorientation means the amount of relative angular deviation from the ideal grain boundary structure at seed joints, measured by the X-ray rocking curve (XRC). $\Sigma 5$ and $\Sigma 25$ grain boundaries with high grain boundary energy are not sensitive to misorientation. Dislocations are observed around $\Sigma 5$ grain boundaries with misorientation (1.15−1.65°), but dislocations are hardly observed around a $\Sigma 25$ grain boundary with misorientation (2.14°). The $\Sigma 3$ grain boundary energy is much smaller than the $\Sigma 5$ and $\Sigma 25$ grain boundary energies, but the effect of misorientation on the grain boundary energy is largest for the $\Sigma 3$ grain boundary, because the non-perfect $\Sigma 3$ grain boundary with a small angle deviation is close to the energy minimum and can reduce the grain boundary energy to generate dislocations as shown by the green point in Fig. 1.9. This situation about the non-perfect $\Sigma 3$ grain boundary is different from the $\Sigma 25$ grain boundary which is categorized as large-angle grain boundary. The $\Sigma 5$ grain boundary is more sensitive to misorientation comparing to the $\Sigma 25$ grain boundary.

1.3.3.3 Random grain boundaries

A grain boundary with a larger Σ value than 27 is categorized in a random grain boundary which belongs to the large angle grain boundary. A random grain boundary is neither symmetric nor straight, which has many zig-zag segments along the random grain boundary observed by transmission electron microscopy (TEM) [87]. Random grain boundaries have separating crystallites without common lattice points [88]. The evolution of random grain boundaries and all other non-twin grain boundaries is mainly influenced by the shape of the growing interface and the surface energy of the adjacent grains slightly influences the growth angle of the grain boundaries [89]. Generally, the fraction of random grain boundary increases and that of $\Sigma 3$ grain boundary decreases with the ingot height (see Section 5.4.2).

Generally, grain boundaries can be source of dislocations depending on the shape or smoothness [90]. Random grain boundaries together with local high stress possibly act as origins of dislocations in Si multicrystalline ingots as shown in Fig. 6.4. Thus, one of origins of dislocation formation in Si multicrystalline ingots may be random grain boundaries formed between grains with random orientations because of their unique grain boundary structure such as many zig-zag segments with local high stress and the decrease of grain boundary energy due to introduction of dislocations. However, this point should be made clear by the further works because random grain boundaries are basically categorized as large-angle grain boundaries. On the other hand, random and high Σ grain boundaries can act as obstacle of dislocation moving one grain to another and can block defect propagation, while low Σ grain boundaries such as $\Sigma 3$ have no blocking effect and dislocation can easily pass through $\Sigma 3$ grain boundaries [90]. The random grain boundaries have strong effects to prevent movement of dislocations and dislocation clusters because of their free surface characteristics (see Section 6.4.4).

The electron beam induced current (EBIC) method can measure the recombination strength of grain boundaries. An electron beam locally induced charge carriers in the sample and a small current is measured at the ohmic contacts in dependence on the recombination strength of crystal defects. Although Σ3 grain boundaries are originally electrically inactive and cannot be detected by EBIC at all, the random grain boundaries can be detected by EBIC [91,92]. Random grain boundaries generally have higher electrical activity than CSL grain boundaries [88]. Thus, the random grain boundaries are electrically active because they have more recombination centers caused by more crystal defects, higher impurity levels and larger grain boundary energy.

1.3.3.4 Twins

Twin boundary is defined as a grain boundary with a reflection symmetry or same grain boundary planes. Fig. 1.10 shows the category of grain boundaries by rotation axis. Grain boundaries are categorized as tilt, twist and mixed grain boundaries. The tilt grain boundary has the rotation axis parallel to the grain boundary plane. The tilt grain boundaries are categorized as the symmetric (same rotation angles: mirror symmetry) and asymmetric (different rotation angles) tilt boundaries. 25 types of Σ3 grain boundaries with different boundary planes are reported including symmetric and asymmetric tilt grain boundaries [93]. The dislocation density of an asymmetric tilt boundary is higher than that of a symmetric tilt boundary [73]. The twist grain boundary has the rotation axis perpendicular

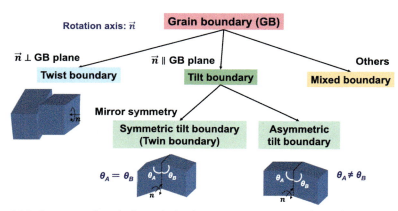

Fig. 1.10 Category of grain boundaries by rotation axis. Grain boundaries are categorized as tilt, twist and mixed grain boundaries. The tilt grain boundary has the rotation axis parallel to the grain boundary plane. The tilt grain boundaries are categorized as the symmetric (same rotation angles) and asymmetric (different rotation angles) tilt boundaries. The tilt grain boundaries are categorized as the symmetric (same rotation angles) and asymmetric (different rotation angles) tilt boundaries. Only symmetric tilt boundaries are twin boundaries. *(Referred from Discussion with K. Kutsukake, 2017.)*

to the grain boundary plane. Only mirror symmetric tilt boundaries are twin boundaries in the strict sense as shown in Table 1.1. As the <111> axis has 3-fold symmetry, the twist processes with a rotation axis of <111> and a rotation angle of ±120°, ±240° and ±360° (±0°) make only the same folded structures, and a rotation angle of ±60° and ±180° makes mirror reflective structures. For example, as shown in Fig. 1.11, the {111} twist grain boundary with a rotation axis of <111> and a rotation angle of ±180° or ± 60° is equivalent with the {111} tilt grain boundary with a rotation axis of <110> and a rotation angle of 70.53°. The {111} twin grain boundaries described by a tilt process can equally well be described as a twist grain boundary [73]. Fig. 1.12 shows a simple tilt process to make an {111} Σ3 grain boundary by contacting the two crystals with the {111} face 1 and the {111} face 2. As shown in Fig. 1.12, the 70.53° can be obtained by 35.27°× 2 because the angle between the <100> axis orthogonally intersected to the <110> rotation axis and the {111} plane parallel to the <110> axis is 35.27°. The tilt and twist operations for the twining are different from the geometric or mathematical point of view even though both twins have the exact same atomic arrangement [79]. The different tilt and twist processes have possibility to form different crystal defects near each twin boundary even though the final twin structures are same [73,79].

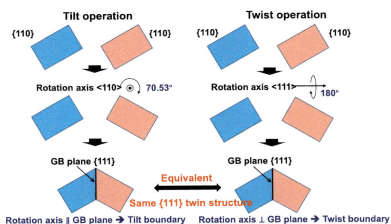

Fig. 1.11 Equivalent twin grain boundaries using the tilt and twist processes. The twist process to prepare the {111} twist grain boundary has a rotation axis of <111> and a rotation angle of ±180°. The tilt process to prepare the {111} tilt grain boundary has a rotation axis of <110> and a rotation angle of 70.53°. *(Referred from Discussion with K. Kutsukake, 2017.)*

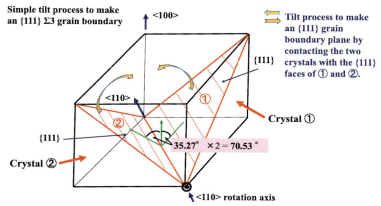

Fig. 1.12 Simple tilt process to make an {111} Σ3 grain boundary by contacting the two crystals with the {111} face 1 and the {111} face 2. The 70.53° can be obtained by 35.27°×2 because the angle between the <100> axis orthogonally intersected to the <110> rotation axis, ⊙ and the {111} plane parallel to the <110> axis is 35.27°.

For example, the {111} twist grain boundary with a rotation axis of <111> and a rotation angle of 60° is expected as a cross-grid of screw dislocations, and the {111} tilt grain boundary with a rotation axis of <110> and a rotation angle of 70.53° is expected as an array of edge dislocations parallel to <110> [73]. This point should be accurately confirmed by some experiments.

As a particular case of fivefold twin centers in the diamond structure, three convergent {111} Σ3 twin boundaries (t_1, t_2, t_3) interact to form an unusual fivefold twin center which includes one single atomic column and a {111}/{110} Σ27 twin boundary [94]. t_1 and t_3 symmetrically orient at 70.53° to t_2. In this case, the structure looks like following the reaction of Σ3 = Σ3 + Σ3 + Σ27 or Σ3 + Σ3 = Σ3 + Σ27 or Σ3 + Σ3 + Σ3 = Σ27.

The twin boundary is generally involved in the CSL grain boundary with several coincidence parameters such as $\Sigma = 3^n = 3, 9, 27, 81-$ [95]. Successive twinning during growth by rotation around different axes creates Σ9, Σ27 and random grain boundaries [96]. The Σ3 grain boundary has the lowest order of the coincidence parameter and the highest coherency. The twin boundaries are easily formed on the closed packed {111} plane. There are Σ3 grain boundaries with different boundary planes such as {111}, {112} and {110} and the {111} Σ3 grain boundary has the lowest energy minimum as shown in Fig. 1.9. These Σ3 grain boundaries are common twin boundaries. However, even for the Σ3 grain boundaries, there are non-twin boundaries which have the rotation axes perpendicular to the grain boundary planes and are formed by the twist processes as shown in Table 1.2.

As shown in Fig. 1.6, the $\Sigma 3$ grain boundary shown by red lines are dominant for the Si multicrystalline ingot grown in a quartz crucible. In the typical multicrystalline ingot, more than 50 % of total grain boundaries are $\Sigma 3$ grain boundaries especially in the lower part of the ingot as shown in Fig. 5.6. Voigt et al. [95] reported the formation of twins of high order and random grain boundaries from twins with low order. The $\Sigma 3$ twins are easily built during growth and multiple twinning leads to the formation of twins of higher order. Even a number of random grain boundaries are produced by multiple twinning in combination with dislocation networks in grain boundaries. For $\Sigma 3$ and $\Sigma 9$ grain boundaries, there exist grain boundary planes associated with a low grain boundary energy. The energy of the Si {111} twin boundary is about $0.051-0.058 \, \mathrm{J \, m^{-2}}$ [4]. Nucleation of twins is thermodynamically unfavorable since it increases the total system energy by adding new grain boundaries. However, the formation of twin boundary can release stress and the twinning has tendency to relax stress developed in a crystal during growth [82,90] (see Section 1.5). When dislocations cross a twin boundary, the crossed plane is a {211} twin boundary plane [97].

1.3.3.5 Twin formation

Stokkan [96] studied origins of twin boundaries denoted as only $\Sigma 3$ grain boundaries in detail. Two origins of twin boundaries are a tri-junction (TJ) between three grain boundaries and a $\Sigma 9$ grain boundary. The former one occurs as a reaction to change grain boundary configuration. The latter one is its straight grain boundary segments of alternating direction caused by reduction of total Gibbs free energy at the $\Sigma 9$ grain boundary. The $\Sigma 9$ grain boundary can easily reduce its energy by dissociating and forming triangles as low energy secondary grains that contain two {111} twins and one {211} twin, and the $\Sigma 9$ grain boundary contains a high density of such twinned triangles. The orientation of the grain boundary essentially configures toward low energy in the CSL relationship. A sudden small increase in the grain boundary energies is compensated and reduced by the small increase in total Gibbs free energy caused by formation of twin boundaries. Thus, the introduction of twin boundaries can slightly reduce total grain boundary energies.

For facet-facet grooves, some grain boundaries follow the bisector rule, in which a grain boundary appears by following the bisector of the angle at the

SSL–TJ between facets of the groove [48,98]. In this case, both facets on the inner surface of the groove have a same growth rate because the same supercooling occurs at the bottom and on the facets of the groove [48,98]. The grain boundary direction of a non $\Sigma3$ can also follow the bisector angle of the two facets if the growth rates of the two facets are the same in a facet-facet groove [99]. However, even for facet-facet grooves, when the two facets have different growth velocities, the bisector rule is not satisfied. When a groove is formed from a random grain boundary between two Si (100) and (110) sheet crystals, the two facets on its inner surface have different growth rates and the bisector rule of the two grains is not satisfied [99].

For rough-facet grooves, grain boundaries do not follow the bisector rule [48]. Fujiwara et al. [100] directly observed formation of the {111} $\Sigma3$ twin boundary. At first, a new grain with a random grain boundary is formed along an {111} inner face of a groove from the SSL–TJ, and the {111} inner face is changed to a new $\Sigma3$ grain boundary plane by the growth of the new grain on the inner face. The random grain boundary leads to the formation of a new groove as shown in Fig. 1.3. The groove deepens during growth and finally a melt inside the groove crystallizes. Finally, the formation of the new grain with the new {111} $\Sigma3$ boundary is completed on the {111} inner facet of the deep groove. Such crystallization of the melt inside the groove and the formation of the new grain easily occur because the supercooling of the melt inside the groove is larger than at the planar interface and the formation energy of a {111} $\Sigma3$ boundary is quite small. Thus, such a deep groove acts as a trigger for the formation of the {111} $\Sigma3$ twin boundary based on the reaction as $R = \Sigma3 + R$ as shown in Fig. 1.3. At the SSL–TJ, the formation energy of twins is low and the twin nucleation easily occurs at a much lower supercooling. Such twin formation of Si at the SSL–TJ was explained by a calculation model [101]. As shown in Fig. 1.3, the facet-facet groove can be changed to the rough-facet groove and the depth of the groove is reduced. Such transformation occurs due to the rapid growth by increased supercooling as the depth of the groove increases during growth [99]. Lin and Lan [50] studied the bisector rule for twin growth from grooves in detail. The twinning is most likely formed from a non-Σ grain boundary groove such as a random grain boundary groove. However, there are less probable for twinning to occur from the $\Sigma3$ grain boundary grooves because new grains are difficult to be formed on the facets due to their much smaller supercooling in the shallower facet-facet grooves as shown in Fig. 1.3. As the growth rate becomes higher, grain boundaries form in the direction based on the bisector rule because a smaller facet becomes larger by a larger supercooling at SSL-TJ. Twins as symmetric tilt grain boundaries such as some $\Sigma 3$, $\Sigma 9$ and $\Sigma 27$ grain boundaries also follow the bisector rule. However, most of $\Sigma 3$ grain boundaries are difficult to follow the bisector rule for the facet-facet groove because a familiar $\Sigma 3$ grain boundary has originally a common stable {111} facet and no new grain grows on/along the {111} facet of the groove [82,100].

Stamelou et al. [49] studied the twin nucleation in detail. Twin nucleation takes place at the {111} facets inside a groove formed by a grain boundary or for the {111} facets situated at edges of samples because the supercooling is higher in the groove and at the edges comparing with that on the solid-liquid interface. Inside the groove with a grain boundary and at the edge, the supercooling is always lower than 1 K. The estimated values are 0.2−8 K [48,50,82,102]. The experimental supercooling at the edges is slightly higher than inside the groove and a higher nucleation probability of twins is expected at the edges. On the other hand, a high supercooling is required when twinning occurs in the middle of a facet. Thus, twin nucleation is commonly observed at the SSL-TJ and at edges or borders of an ingot where the Si melt is in contact with the crucible wall [82]. The twinning probability can be calculated as a function of the supercooling [50,102].

When an <110> oriented seed is used for the ingot growth, dislocations originating from the seed propagate upward vertically being aligned along the <110> growth direction. Sometimes, their propagation is efficiently blocked by formation of a horizontally arranged successive multi-twin boundaries in a $\Sigma 3$ <111> (the rotation axis: <111>) relation between them. All the horizontal and diagonal twin boundaries initiating at the edge of the sample are a $\Sigma 3$ <111> type. The growth dislocations after the intersection with the $\Sigma 3$ <111>/{111} (the grain boundary plane:{111}) boundary manage to cross-slip on the two inclined {111}. The twin boundary laterally grows along the solid-liquid interface with macro steps from an edge of a crystal toward the center of the sample with a high supercooling and such twin growth is repeated to form the successive multi-twin boundaries when the supercooling at the edge is again high enough to form new two dimensional twin nucleus [82].

The twin formation is strongly related to an abrupt change of the growth rate during growth. The twin boundaries always appear just when the local growth rate is markedly increased and rarely appear at a constant growth rate because the thermal stress becomes little bit larger at the moment [82,103]. For the twin formation, the significant increase of the growth rate is more effective than the high growth rate. Twins can also appear on facet planes by the two-dimensional nucleation without using SSL-TJ. As shown in Section 4.2, the parallel twins in dendrite crystals are shown to follow this mechanism of the two-dimensional nucleation and the following step growth. A numerical mode of twinning is proposed for columnar grain structure [104]. The formation mechanism of the parallel twins in a dendrite crystal is also explained.

Möller reported the relation between C precipitates and formation of twins [105]. In a Si mulricrystalline ingot, C can precipitate in high concentration together with the formation of twin boundaries. The formation of a twin is induced by the formation of a single C layer on a strained (111) twin plane with a SiC structure. The planar precipitation occurs by fast diffusion of C in the melt toward the growing interface and accumulation of C atoms at (111) steps of the facetted interface. Once a layered SiC structure has formed at some nucleus, the next layer on top of the (111) plane can be either in the regular or twinned orientation, which depends on their energetic difference and the crystallographic orientation of the interface. A numerical mode of heterogeneous twinning is proposed by extending the previous model [101,106].

1.3.3.6 Coherent grain boundaries

The coherency is an important factor to characterize grain boundaries. The most coherent one is a symmetric $\Sigma 3$ grain boundary and the most incoherent one is called as a random grain boundary. The $\Sigma 3$ grain boundary with a tilt rotation is a kind of twin boundaries, which have a reflective mirror symmetry between the configured grains such as an {111} $\Sigma 3$ grain boundary. The {111} $\Sigma 3$ grain boundary is a coherent grain boundary with the extremely low interfacial energy [77]. It has the lowest energy at the {111} plane and usually has {111} grain boundary planes [107]. According to the strict definition of Randle [67], the coherent twin has a $\Sigma 3$ misorientation and {111} grain boundary planes and it is immobile. The incoherent twin has a $\Sigma 3$ misorientation and grain boundary planes other than {111} such as {112} which are tilt and twist types on rational planes and it is very mobile. The coherent twin has very few crystal defects as shown by the atomic arrangement near a $\Sigma 3$ grain boundary in Fig. 1.6. $\Sigma 3$ grain boundaries are typically electrically inactive. Twin boundaries are often associated with low dislocation density as shown in Figs. 4.14 and 4.15 [108].

The incoherent $\Sigma 27$ <110> grain boundary is more distorted in comparison to the incoherent $\Sigma 9$ <110> grain boundary and the coherent $\Sigma 3$ <111> grain boundary. Areas in which $\Sigma 27$ <110> grain boundaries are presented are more distorted and dislocations are emitted at the encounter of a $\Sigma 27$ <110> grain boundary with either a $\Sigma 3$ <111> grain boundary or a $\Sigma 9$ <110> grain boundary [82]. Thus, the types of grain boundaries such as coherent or incoherent, low Σ or high Σ, symmetry or asymmetry have an impact on the distortion of the formed grain boundary and on the emission of dislocations [82].

1.3.4 Dislocations

1.3.4.1 Generation of dislocations during growth

Dislocations are lattice defects in Si crystals, which are formed and multiplied during crystal growth and cooling by plastic deformation to release residual thermal stress in the crystal. Dislocations are mainly generated at grain boundaries appeared at the growing interface during crystal growth. The residual strain is highest in the peripheral areas of an ingot like top, bottom and central side walls, while the ingot center as well as the ingot corners shows low residual strain, resulting in the same distribution of dislocations [109]. Klapper [110] categorized dislocations as growth dislocations and post-growth dislocations. The growth dislocations or grown-in dislocations connect with the growth front and proceed with it during growth. They tend to minimize their elastic strain energy by aligning and propagating perpendicular to the growth interface [110,111]. Thus, the growth dislocations usually develop the as-grown geometry with preferred minimum-energy direction during keeping their straight lines. Inclusions with accumulated stress around them generate dislocations, especially large inclusions (>50 µm) usually emit bundles of dislocations. The post-growth dislocations are generated behind the growth front during growth or during cooling. The post-growth movement of dislocations is induced by thermal stress (dislocation glide) or by the absorption of interstitials and vacancies (dislocation climb). The as-grown geometry can be destroyed by thermal stress during cooling to room temperature.

For a multicrystalline ingot, dislocations are mainly introduced by the thermal and mechanical stress caused by the non-uniform temperature distribution in the ingot during growth and the expansion force generated between the ingot and the crucible wall (a Si melt expands by 11% during solidification). The dislocation density induced by the thermal stress during the cooling process of crystal growth is in the order of $10^3 \, \text{cm}^{-2}$ [109].

The dislocations present in the ingot are grown-in dislocations. Once dislocations are nucleated, they extend over long distances in the growth direction and locally remain. The dislocation density and distribution are not significantly changed before and after the thermal treatment such as annealing, but slightly increase by a tensile stress [112]. The dislocation recombination or annihilation due to thermally activated cross glide and climb processes are not easy to occur due to the high Peierls barrier of Si [112]. Two dislocations can annihilate only when they have counter sighs for their Burgers vectors and exist within the same slip plane.

Especially, structure-related dislocations appear in Si multicrystalline ingots, which are based on coincident structures. A small angle grain boundary or sub-grain boundary with a small angle deviation and a non-perfect $\Sigma 3$ grain boundary with a small angle deviation can generate such dislocations to minimize its energy as source of them. Dislocation are emitted from some locally stressed parts of the grain boundary, specifically kinked areas in the boundary plane and emitted dislocations appear in arrays in slip planes [74]. The emitted dislocations interact with dislocations emitted from other parts of the grain boundary, with other defects such as stacking faults and with growth dislocations, resulting in dislocation pile-up near the sub-grain boundary [74]. The structure-related dislocations are especially important for the mono-like cast method which uses several single-crystal seed plates with unavoidable small misorientations between adjacent seed plates.

One of the most frequent dislocation types in a Si crystal is the $60°$ dislocation which is electrically active [113]. The $60°$ dislocation has a burgers vectors of b = 1/2 <110> a = 0.37 nm (a: lattice constant of Si crystal = 0.54 nm, <110> = $\sqrt{2}$) and gliding of the dislocation is parallel to it [111]. Dislocations can easily move and multiply on the preferred slip systemby thermal stress. Dislocations usually glide in {111} planes under stresses during growth and cooling because these planes are the common slip planes in Si crystal. Moving $60°$ dislocations generate high density of extended defects in their slip plane, which are terminated on dislocation lines and stretch along all path of dislocation [113]. These defects are agglomerates of intrinsic point defects and their density is much higher than that of dislocations [113]. The most preferred slip system of dislocations on fcc-type crystals such as a diamond structure is of the type{111} <110> which means {111} slip or glide planes with 1/2 <110> Burgers vectors or <110> slip direction [97,111]. The dislocation axis is expected to follow low energy orientation, which are <110> for perfect pure screw and $60°$ dislocations.

Near (111) wafer surface, the dislocation lying on (1-1-1) slip planes are arranged in arrays which are parallel to [2−1-1] direction and normal to [0-11] direction [114]. The {111} slip planes incline to the surface in mainly <112> 30° orientation [114]. For the (111) system, twelve slip systems are available for dislocation in Si. There are four {111} slip planes such as (-1-11), (1-11), (-111) and (111) with three <110> slip directions each. Four equivalent slip planes or slip systems exist for the (100) system and three equivalent slip planes or slip systems exist for the (111) system [115]. Dislocations are much free to move in the (111) system than in the (100) system because two slip systems are simultaneously activated in the (100) system and cause dislocations to intercept each other, but such interception of dislocations does not appear in the (111) system [115]. The dislocation density and the stress exhibit the lowest values in <110> directions because these directions are the slip directions in Si and the stress is easily released by motion of dislocations [109,116]. The dislocation generation in <110> directions is not much affected by the thermal process such as cooling or annealing comparing with the <100> and <112> directions, and the <111> directions can reduce the dislocation density by 50% comparing with the <100> and <112> directions [116]. For two adjacent grains, if several <110> directions are common and the same slip systems are activated during growth, dislocations can move from one grain into another. The {111} grains, in which the growth direction is closed to <111>, tend to form dislocations due to the higher number of potential slip planes. More activated slip systems result in higher dislocation densities [97]. Dislocation clusters generated during growth have orientations determined by the slip systems [117]. Dislocations originally aligning along {111} slip planes rearrange in sub-grain boundaries to lower the crystal energy [118].

A dislocation near a free surface is attracted to the surface due to image force. During crystal growth, a dislocation line close to the solid-liquid interface minimize its energy by glide and cross-slip and penetrate the interface at an angle of near 90° toward the surface to shorten the length of the dislocation line [117]. The exact angle depends on the angle between the interface and the glide plane. The solid-liquid interface is faceted with many {111} facets of the order of 1 mm in size [23]. The angle between the {111} facet on the interface and the other {111} facet planes is always 70.53°. Thus, cross-slip between the {111} planes do not enable a dislocation to penetrate the interface at angle of 90° without stress. During an annealing process of a grown ingot, dislocation segments of opposite sigh annihilate to minimize energy and remaining dislocations toward low energy configurations of dislocation structure such as aligning perpendicular to the glide plane, which have undergone recovery, i.e., movement of dislocations and sub-grain boundaries in order to minimize the stored energy [117].

Ervik et al. [119] studied the effect of high-temperature annealing for Si multicrystals with high density of dislocations. When much large amount of dislocations is produced by strong deformation such as bending, dislocations tend to reduce the internal energy by rearranging into low-energy configurations during annealing. High temperatures are needed to initiate such recovery processes, but the high-temperature annealing does not reduce dislocation density because dislocations are already in low-energy configurations. Stress can enhance the dislocation density reduction when it is applied at high temperature. In a Si multicrystal, applied shear stress varies from grain to grain, and grain boundaries act as strong barriers to dislocation motion. After annealing, polygonised structures appear even in Si multicrystals, where dislocation etch-pits are aligned in arrays perpendicular to {111} planes.

1.3.4.2 Effects of shear stress on formation of dislocations

Essentially, Si has a decisive advantage in its resistance to thermal stress-induced slipping because Si exhibits a small tendency to slip during growth [120]. Dislocation emissions can occur not only thermal stress but also shear stress irregularities introduced by grain boundary topological imperfections [118]. The shear stress on {111} slip planes strongly depends on the multicrystalline structure such as the grain orientation and the grain-boundary character due to the anisotropy of elastic coefficient of Si [121]. The resolved shear stress developing in the {111} slip planes, τ can be obtained by the applied load, σ and the Schmid factor, m as follows,

$$\tau = m\sigma, \tag{1.33}$$

where m is expressed as follows,

$$m = \cos\varphi \, \cos\phi, \tag{1.34}$$

where φ is the angle between the applied load direction and the direction normal to the slip plane such as <111>, and ϕ is the angle between the applied load direction and the slip direction such as <110>. Dislocations are expected to be first generated in a strained grain with the higher maximum Schmid factor because τ is the highest in it [118,122].

Inhomogeneous distribution of dislocations in different grains is related to the crystal orientation of each individual grains and in-plane stress is mainly applied [90]. Dislocations generate on only one side of the grain boundary as shown in the right lower figure in Fig. 1.13. This reason is made clear by the following analysis of stress distribution in a Si ingot which depends on the combination of grain orientation on both sides of grain boundaries. The distribution of shear stress near artificially manipulated grain boundaries can be calculated by three-dimensional finite element analysis using a bi-crystal model as shown in Fig. 1.13, because shear stress around a grain boundary strongly depends on structure of the grain boundary. The cylindrical sample used has two grains and a grain boundary. The growth direction is the z-direction which is perpendicular to the grain surface, and the grain boundary plane is perpendicular to the x-direction shown in Fig. 1.13 [121–123]. The diameter and height of the sample are 10 and 1 cm, respectively. A Si ingot grown inside a silica crucible has large compressive stress because Si expands by 11% in volume during crystallization. As a boundary condition, the displacement of 0.0001–0.01% toward the sample center is given to all the nodes at the sample edge, and it corresponds to external compressive forces caused by expansion of Si during solidification [121–123]. The local shear stress around the grain boundary depends on the relative orientations between the two grains owing to the anisotropic elastic constants in Si crystal [124]. The reverse rotation angles

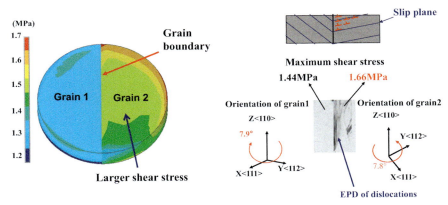

Fig. 1.13 Distribution of shear stress near artificially manipulated grain boundaries calculated by three-dimensional finite element method. The cylindrical sample used has two grains and its growth direction is the z-direction. *(Referred from Fig. 6 in I. Takahashi et al. J. Cryst. Growth* **312** *(2010) 897.)*

of grain 1 and grain 2 are 7.9 and 7.8 in the axis of z <110>, respectively. The angle rotated around the z-axis can partially determine the character of the grain boundary.

The grain boundary plane is assumed to be flat and the bi-crystal sample with two grains is treated as continuous elastic material even through 5 parameters are needed to describe the geometry of a grain boundary [125]. Therefore, the characters of the grain orientation and grain boundary are expressed by the discontinuous elastic modulus [123]. In a Si crystal, there are 12 equivalent types of the {111} slip plane in the <110> direction. The shear stress on {111} face in <110> direction is calculated to be 1.35−1.73 MPa, which depends on the rotation angle [121]. The shear stress also depends on the growth direction. Among the three growth directions of <100>, <110> and <111>, the shear stress is smaller when the growth directions are <100> and <110> [121]. The shear stress is smallest in <110> directions because they are the slip directions and the stress is easily released by motion of dislocations [109]. This result is preferable for the dendritic cast growth (see Section 4.2). The maximum shear stress around the grain boundary is adopted among the 12 types originated from 8 equivalent {111} slip planes and 3 equivalent <110> orientations as shown in Fig. 1.13 [121,123]. The grain 2 has larger shear stress comparing to the grain 1. The grain 2 with larger shear stress corresponds to the grain in which many dislocations generate as shown in Fig. 6.5. Thus, the grain with larger shear stress corresponds to the grain with generated dislocations [126]. The shear stress concentrates near the grain boundary. Unilateral emission of dislocations from a central grain boundary can be also observed for bi-crystal Si ingots separated by $\Sigma 9$ and $\Sigma 27$ grain boundaries [118]. Such semi-coherent grain boundaries tend to generate dislocations as a result of their re-arrangement using straight segments, which is often observed for the mono-like cast growth (see Section 6.4). Thus, the reason why dislocations appear on only one side of the grain boundaries can be known by this calculation based on the anisotropic mechanical properties of Si crystals.

1.3.5 Other crystal defects

Kivambe et al. [127] studied stacking faults in Si multicrystals. Stacking faults lie on {111} planes, appear as parallel straight lines to originate from a small angle grain boundary and end inside a grain or in a grain boundary. Stacking faults form strong barrier to movement of dislocations and formation of sub-grain boundaries. Dislocations are frequently observed to emerge from their edges of stacking faults. Growth of stacking faults is associated with generation of dislocations. Stacking faults play an important role in formation of the dislocation microstructure in multicrystalline Si crystals.

Stacking faults can be formed in a Si crystal by condensation of vacancies near the growing interface. The equilibrium concentration of vacancies near the interface are on the order of 1×10^{15} cm^{-3} at the melting point [78]. On the view point of the atomic arrangement of stacking faults, although the introduction of stacking faults does not alter the nearest- and second-neighbor coordination of the individual atoms from that of a prefect crystal, it does modify the number of third neighbors [4].

1.4 Impurities and their activities in defects

1.4.1 O, C and N atomic impurities

O is one of main impurities in a Si crystal for solar cells. Main origin of O contamination is silica crucible and O is dissolved in a Si melt by the reaction between the crucible and the melt. O atoms exist in a Si ingot as interstitial oxygen (O$_i$) atoms to form covalent bonds which are electrically neutral. O$_i$ atoms can be measured by the Fourier transform infrared spectrometry (FTIR) analysis. The O$_i$ concentration is determined from an intensity of absorption peak at 607 cm^{-1} with a conversion factor of 8.2×10^{16} cm^{-2} [128,129]. The equilibrium segregation coefficient of O in a Si ingot solidified from a Si melt (amount of the impurity in Si crystal/amount of the impurity in Si melt) is 1.25 or 1.4 [130–132]. The solubility limit of O in a Si melt is 2.2×10^{18} cm^{-3} at the melting point of Si (Mp) [131]. The O concentration in an ingot grown by the cast method is measured as 1.3×10^{17} cm^{-3} in the bottom part, 1.0×10^{17} cm^{-3} in the middle part and 0.5×10^{17} cm^{-3} in the upper part [129]. It typically decreases from bottom to top of the Si ingot because the segregation coefficient of O in Si is 1.25 which is larger than 1.0. The average of the O concentration is 0.9×10^{17} cm^{-3}. The dissolution rate of O from a Si$_3$N$_4$ coated quartz crucible is estimated to be smaller than 3.5×10^{12} atoms cm^{-2} s^{-1} and that from an uncoated quartz crucible is 1×10^{15} atoms cm^{-2} s^{-1} [129,133]. The diffusion coefficient of O in Si is $D_0 = 0.17\exp(-2.54\text{eV}/k_B T)cm^2\ s^{-1}$ over a wide temperature range, where k_B is Boltzmann constant (1.380×10^{-23} J K^{-1} or 8.617×10^{-5} eV K^{-1}) [134]. The diffusion coefficient of O in SiO$_2$ glasses at $1000\,^\circ$C is from 3×10^{-14} cm^2 s^{-1} to 3×10^{-15} cm^2 s^{-1} and about 10^{-18} cm^2 s^{-1} in quartz [135]. The O concentration in a Si ingot increases with the dissolution rate. The evaporation coefficient of O is estimated as 3.9×10^{-5} cm s^{-1} [129].

The evaporated rate of O can be obtained as a product of the evaporation coefficient and the O concentration in a Si melt. The O concentration in a Si ingot decreases with the evaporation coefficient (see Section 3.3.4).

Carbon (C) is incorporated into a Si melt through the vapor transport of CO and CO_2. C atoms exist in a Si ingot as substitutional carbon (C_s) which can be also measured by the FTIR analysis. C is distributed homogeneously in a cast-grown Si multicrystalline ingot as isolated C_s [136]. The C_s concentration is determined from an intensity of absorption peak at 1106 cm^{-1} with a conversion factor of 3.14×10^{17} cm^{-2} [129,137]. The C_s-oxygen dimer (O_{2i}) pair is generated by the reaction $C_iO_i + VO = C_sO_{2i}$, where VO is vacancy (V) trapped by O [138]. The interstitial carbon (C_i) is mobile and a fast diffuser, which is electrically active [138]. The equilibrium segregation coefficient and solubility limit of C in a Si melt is 0.07 and about 4×10^{18}–5×10^{18} cm^{-3} at Mp [131,132,138–140], respectively. The reported solubility limits of C in a Si melt are listed in Ref. [141]. The solubility limit of C in a Si solid is 3.5×10^{17} cm^{-3} at the melting point of Si [105,132]. The C_s concentration increases with the ingot height because of the small segregation coefficient of 0.07 [140,142], but it is largely varied by convection in the melt [142]. Carbon monoxide (CO) gas in the atmosphere dissolves into a Si melt through the melt surface with the incorporation rate of C, which is estimated as 1.1×10^{14} atoms cm^{-2} s^{-1} [129]. The C concentration in a Si ingot increases with the incorporation rate. The incorporation rate of C is affected by the CO concentration above the melt surface and the diffusivity of CO in Ar gas, which is inversely proportional to the pressure [143]. C diffuses substitutionary, and the diffusion coefficient of C_s in Si is $D_C = 0.33$ $exp(-2.92 \pm 0.25$ $eV/k_BT)cm^2$ s^{-1} [144]. The diffusion coefficient of C_i in Si is $D_C = 0.44 - 0.86$ $exp(-0.87$ $eV/k_BT)$ cm^2 s^{-1} [138]. O and C can precipitate at dislocations and grain boundaries and change their electrical behaviors (see Section 1.4.2 and 2.2).

The equilibrium segregation coefficient and solubility limit of nitrogen (N) in a Si melt is 0.0007 [132,145] and about 5.7×10^{18} cm^{-3} at Mp [131,140], respectively. The solubility limit of N in a Si solid is 4.5×10^{15} cm^{-3}, which depends on the growth rate [132,140]. N atoms can form several defects by reactions as interstitial N (N_i) and few substitutional N (N_s), and the binding energy of N_i-N_i is 3.86–4.30 eV [146]. The N concentration in a Si crystal can be quantitatively measured by a Photoluminescence (PL) method. In this method, N impurities are transformed into radiative centers responsible for the A-line at 1.1223 eV by forming an isoelectronic Al-N pair using Al ion implantation and the detection limit for N is around 2×10^{13} cm^{-3} [147]. The migration activation energies of the N monomer and dimer are estimated to be 0.4 and 2.5 eV, respectively [138]. The N diffusivity as the form of monomer is determined as $D_N = 0.25 \exp(-1.25 \text{ eV}/k_B T) \text{ } cm^2 \text{ } s^{-1}$ and the N diffusivity as the form of dimer is determined as $D_N = 2700 \exp(-2.8 \text{ eV}/k_B T) \text{ } cm^2 \text{ } s^{-1}$, which is three order of magnitude larger than that of O at 1100 °C [138,148]. For the cast method, the convection is moderate in the Si melt and the N concentration in the ingot is usually determined by the segregation effect. The N concentration slightly increases with the ingot height at the low growth rate of 0.2 cm h^{-1}, but it is largely varied by convection in the melt [140,142]. As the segregation coefficients of C and N are very small, these elements are less incorporated in the solid and enriched in the growing interface. When the enrichment exceeds the solubility limit in the melt, the silicon carbide (SiC) and silicon nitride (Si$_3$N$_4$) precipitates appear in the melt near the growing interface. The thermal conductivities of β-SiC and Si$_3$N$_4$ are 15−70 W m^{-1} K^{-1} at 1,700K and 10−40 W m^{-1} K^{-1}, respectively [149]. For a given particle size, the critical growth rate to capture these particles on the solid-liquid interface is higher for Si$_3$N$_4$ particles than for SiC particles because the critical growth rate is inversely proportional to the thermal conductivities [149]. The critical growth rate becomes higher under convective conditions and becomes lower as the particle size becomes larger [149]. These particles are easily removed from the growing interface as the critical growth rate becomes higher.

The Si$_3$N$_4$ coating materials generally contain a significant amount of O due to an oxidation treatment or silica powder [150]. A non-wettable SiO$_2$ layer exists on Si$_3$N$_4$ particles (see Section 3.4.4). The O can dissolve from the quartz crucible wall through the Si$_3$N$_4$ coating layer to the Si melt during growth, and the O concentration in a Si ingot is higher near the crucible wall which has a 2 6 mm thick boundary layer [151,152]. The diffused O concentration in the ingot rapidly decreases within the distance of the order of 1 mm from the crucible wall, and the O contamination strongly depends on the contact condition between Si and SiO$_2$ [150]. The O concentration gradually decreases with the ingot height except for the initial part grown from a melt with high O content [150]. The O concentration in the Si melt increases as the crucible rotation rate increases and the O concentration in the Si ingot distributes inhomogeneously in the radial direction [153].

These light-element impurities can be origins of crystal defects and incorporated through the atmosphere over the Si melt [154]. The light-element impurities precipitate first, and these precipitates generate the small angle grain boundaries due to their slight misorientation and the strain field around the precipitates [154]. The solubility limits of light-element impurities (C, N, O) in Si melt and in Si crystal are more precisely listed in Appendixes.

1.4.2 O precipitates

During thermal processing, the supersaturated O_i atoms in a Si single ingot can aggregate into O precipitates in a wide temperature range and further induce secondary lattice defects [43,138]. The content of the O precipitates is almost amorphous or crystalline SiO_x ($x = 1-2$), and its maximum size is 1 µm and its volume is 1.25 times larger than the Si atomic volume in the lattice [43,155]. The O precipitates are decorated by metal impurities in the Si single ingot, which grow larger by thermal processing and generates crystal defects [156]. In a Si multicrystalline ingot, the O precipitates are most likely formed in the bottom part of an ingot where the O concentration is higher [157]. The O or SiO_2 precipitates exerts compression on the Si lattice because of its larger volume than Si. The O precipitates can act as preferential nucleation sites for precipitates of harmful metallic impurities, so by controlling the location of the O precipitates, metallic impurities can be confined to inactive regions of wafers as internal gettering [158]. Thus, the O precipitates can act as the gettering sites for harmful metal contaminants [138]. The O precipitates which are locking dislocations in place cause disordered distribution of dislocations [159]. The amount of as-grown O precipitates is lower as the C concentration decreases using the liquinert crucible because C can act as nucleation sites of the O precipitates [160]. Usually, C enhances O precipitation, exhibiting an increase of the precipitate density and simultaneously decreased C concentration in the Si matrix [138]. C-O complexes are proposed to be the heterogeneous nuclei of O precipitates. SiO_2 forms at high temperature ($700-800\,°C$) in a Si ingot, and its thermal dilatation is about five times smaller than that of Si [161]. The formation enthalpy of SiO_2 is $\Delta H = -909\ kJ\ mol^{-1}$ [135] (see Section 3.3.4).

Möller et al. [157] studied O precipitates in detail. Two types of SiO_2 precipitates such as spherical and plate-like ones are observed by TEM. The spherical defects have amorphous structure with sizes between 0.01 and 0.1 µm which are formed at high temperature. The plate-like defects on {111} planes have a lateral extension up to 0.1 µm and a thickness of one or two atomic layers. The transition from plate-like to spherical precipitates is observed at high temperature. No precipitates are observed below the O concentration of 5×10^{17} cm^{-3} because the number of precipitates per unit volume decreases as the initial O concentration decreases for constant annealing time and temperature. The precipitation of O is enhanced in Si multicrystals because of a high density of nucleation sites. Small O clusters and C can enhance the formation of O precipitates because the presence of the clusters and C additionally increases the number of nucleation sites for precipitation. C impurities in MCZ Si crystals which act as heterogeneous nucleation sites for O precipitates can be largely reduces to lower than 1×10^{14} cm^{-3} by preventing back-diffusion of CO [162,163].

O can precipitate at dislocations and grain boundaries and change their mechanical behaviors. In Si single crystals, softening of crystals occurs rapidly in crystals with higher O concentration than 5×10^{17} atoms cm^{-3} on annealing at 1050 °C due to precipitation of supersaturated O atoms [164]. After annealing process, the dislocation density decreases as the O concentration increases because of the effect of O_i atoms on the locking of dislocation multiplication [165]. O atoms in a dislocated crystal effectively lock dislocations and result in the strengthening of the crystal which becomes more effective with the O concentration [164].

Murphy et al. [158] studied the O precipitates in detail. O-related defects can act as recombination centers, including thermal donor defects formed upon low temperature annealing, B-O complexes formed upon illumination, and O precipitates with associated defects such as decorated dislocations and stacking faults. The lifetime of wafers with mostly unstrained O precipitates is extremely high (up to 4.5 ms), but the lifetime falls rapidly as the relative density of strained O precipitates increases by longer growth time or annealing time because such strained precipitates are surrounded by dislocations and stacking faults during annealing for longer time. Recombination at unstrained O precipitates is week, but recombination at strained O precipitates is much stronger, which depends on the density of the precipitates. The recombination activity at strained O precipitates with other extended defects is approximately 3—4 times greater than that at unstrained O precipitates (see Sections 2.4.3 and 2.5.3).

1.4.3 Metallic impurities and precipitates

The top part of a Si ingot grown by the unidirectional solidification cast method contains more impurities than its middle part because the segregation coefficients of these impurities are much less than one. A Si ingot grown using a cast furnace contains a large variety of metallic impurity species such as Fe, nickel (Ni), copper (Cu), cobalt (Co), chromium (Cr), molybdenum (Mo), aluminum (Al) and manganese (Mn). Cr, Cu, Ni, and Fe readily form recombination-active precipitates. A standard crucible is an infinite source of Fe impurities and a Si_3N_4 coating behaves like a finite Fe source [166]. The sintering of the Si_3N_4 coating prior to growth leads to remarkable increase for concentrations of transition metal impurities such as tantal (Ta), zirconium (Zr), neodymium (Nd), Fe, Cr, silver (Ag), scandium (Sc), thorium (Th), cerium (Ce), antimony (Sb) and calcium (Ca) which originally present in the Si_3N_4 coating layer [167]. The source of their excess impurity concentrations is diffusion from silica crucible [167] (see Section 3.3.3).

The equilibrium segregation coefficient of Fe between solid and liquid Si is 8×10^{-6} [43,168,169], and the effective segregation coefficient of Fe is 2×10^{-5} [166]. Si is purified during crystal growth because the segregation coefficient of Fe is quite small. The equilibrium segregation coefficients of main impurities in Si are Cu (4×10^{-4}), tin (Sn (0.016)), Ni (8×10^{-6}), Co (8×10^{-6}), gold (Au (2.5×10^{-5})), zinc (Zn (0.001)), Mn (1×10^{-5}), cadmium (Cd (1×10^{-6})), titanium (Ti (3.6×10^{-6})), Cr (1.1×10^{-5}), lithium (Li (0.01)), Ag (1×10^{-6}), Ge (0.56), B (0.7–0.8), Al (2×10^{-3}), Ga (0.008), In (4×10^{-4}), P (0.35), As (0.3), Sb (0.023), O (0.2–1.4) and C (0.034–0.3) [43,168–169]. Among them, the equilibrium segregation coefficients of dominant impurities and dopants in Si are listed in Table 1.3 [43,130–132,137–140,145,168–169]. The equilibrium segregation coefficients of all impurities and dopants in Si are more precisely listed in Appendixes together with each reference. The heats of segregation for larger solute atoms such as Fe in a matrix of smaller atoms

Table 1.3 Equilibrium segregation coefficients of main impurities in Si.

Impurities	Segregation coefficients	Impurities	Segregation coefficients
O	1.25 [167], 1.4 [131]	B	0.716 [167], 0.8 [34]
C	0.3 [167], 0.07 [131]	P	0.35 [34]
N	0.0007 [129]	Al	$2 \times 10-3$ [34]
Fe	8×10^{-6} [34,167]	Ga	0.008 [34]
Cu	4×10^{-4} [34]	In	$4 \times 10-4$ [168]
Ni	8×10^{-6} [167]	As	0.3 [34], 0.8 [43]
Co	8×10^{-6} [34]	Sb	0.023 [34]
Cr	1.1×10^{-5} [167]	Ge	0.33 [34], 0.56 [43]
Cd	1×10^{-6} [43]	Sn	0.016 [34]

Basic growth and crystallographic quality of Si crystals for solar cells 45

such as Si are larger than the opposite case [170]. Co is prominent in Si_3N_4 coating materials which behaves like a finite Co source [166]. The inductively coupled remote plasma mass spectroscopy (ICP-MS) is very well suited for the rapid analysis of Si crystals, but it cannot be used for the analysis of volatile components such as B or As, which are lost during the sample evaporation step [135]. The detection limits of ICP-MS for Fe and Co are in the orders of 2×10^{13} and 2×10^{12} cm^{-3}, respectively [166]. In addition to these impurities, the applied phosphoryl chloride ($POCl_3$) diffusion delivers O and P which diffuse into Si wafers. A high O flow produces a large amount of Si self-interstitials due to amorphous SiO_2 formation with a volume expansion of 130% [172]. V and O form VO_2 complexes acting as O precipitation centers and gettering sites for Fe [172]. These impurities exist not only as single impurity atoms, but also complexes of atoms. The formation enthalpies, ΔH, of metal oxides are $\Delta H = -1,675$ kJ mol^{-1} for Al_2O_3, $\Delta H = -1,130$ kJ mol^{-1} for Cr_2O_3, $\Delta H = -1,118$ kJ mol^{-1} for Fe_3O_4, $\Delta H = -241$ kJ mol^{-1} for NiO and $\Delta H = -155$ kJ mol^{-1} for CuO [135].

As shown in Section 1.3.3, the concentration of metallic impurities is high at high-Σ grain boundaries and at random grain boundaries, but low at $\Sigma3$ grain boundaries. Precipitation of metallic impurities occurs at dislocations, grain boundaries and other crystal defects. Metallic impurities precipitated at such defects enhance the recombination activities of these defects. Especially, fast diffusing metals such as Fe, Cu, Cr and Ni segregate or precipitate at dislocations and grain boundaries after crystallization [157]. The influence of Fe, Cu and Ni atoms is qualitatively identical, and recombination activity of dislocations remains low both for starting samples and after diffusion of these atoms [173]. Cu and Ni can more easily form silicide precipitations than Fe because of their faster diffusivity [135]. Cu contaminants form precipitates homogeneously and heterogeneously in regions with low and high defect densities, respectively, because they mostly segregate at the defects [174]. Cu preferably precipitates on grain boundaries as hetero nucleation sites [175]. These Cu precipitates increase the recombination activity of defects and degrade the carrier lifetime [174].

The diffusion coefficient, D, of an impurity in Si can be expressed as follows,

$$D(T) = D_0 \exp(-\Delta H_m eV/k_B T) \ cm^2 \ s^{-1}, \qquad (1.35)$$

where D_0 is the pre-exponential factor and ΔH_m is the migration barrier [176]. Parameters describing the diffusivity of some transition metals in intrinsic Si is listed in Ref. [176]. The interstitial copper (Cu_i) is always positively charged, implying a shallow donor level in the upper half of the bandgap, and its effect on the excess carrier recombination rate is negligible [176]. The Cu_i intrinsic diffusion coefficient is determined as $D_{Cu}(T) = (3.0 \pm 0.3) \times 10^{-4} \exp(-0.18 \pm 0.01 \text{ eV}/k_B T) \text{ } cm^2 \text{ } s^{-1}$ in the temperature range of 538–1446 °C [177]. The diffusion coefficient of Cu in Si at 1100 °C is 1×10^{-4} $cm^2 s^{-1}$ [178]. Ni precipitates or segregates in defects such as grain boundaries, creates highly recombination–active precipitates and increases the recombination activity of the defects with annealing temperature [179,180]. Ni diffuses mainly via interstitial sites, but interstitial nickel (Ni_i) is neutral and does not affect the electrical properties and only substitutional nickel (Ni_s) or its complexes can be detected by electrical methods [176]. The diffusivity of Ni_i is measured as $D_{Ni}(T) = (1.69 \pm 0.74) \times 10^{-4} \exp(-0.15 \pm 0.04 \text{eV}/k_B T) cm^2 s^{-1}$ in the temperature range of 665–885 °C [179]. The diffusion coefficient of Ni in Si at 1100 °C is 4×10^{-5} $cm^2 s^{-1}$ [178]. Usually, the effective diffusivity of donors in p-type material is lower than their diffusivity in intrinsic material due to acceptor–donor pairing [177]. The diffusivities of dominant impurities in solidus Si are shown in Fig. 1.14, which are described on the basis of experimental data [171]. The activation energy of Ni with -0.15 ± 0.04 $eV/k_B T$ is comparable with that of Cu_i with -0.18 ± 0.01 $eV/k_B T$ [177,179]. The solubilities of Cu in Si are 5×10^{17} cm^{-3} at 1000 °C [159] and 9×10^{17} cm^{-3} at 1100 °C [178], and the solubility of Ni in Si at 1100°C is 5×10^{17} cm^{-3} [178] The solubilities of dominant impurities in solidus Si are shown in Fig. 1.15, which are thermodynamically estimated using experimental data [171]. Ni_i has both high solid solubility and diffusivity, and dissolved Ni_i easily precipitate during rapid cooling due to the small lattice mismatch between Si and nickel silicide ($NiSi_2$) with cubic lattice symmetry and Ni_i–Si distance of 2.438 Å [176,179]. $NiSi_2$ starts to transform into NiSi at temperature below 800 °C [179].

In combination with dislocations and grain boundaries, O precipitates enhance the precipitation of metals at these defects [157]. Ni is known to precipitate at O precipitates and extended defects [179]. Ni precipitates in grain boundaries for the annealing at low temperature under slow cooling and their recombination activity increases with the annealing temperature [180] (see Sections 2.4.2 and 3.3.3). Fe, Cr and Ti cause a reduction in the diffusion length, but Ni does not reduce the diffusion length significantly, even at the highest concentration used in the feedstock (200 ppm wt) [181].

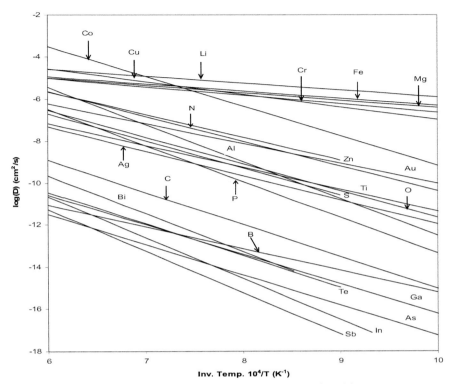

Fig. 1.14 Diffusivities of dominant impurities in solidus Si. *(Referred from Fig. 12.11 in K. Tang et al. in: K. Nakajima and N. Usami (Eds.), Chapter 12, Springer, Berlin Heidelberg, 2009.)*

Smaller amount of Ti (10 ppm wt) can dramatically reduce the efficiency of solar cells along the entire ingot among Fe, Cr and Ni, and a few amount of Ti (0.1 ppm wt) can more strongly affect the solar cell performance comparing with Fe (11 ppm wt), Cr (8 ppm wt) Ni (13 ppm wt) and Cu (8 ppm wt) [181]. Cr has intermediate diffusivity and can occupy both interstitial and substitutional positions in the Si lattice [176].

Transition-metal impurities gettered by grain boundaries act as recombination centers of carriers. Gindner et al. [172] studied the gettering effect on grain boundaries. During P gettering, mobile impurities are effectively removed from crystal defects, but still many of them remain inside the defects or are internally gettered there, which results in highly recombination active. This amount of impurities requires another gettering process such as the hydrogenation and/or hydrogen (H) passivation. As the

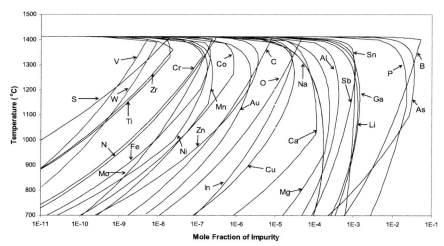

Fig. 1.15 Solubilities of dominant impurities in solidus Si. *(Referred from Fig. 12.1 in K. Tang et al. in: K. Nakajima and N. Usami (Eds.), Chapter 12, Springer, Berlin Heidelberg, 2009.)*

crystal structure of Σ3 grain boundaries is less distorted because of the small tilt angle between neighboring grain orientations, impurities are completely removed out of these grain boundaries after P gettering. The best gettering efficacy occurs for wafers with the lowest as-grown lifetime and the highest number of Σ3 grain boundaries because of a facilitated gettering effect. The initial as-grown lifetime alone is not a good indicator for predicting the gettering efficacy of Si multicrystalline wafers. Their local defect structures should be considered. Lattice defects like dislocations and grain boundaries can enhance impurity diffusion and precipitation of impurities since activation energy for migrating atoms can be lowered in these regions depending on the type of foreign atom [172]. On the other hand, dislocations and grain boundaries also act as internal gettering sites with dangling bonds and vacancies for metallic impurities.

1.4.4 Fe impurities and precipitates

Fe is a dominant and very common metallic impurity in Si crystals and is difficult to eliminate completely on a production line. Fe is mobile at room temperature (RT) and can diffuse quickly above 100 °C [182]. The principle peaks of X-ray fluorescence spectra associated with Fe are at 6.403 keV ($K_{\alpha 1}$) and 7.057 keV ($K_{\beta 1}$) [135]. Fe occupies in an interstitial lattice site of Si crystal and diffuses interstitially [182]. Fe behaves as donor, and Fe atoms are electrically neutral in *n*-type Si [183]. As substitutional iron (Fe$_s$) is unstable in Si, Fe is presented as interstitial iron (Fe$_i$) which is positively charged in p-type Si at RT and readily forms pairs or complexes with negatively charged defects such as shallow acceptors [176,182]. In B-doped p-type Si, the total dissolved Fe concentration is given by a sum of three species such as neutral interstitial iron (Fe$_i^0$), interstitial ionized iron (Fe$_i^+$) and Fe paired with an acceptor such as B (FeB) [135]. At temperatures above 100—200 °C, the pairs dissociate and the Fe$_i^+$ concentration increases [182]. At sufficiently high temperatures above 600 K, Fe becomes mostly neutral in both n-type and p-type Si because the Fermi level moves to the mid-gap position [182]. The level of Fe$_i$ is known to be $E_v + 0.40$ eV, and the levels of Fe$_i^+$ and Fe$_i^0$ are $E_v + 0.36$ eV and $E_v + 0.45$ eV, respectively, where E_v is the valence level in the band structure [182]. The level of Fe$_i$ determined from deep-level transient spectroscopy (DLTS) varies from $E_v + 0.39$ eV to $E_v + 0.45$ eV [181]. Fe$^+$ forms neutral Fe-acceptor pairs with negatively charged substitutional acceptor impurities such as B, Al, Ga and In Ref. [183]. In B-doped Si crystal, the Fe$_i$ significantly reduces the effective minority carrier lifetime and reacts with substitutional boron (B$_s$) forming an Fe-B pair even at RT, which acts as an effective lifetime killer [182—184]. FeB pairs have two energy levels such as a donor at $E_v + 0.09$ eV and an acceptor at $E_v - 0.29$ eV, and the concentration of both defects can be measured by DLTS [176]. At a low excitation level, the FeB effect on the excess carrier lifetime is smaller than that of Fe$_i$ [176]. In Ga-doped Si crystal, Fe$_i$ also significantly reduces the effective minority carrier lifetime [183] (see Sections 2.4.2, 3.3.3 and 8.4).

Fe with a high diffusion coefficient interstitially diffuses into Si wafers at high processing temperatures, resulting in fast contamination of large wafer areas. Fe$_i$ is accumulated at dislocations when a wafer is annealed at 600 °C and Fe$_i$ is diffused uniformly on the wafer when annealed at 800 °C and 1000 °C [185]. Thus, the threshold temperature of trapping Fe$_i$ at dislocations is between 600 °C and 1000 °C (see Section 6.5.1). The diffusion coefficient of Fe$_i$ in Si measured by DLTS is $D_{Fe} = 9.5 \times 10^{-4} \exp(-0.65 \text{ eV}/k_B T)$ cm^2 s^{-1} in the temperature range of 800—1070 °C [184]. It is approximately $D_{Fe} = 3 \times 10^{-15}$ $cm^2 s^{-1}$ at 20°C and it is between 10^{-14} to 10^{-15} $cm^2 s^{-1}$ in SiO$_2$ at 1000°C [135] The best fit equation of many reported data about the diffusion coefficient of Fe$_i$ in Si is $D_{Fe} = (1.0 + 0.8 \text{ to} - 0.4) \times 10^{-3} \exp(-0.67 \text{ eV}/k_B T)$ cm^2 s^{-1} over a wide range of temperatures from 0°C to 250°C [182]. This expression agrees well with the previously reported one, namely $D_{Fe} = 1.3 \times 10^{-3} \exp(-0.68 \text{ eV}/k_B T)$ cm^2 s^{-1} over a range of temperatures from 20 °C to 1250°C by Weber, which is obtained from a least squares fit of all diffusivity results for Fe in Si [178]. The diffusion coefficient of Fe in Si at 1100 °C is 4×10^{-6} cm^2 s^{-1}, and the diffusion coefficients of other dominant impurities are listed in Ref. [178].

The solubility, S, of an impurity in Si can be described as follows at lower than the eutectic temperature,

$$S = 5 \times 10^{22} \exp(-\Delta G/k_B T) \ cm^{-3}, \tag{1.36}$$

Where ΔG is the excess Gibbs free energy for impurity introduction in Si [176]. Parameters of ΔG for some transition metals in intrinsic Si is listed in Ref. [176]. The solubility of Fe in Si increases with temperature below the eutectic temperature, passes through a maximum point, and decreases above the eutectic temperature of 1207°C for the Si-rich FeSi$_2$ phase [182,186]. The linear fit equation of many reported data about the solubility of Fe in Si is $S_{Fe} = (8.4 + 5.4 \text{ to} - 3.4) \times 10^{25} \exp(-(2.86 \pm 0.05 \text{eV})/k_B T) \ cm^{-3}$ in a temperature range from 800 °C to 1200 °C [182]. This expression agrees well with the previously reported one such as $S_{Fe} = 1.8 \times 10^{26} \exp(-(2.94 \pm 0.05 \text{ eV})/k_B T) \ cm^{-3}$ by Weber [178]. The solubilities of Fe in Si are 4×10^{14} cm^{-3} at 1000 °C [159] and 3×10^{15} cm^{-3} at 1100 °C [178]. The solubilities of other dominant impurities are listed in Ref. [178]. The diffusivity at 1100 °C is in the order of Fe < Ni < Cu and the solubility at 1100 °C is also in the order of Fe < Ni < Cu, and the diffusivity and solubility of Cu is about 2 order of magnitude larger than those of Fe.

Fe is detrimental for Si devices because Fe, its complexes and precipitates introduce deep levels in the band gap or provide recombination centers, reducing the minority carrier lifetime or the minority carrier diffusion length [135]. Fe forms several types of silicide such as FeSi and FeSi$_2$. The solubility of Fe is much lower than that of Ni or Cu, and the density of silicide precipitates is low in a crystal [135]. Therefore, it is difficult to find Fe-silicide precipitates in a wafer, but it is easily to observe agglomeration of Fe at extended defects such as dislocations after thermal diffusion of Fe. The agglomerated Fe is not Fe-silicide, but rather an Fe-silicate, primarily with Fe^{3+} [135]. Areas with high Fe concentration are formed at the top of an ingot because of segregation of Fe in a Si melt and at the side of the ingot close to the crucible wall because of high solid-diffusion from the Si$_3$N$_4$ coated wall [187]. Fe is dissolved in the interstitial state and precipitated at grain boundaries and dislocations. Fe atoms easily diffuse and form precipitates even onto the {110} Σ3 and

{112} Σ3 grain boundaries during slow cooling [91]. Schön et al. [188] studied Fe and Fe-silicide precipitates in Si multicrystals in detail. The total amount of Fe within a grain is estimated to be higher than the amount of Fe at a grain boundary after crystal growth even though the density of Fe is three orders of magnitude higher at the grain boundary. Supersaturated Fe tends to form energetically favorable large precipitates during a slow cooling process. With increasing cooling rate, an increasing fraction of the supersaturated Fe is no longer able to reach existing precipitates, and the spatial distribution becomes more homogeneous. A low density of large Fe precipitates is favored over a high density of small ones although the recombination activity increases with the size of the precipitates. The size of precipitates ranges from 0.2 nm (0.0002 μm: few Fe atoms) to 100 nm (several 10^8 atoms). The total Fe concentration mainly increases the mean size of the precipitates and affects only slightly the precipitate density. Precipitation of Fe in area of high defect densities leads to regions of low minority carrier diffusion lengths [135] (see Sections 2.4.2 and 5.4.3).

Chen et al. [93] studied Fe contamination effects on Σ3 grain boundaries. The {111}, {110} and {112} Σ3 grain boundaries are not electrically active, but the {110} and {112} Σ3 grain boundaries become electrically active when they are contaminated by Fe. The many zig–zag segments of the {112} Σ3 grain boundaries have disorder which introduces extra defects such as dislocations as recombination active sites. Due to grain boundary dislocations, Fe is preferentially gettered at irregular parts of the [108] and {112} Σ3 grain boundaries and exists as very small clusters. Fe impurities is the dominant factor for the strong recombination activity. Fe impurities segregated on grain boundaries act also as nucleation sites for precipitates caused by the strain energy in highly disturbed regions of grain boundaries. The denuded zones appear around the Σ3 and random grain boundaries because Fe atoms are gettered at these grain boundaries [189]. The {110} Σ3 grain boundaries have smaller number of zig–zag segments comparing with the {112} Σ3 grain boundaries. Basically, there are no extra defects along the {111} straight boundary plane.

In cooling processes of a bulk crystal, the impurities such as Fe, Cr and Cu are removed from the crystal and depleted in the crystal below 650 °C owing to dislocation gettering and a large part of the elements is dissolved in the crystal above 650 °C [132]. Dislocations form immediately after solidification and their density increases during cooling down to temperatures of about 800 °C [132]. The contamination level of Fe should be reduced to the low level of $10^{10}–10^{11}$ cm^{-3} [135] (see Section 3.3.3).

1.5 Strain and stress

Basically, a Si ingot grown in a crucible has residual stress because Si expands by 11% during solidification. The distribution of thermal stress depends on the temperature field. The residual strain is estimated to be highest in the periphery of ingot such as top, bottom and side because the thermal gradient is high in this area [109]. The strain shows a 4-fold crystallographic symmetry depending on different slip planes of {111} [109]. By combining EBSD and a scanning infrared polar-iscope (SIRP), the absolute values of residual strain and a two-dimensional distribution of residual strain can be quantitatively characterized in multicrystalline crystals [190]. The residual strain increases at grains with multi-twin grain boundaries, sometimes together with the $\Sigma 3$, $\Sigma 9$ and Σ 27 grain boundaries, in the vicinity of small angle boundaries, reaching the order of 10 MPa in stress [190]. The strain does not directly result in the electrical activity, but the strain acts as the driving force to form twin boundaries and small angle grain boundaries [71]. The radial temperature gradient affects the shape of the growing interface and the thermal stress in the ingot. The flat interface with little stress can be obtained by the almost flat radial temperature gradient. The slightly convex interface with little stress in the growth direction is obtained when the melt temperature is lower in the center than near the crucible wall.

The stress in the ingot associated with plasticity cannot be removed completely by post-growth annealing [191]. The stress is an origin of generation of crystal defects such as dislocations. As a special case, in Si edge-defined film-fed growth (EFG) ribbon sheets, the residual stresses are locked in very high residual stress areas and block dislocation motions [191]. In this case, high residual stresses seem to be beneficial to the solar cell efficiency.

On the other hand, crystal defects have their own strain depending on their structure. Chen et al. [71] studied the relation between strain and grain boundaries. A certain amount of strain remains near small angle grain boundaries. Some amount of strain also exists even around the $\Sigma 3$ and random grain boundaries. The heavily strained area appears at the multi-twin grain boundaries, but the strain level at the $\Sigma 3$, $\Sigma 9$ and Σ 27 grain boundaries is smaller than at the multi-twin grain boundaries [190]. However, the strain increase is also found around the random and Σ 27 grain boundaries and at the junction of the $\Sigma 3$ and Σ 27 grain boundaries due to grain boundary combinations [190]. The {112} $\Sigma 3$ grain boundaries have zig-zags and are surrounded by twin boundaries to reduce the large strain field of the {112} Σ3 grain boundaries [93]. Large strain is confined in regions containing multi-twin boundaries which are promoted by fast crystal growth. During the fast solidification process, strain can be relaxed to the value of the critical stress of the twin deformation, but residual strain still remains. Large residual strain inside grains acts as the driving force for formation of twin boundaries and small angle grain boundaries.

Strain can be also relaxed to the value of the critical stress of dislocation alignment in small angle grain boundaries. Small angle grain boundaries possess both strong electrical activity and large strain. However, such strain is not directly related to electrical activity which is mainly determined by grain boundary character and impurity. A long lifetime is detected at twin boundaries with large strain [191].

1.6 Appendixes

1. Equilibrium segregation coefficients (amount of the impurity in Si crystal/amount of the impurity in Si melt) of all impurities and dopants in Si crystal used in the text, arranging in order of atomic number

    * Effective segregation coefficient, k_e

    Li (atomic number 3): 0.01 [169].

    B (5): 0.7 [53], 0.716 [168,192],0.8 [53,168,169].

    C (6): 0.034−0.3 [168], 0.30 ± 0.16 [193], 0.07 [132], 0.07 ± 0.01 [130,139].

    N (7): 4×10^{-4} [132], 7×10^{-4} [130], O (8): 0.2−1.25 [168], 0.085 ± 0.08 [193], 0.5 [169], 1.25 ± 0.17 [130], 1.4 [132]

    Al (13): 0.002 [168,169].

    P (15): 0.35 [168,169].

    S (16): 1×10^{-5} [169]

    Ti (22): 3.6×10^{-6} [43,168,194]*

    V (23): 4×10^{-6} [168,194]*

    Cr (24): 1.1×10^{-5} [43,168,194]*

    Mn (25): ∼1.0×10^{-5} [168,169]

    Fe (26): 6.4×10^{-6} [194]*, 8×10^{-6} [168,169], 2.2×10^{-5} [195]*

    Co (27): 8×10^{-6} [168,169]

    Ni (28): 8×10^{-6} [43,168], 3.2×10^{-5} [194]*

    Cu (29): 4×10^{-4} [168,169], 6.4×10^{-4} [194]*

    Zn (30):1×10^{-5} [169], 0.001 [43].

    Ga (31): 0.008 [53,169].

Ge (32): 0.33 [169], 0.56 [43].
As (33): 0.3 [168,169], 0.8 [43].
Ag (47): 1×10^{-6} [43]
Cd (48):1×10^{-5} [43]
In (49): 4×10^{-4} [34,169]
Sn (50): 0.016 [169].
Sb (51): 0.023 [169].
Ta (73): 1×10^{-7} [168,169]
Au (79): 2.5×10^{-5} [169]
Bi (83): 7×10^{-4} [169]

The relation between the effective segregation coefficient, k_e, and the equilibrium segregation coefficient, k_0, can be expressed as follows,

$$k_e = \frac{k_0}{k_0 + (1 - k_0)\, exp(-v\delta/D)}, \tag{A-1}$$

where v is the crystal growth rate, δ is the diffusion-layer thickness and D is the diffusion coefficient [168,192]. k_e depends on the crystal growth rate and the diffusion-layer thickness, and it increases toward to 1 as the crystal growth rate increase. When the diffusion-layer thickness is 0, $k_e = k_0$.

2. Solubility limits of light-element impurities in Si melt and in Si crystal used in the text, arranging in order of atomic number

  * Mp: the melting point of Si.

(2.1) In Si melt

C (6): $3.0 \pm 0.3 \times 10^{18}$ cm^{-3} at Mp [196], $4.5 \pm 0.5 \times 10^{18}$ cm^{-3} at Mp [130], 5×10^{18} cm^{-3} at 1685 K [131], 9.0×10^{18} cm^{-3} at Mp [197]

N (7): 5.7×10^{18} cm^{-3} at 1685 K [131,140], 6×10^{18} cm^{-3} [130]

O (8): $2.2 \pm 0.15 \times 10^{18}$ cm^{-3} at Mp [130], 2.2×10^{18} cm^{-3} at 1685 K [131], 4.4×10^{18} cm^{-3} at Mp [197]

(2.2) In Si crystal

C (6): $3.2 \pm 0.3 \times 1017$ cm^{-3} at Mp [130], 3.5×1017 cm^{-3} at Mp [132,139], N (7): 4×1015 cm^{-3} at Mp [140], $4.5 \pm 1.0 \times 1015$ cm^{-3} at Mp [130,132]

O (8): $2.75 \pm 0.15 \times 1018$ cm^{-3} at Mp [130].

3. List of physical properties of Si

  (1) Atomic number: 14

  (2) Atomic weight: 28.1

(3) Lattice constant: 5.4307 Å [4].

(4) Density: 2.3290 g cm^{-3} at RT [198].

(5) Melt density: 2.57 g cm^{-3} at Mp [198].

(6) Melting point (Mp): 1414 °C (1687 K) [198].

(7) Boiling point: 2355 °C (2628 K) [198].

(8) Heat content: 19.789 J mol^{-1}K^{-1} at 25 °C [198].

(9) Heat of fusion: 50.21 kJ mol^{-1}[198]

(10) Heat of vaporization: 359 kJ mol^{-1}[198]

(11) Vapor Pressure: 1 Pa at 1908 K [198].

(12) Thermal conductivity: 149−168 W m^{-1}K^{-1} at 300 K [198,199].

(13) Thermal expansion coefficient: 2.6 μm m^{-1}K^{-1} at 25 °C [198].

(14) Linear expansion coefficient: 2.3−2.6 × 10^{-6} K^{-1} at RT [198,199].

(15) Young's modulus: 185−188 GPa (× 10^3 N mm^{-2}) [198,199].

(16) Bulk modulus: 100 GPa [198].

(17) Modulus of rigidity: 52 GPa [198].

(18) Poisson ratio: 0.28 [198].

(19) Electrical resistivity: 10^3 Ω m at 20 °C [198].

(20) Energy gap: 1.12 eV at 300 K [198].

(21) Crystal-melt interfacial average free energy: 0.37 J m^{-2} [5].

(22) Equilibrium vacancy concentration in Si: 3.5 × 10^{14}−1.5 × 10^{15} cm^{-3} at Mp [200].

(23) Equilibrium interstitial Si concentration in Si: 2.4 × 10^{14}−1.2 × 10^{15} cm^{-3} at Mp [200].

(24) Diffusion coefficient of vacancy in Si: 3.74 × 10^{-5}−4.48 × 10^{-5} cm^2 s^{-1} at Mp [200].

(25) Diffusion coefficient of interstitial Si in Si: 5.0 × 10^{-4} cm^2 s^{-1} at Mp [200].

References

[1] J. Czochralski, Z. Phys. Chem. 92 (1917) 219.

[2] L. Arnberg, M. Di Sabatino, E. Øvrelid, Electron. Mater. Solid. 63 (2011) 38.

[3] T.F. Ciszek, G.H. Schwuttke, K.H. Yang, J. Cryst. Growth 46 (1979) 527.

[4] L.F. Mattheiss, J.R. Patel, Phys. Rev. B 23 (1981) 5384.

[5] P.A. Apte, X.C. Zeng, Appl. Phys. Lett. 92 (2008) 221903.

[6] W. Kossel, Naturwissenschaften 18 (1930) 901.

[7] I.V. Markov, Crystal Growth for Beginners, World Scientific, Singapore, 1995, pp. 1−62 (Chapter 1).

[8] J. Lipton, W. Kurz, R. Trivedi, Acta Metall. 35 (1987) 957.

56 Crystal Growth of Si Ingots for Solar Cells Using Cast Furnaces

[9] K. Nakajima, K. Kutsukake, K. Fujiwara, N. Usami, S. Ono, I. Yamasaki, in: Proceedings of the 35th IEEE Photovoltaic Specialists Conference, Hawaii Convention Center, Honolulu, Hawaii, USA, June 20-25, 2010, pp. 817—819.

[10] M.G. Tsoutsouva, T. Duffar, C. Garnier, G. Fournier, Cryst. Res. Technol. 50 (2015) 55.

[11] T. Aoyama, K. Kuribayashi, Acta Mater. 48 (2000) 3739.

[12] C.F. Yang, M.G. Tsoutsouva, H.P. Hsu, C.W. Lan, J. Cryst. Growth 453 (2016) 130.

[13] A. Appapillai, E. Sachs, J. Cryst. Growth 312 (2010) 1297.

[14] I. Brynjulfsen, L. Arnberg, J. Cryst. Growth 331 (2011) 64.

[15] L. Alphei, A. Braun, V. Becker, A. Feldhoff, J.A. Becker, E. Wulf, C. Krause, F.-W. Bach, J. Cryst. Growth 311 (2009) 1250.

[16] K.E. Ekstrøm, E. Undheim, G. Stokkan, L. Arnberg, M.D. Sabatino, Acta Mater. 109 (2016) 267.

[17] K. Fujiwara, Y. Obinata, T. Ujihara, N. Usami, G. Sazaki, K. Nakajima, J. Cryst. Growth 266 (2004) 441.

[18] W.B. Hillig, D. Turnbull, J. Chem. Phys. 24 (1956) 914.

[19] K. Fujiwara, W. Pan, N. Usami, K. Sawada, M. Tokairin, Y. Nose, A. Nomura, T. Shishido, K. Nakajima, Acta Mater. 54 (2006) 3191.

[20] W. Miller, J. Cryst. Growth 325 (2011) 101.

[21] K.A. Jackson, Growth and Perfection of Crystals, Wiley, New York, 1958, p. 319.

[22] W. Obretenov, D. Kashchiev, V. Bostanov, J. Cryst. Growth 96 (1989) 843.

[23] K. Fujiwara, K. Nakajima, T. Ujihara, N. Usami, G. Sazaki, H. Hasegawa, S. Mizoguchi, K. Nakajima, J. Cryst. Growth 243 (2002) 275.

[24] G. Gilmer, J. Cryst. Growth 49 (1980) 465.

[25] W. Obretenov, V. Bostanov, E. Budevski, R.G. Barradas, T. Van der Noot, J. Electrochem. Acta 31 (1986) 753.

[26] K.M. Beatty, K.A. Jackson, J. Cryst. Growth 211 (2000) 13.

[27] K. Fujiwara, K. Maeda, N. Usami, G. Sazaki, Y. Nose, A. Nomura, T. Shishido, K. Nakajima, Acta Mater. 56 (2008) 2663.

[28] T.-J. Liao, Y.S. Kang, C.W. Lan, J. Cryst. Growth 499 (2018) 90.

[29] G.Y. Chen, C.W. Lan, J. Cryst. Growth 474 (2017) 166.

[30] K.A. Jackson, J. Cryst. Growth 264 (2004) 519.

[31] J. Friedrich, L. Stockmeier, G. Müller, Acta Phys. Pol. 124 (2013) 219.

[32] H. Fu, X. Geng, Sci. Technol. Adv. Mater. 2 (2001) 197.

[33] T. Taishi, X. Huang, M. Kubota, T. Kajigaya, T. Fukami, K. Hoshikawa, Jpn. J. Appl. Phys. 39 (2000) L5.

[34] X. Yu, X. Zheng, K. Hoshikawa, D. Yang, Jpn. J. Appl. Phys. 51 (2012) 105501.

[35] K.A. Jackson, Mater. Sci. Eng. 65 (1984) 7.

[36] M. Tokairin, K. Fijiwara, K. Kutsukake, N. Usami, K. Nakajima, Phys. Rev. B 80 (2009) 174108.

[37] W.W. Mullins, R.F. Sekarka, J. Appl. Phys. 35 (1964) 444.

[38] K. Fujiwara, M. Ishii, K. Maeda, H. Koizumi, J. Nozawa, S. Uda, Scripta Mater. 69 (2013) 266.

[39] K. Fujiwara, R. Gotoh, X.B. Yang, H. Koizumi, J. Nozawa, S. Uda, Acta Meter 59 (2011) 4700.

[40] M. Tokairin, K. Fujiwara, K. Kutsukake, H. Kodama, N. Usami, K. Nakajima, J. Cryst. Growth 312 (2010) 3670.

[41] K. Fujiwara, M. Tokairin, W. Pan, H. Koizumi, J. Nozawa, S. Uda, J. Appl. Phys. 104 (2014) 182110.

[42] R. Gotoh, K. Fujiwara, X. Yang, H. Koizumi, J. Nozawa, S. Uda, Appl. Phys. Lett. 100 (2012) 021903.

[43] X. Yu, D. Yang, Growth of crystalline silicon for solar cells: czochralski Si, in: D. Yang (Ed.), Chapter 13, Section Two, Crystalline Silicon Growth, Hand Book of "Photovoltaic Silicon Material", Springer, Berlin Heidelberg, 2018, pp. 1—45 (On line).

[44] X. Yang, K. Fujiwara, N.V. Abrosimov, R. Gotoh, J. Nozawa, H. Koizumi, A. Kwasniewski, S. Uda, Appl. Phys. Lett. 100 (2012) 141601.

[45] M. Mokhtari, K. Fujiwara, H. Koizumi, J. Nozawa, S. Uda, Scripta Mater. 117 (2016) 73.

[46] M. Mokhtari, K. Fujiwara, G. Takakura, K. Maeda, H. Koizumi, J. Nozawa, S. Uda, J. Appl. Phys. 124 (2018) 085104.

[47] K. Fujiwara, S. Tsumura, M. Tokairin, K. Kutsukake, N. Usami, S. Uda, K. Nakajima, J. Cryst. Growth 312 (2009) 19.

[48] A. Tandjaoui, N. Mangelinck-Nöel, G. Reinhart, B. Billia, T. Lafford, J. Baruchel, J. Cryst. Growth 377 (2013) 203.

[49] V. Stamelou, M.G. Tsoutsouva, T. Riberi-Bèridot, G. Reinhart, G. Regula, J. Baruchel, J. Cryst. Growth 479 (2017) 1.

[50] H.K. Lin, C.W. Lan, Acta Meter 131 (2017) 1.

[51] K. Maeda, A. Niitsu, H. Morito, K. Shiga, K. Fujiwara, Scripta Mater. 146 (2018) 169.

[52] M. Neuberger, S.J. Wells, Silicon Data Sheet DS-162 48 (1969) 1017.

[53] S. Uda, X. Hung, M. Arivanandhan, R. Gotoh, in: The 5th Inter. Symp. On Advanced Sci. and Tech. of Si Mater. Kona, Hawaii, USA, Nov. 10-14, 2008.

[54] R. Søndenå, H. Haug, A. Song, C.-C. Hsueh, J.O. Odden, in: Proceedings of SiliconPV 2018, the 8th Intern. AIP Conf. On Crystalline Silicon Photovoltaics, 1999, p. 130016.

[55] K. Kakimoto, L. Liu, S. Nakano, Mater. Sci. Eng., B 134 (2006) 269.

[56] K. Nakajima, R. Murai, K. Morishita, K. Kutsukake, J. Cryst. Growth 372 (2013) 121.

[57] K. Nakajima, N. Usami, K. Fujiwara, K. Kutsukake, S. Okamoto, in: Proceedings of the 24th European Photovoltaic Solar Energy Conference, 2009, p. 1219.

[58] K. Nakajima, K. Kutsukake, K. Fujiwara, K. Morishita, S. Ono, J. Cryst. Growth 319 (2011) 13.

[59] K. Nakajima, R. Murai, K. Morishita, K. Kutsukake, N. Usami, J. Cryst. Growth 344 (2012) 6.

[60] K. Nakajima, K. Morishita, R. Murai, N. Usami, J. Cryst. Growth 389 (2014) 112.

[61] K. Nakajima, K. Morishita, R. Murai, K. Kutsukake, J. Cryst. Growth 355 (2012) 38.

[62] K. Nakajima, R. Murai, K. Morishita, Jpn. J. Appl. Phys. 53 (2014) 025501.

[63] E. Scheil, Z. Metallkd. 34 (1942) 70.

[64] Y. Azuma, N. Usami, K. Fujiwara, T. Ujihrara, K. Nakajima, J. Cryst. Growth 276 (2005) 393.

[65] T. Lehmann, M. Trempa, E. Meissner, M. Zschorsch, C. Reimann, J. Friedrich, Acta Mater. 69 (2014) 1.

[66] T. Muramatsu, Y. Hayama, K. Kutsukake, K. Maeda, T. Matsumoto, H. Kudo, K. Fujiwara, N. Usami, J. Cryst. Growth 499 (2018) 62.

[67] V. Randle, Acta Mater. 52 (2004) 4067.

[68] K. Kutsukake, Growth of cryatalline silicon for solar cells: mono-like method, in: D. Yang (Ed.), Chapter 11, Section Two, Crystalline Silicon Growth, Hand Book of "Photovoltaic Silicon Material", Springer, Berlin Heidelberg, 2018, pp. 1—20 (On line).

[69] K. Kutsukake, Results from TEM, Personal communication, 2017.

[70] D.G. Brandon, Acta Metall. 14 (1966) 1479.

[71] J. Chen, B. Chen, T. Sekiguchi, M. Fukuzawa, M. Yamada, Appl. Phys. Lett. 93 (2008) 112105.

[72] J. Chen, T. Sekiguchi, Jpn. J. Appl. Phys. 46 (2007) 6489.

[73] J. Hornstra, Physica 25 (1959) 409.

[74] M.M. Kivambe, G. Stokkan, T. Ervik, B. Ryningen, O. Lohne, Solid State Phenom. 178−179 (2011) 307.

[75] G. Stokkan, S. Riepe, O. Lohne, W. Warta, J. Appl. Phys. 101 (2007) 053515.

[76] S. von Alfthan, K. Kaski, A.P. Sutton, Phys. Rev. B 74 (2006) 134101.

[77] M. Kohyama, R. Yamamoto, M. Doyama, Phys. Status Solidi 138 (1986) 387.

[78] V.A. Oliveira, B. Marie, C. Cayron, M. Marinova, M.G. Tsoutsouva, H.C. Sio, T.A. Lafford, J. Baruchel, G. Audoit, A. Grenier, T.N. Tran Thi, D. Camel, Acta Mater. 121 (2016) 24.

[79] T. Jain, H.K. Lin, C.W. Lan, J. Cryst. Growth 485 (2018) 8.

[80] M. Kohyama, R. Yamamoto, Y. Watanabe, Y. Ebara, M. Kinoshita, J. Phys. C Solid State Phys. 21 (1988) L695.

[81] K. Kutsukake, N. Usami, K. Fujiwara, Y. Nose, T. Sugawara, T. Shishido, K. Nakajima, Mater. Trans. 48 (2007) 143.

[82] M.G. Tsoutsouva, T. Riberi-Beridot, G. Regula, G. Reinhart, J. Baruchel, F. Guittonneau, L. Barrallier, N. Mangelinck-Noël, Acta Mater. 115 (2016) 210.

[83] N. Usami, M. Kitamura, T. Sugawara, K. Kutsukake, K. Ohdaira, Y. Nose, K. Fujiwara, T. Shishido, K. Nakajima, Jpn. J. Appl. Phys. 44 (2005) L778.

[84] A. Otsuki, Interface Sci. 9 (2001) 293.

[85] J. Han, V. Vitek, D.J. Srolovitz, Acta Mater. 104 (2016) 259.

[86] T. Iwata, I. Takahashi, N. Usami, Jpn. J. Appl. Phys. 56 (2017) 075501.

[87] F. Liu, C.-S. Jiang, H. Guthrey, S. Johnston, M.J. Romero, B.P. Gorman, M.M. Al-Jassim, Sol. Energy Mater. Sol. Cells 95 (2011) 2497.

[88] M.D. Sabatino, G. Stokkan, Phys. Status Solidi 210 (2013) 641.

[89] M. Trempa, C. Kranert, I. Kupka, C. Reimann, J. Friedrich, J. Cryst. Growth 514 (2019) 114.

[90] J. Chen, R.R. Prakash, J.Y. Li, K. Jiptner, Y. Miyamura, H. Harada, A. Ogura, T. Sekiguchi, Solid State Phenom. 77 (2014) 205−206.

[91] J. Chen, D. Yang, Z. Xi, J. Appl. Phys. 97 (2005) 033701.

[92] J. Chen, T. Sekiguchi, D. Yang, Phys. Status Solidi 4 (2007) 2908.

[93] B. Chen, J. Chen, T. Sekiguchi, M. Saito, K. Kimoto, J. Appl. Phys. 105 (2009) 113502.

[94] S. Delclos, D. Dorignac, F. Phillipp, F. Silva, A. Gicquel, Diam. Relat. Mater. 9 (2000) 345.

[95] A. Voigt, E. Wolf, H.P. Strunk, Mater. Sci. Eng., A B54 (1998) 202.

[96] G. Stokkan, J. Cryst. Growth 384 (2013) 107.

[97] S. Würzner, R. Helbig, C. Funke, H.J. Möller, J. Appl. Phys. 108 (2010) 083516.

[98] T. Duffar, A. Nadri, Compt. Rendus Phys. 14 (2013) 185.

[99] K.-K. Hu, K. Maeda, H. Morito, K. Shiga, K. Fujiwara, Acta Mater. 153 (2018) 186.

[100] K. Fujiwara, R. Maeda, K. Maeda, H. Morito, Scripta Mater. 133 (2017) 65.

[101] T. Jain, H.K. Lin, C.W. Lan, Acta Mater. 144 (2018) 41.

[102] T. Duffar, A. Nadri, Scripta Mater. 62 (2010) 955.

[103] K. Kutsukake, T. Abe, N. Usami, K. Fujiwara, K. Morishita, K. Nakajima, Scripta Mater. 65 (2011) 556.

[104] A. Nadri, Y. Duterrail-Couvat, T. Duffar, J. Cryst. Growth 385 (2014) 16.

[105] H.J. Möller, Solid State Phenom. 95−96 (2004) 181.

[106] J.W. Jhang, G. Regula, N. Mangelinck-Noël, C.W. Lan, J. Cryst. Growth 508 (2019) 42.

[107] K. Kutsukake, N. Usami, Y. Ohno, Y. Tokumoto, I. Yonenaga, Jpn. Appl. Phys. Express 6 (2013) 025505.

[108] G. Stokkan, Acta Mater. 58 (2010) 3223.

[109] K. Jiptner, B. Gao, H. Harada, Y. Miyamura, M. Fukuzawa, K. Kakimoto, T. Sekiguchi, J. Cryst. Growth 408 (2014) 19.

[110] H. Klapper, Mater. Chem. Phys. 66 (2000) 101.

[111] M. Trempa, C. Reimann, J. Friedrich, G. Müller, A. Krause, L. Sylla, T. Richter, J. Cryst. Growth 405 (2014) 131.

[112] C. Reimann, J. Friedrich, E. Meissner, D. Oriwol, L. Sylla, Acta Mater. 93 (2015) 129.

[113] V. Eremenko, E. Yakimov, N. Abrosimov, Phys. Status Solidi 4 (2007) 3100.

[114] M.M. Kivambe, G. Stokkan, T. Ervik, B. Ryningen, O. Lohne 110 (2011) 063524.

[115] K. Jiptner, Y. Miyamura, H. Harada, B. Gao, K. Kakimoto, T. Sekiguchi, Prog. Photovoltaics Res. Appl. 24 (2016) 1513.

[116] B. Gao, S. Nakano, H. Harada, Y. Miyamura, K. Kakimoto, J. Cryst. Growth 474 (2017) 121.

[117] B. Ryningen, G. Stokkan, M. Kivambe, T. Ervik, O. Lohne, Acta Mater. 59 (2011) 7703.

[118] A. Autruffe, V.S. Hagen, L. Arnberg, M.D. Sabatino, J. Cryst. Growth 411 (2015) 12.

[119] T. Ervik, M. Kivambe, G. Stokkan, B. Ryningen, O. Lohne, Acta Mater. 60 (2012) 6762.

[120] A.S. Jordan, R. Caruso, A.R. VonNeida, J.W. Nielsen, J. Appl. Phys. 52 (1981) 3331.

[121] I. Takahashi, N. Usami, K. Kutsukake, K. Morishita, K. Nakajima, Jpn. J. Appl. Phys. 49 (2010) 04DP01.

[122] I. Takahashi, N. Usami, K. Kutsukake, G. Stokkan, K. Morishita, K. Nakajima, J. Cryst. Growth 312 (2010) 897.

[123] N. Usami, R. Yokoyama, I. Takahashi, K. Kutsukake, K. Fujiwara, K. Nakajima, J. Appl. Phys. 107 (2010) 013511.

[124] H.J. McSkimin, W.L. Bomd, E. Buehler, G.K. Teal, Phys. Rev. 83 (1951) 1080.

[125] A.P. Sutton, R.W. Balluffi, Interface in Crystalline Materials, Oxford Science, New York, 1995, p. 21.

[126] K. Nakajima, K. Kutsukake, K. Morishita, N. Usami, S. Ono, I. Yamasaki, in: Conference Record of 25th European Photovoltaic Solar Energy Conference and Exhibition/5th World Conference on Photovoltaic Energy Conversion, Valencia, Spain, Sept. 6-10, 2010, pp. 1299−1301.

[127] M.M. Kivambe, T. Ervik, B. Ryningen, G. Stokkan, J. Appl. Phys. 112 (2012) 103528.

[128] A. Baghdadi, W.M. Bullis, W.C. Croarkin, Y.Z. Li, R.I. Scace, R.W. Series, P. Stallhofer, M. Watanabe, J. Electrochem. Soc. 136 (1989) 2015.

[129] K. Kutsukake, H. Ise, Y. Tokumoto, Y. Ohno, K. Nakajima, I. Yonenaga, J. Cryst. Growth 352 (2012) 173.

[130] Y. Yatsurugi, N. Akiyama, Y. Endo, J. Electrochem. Soc. 120 (1973) 975.

[131] C. Reimann, M. Trempa, T. Jung, J. Friedrich, G. Müller, J. Cryst. Growth 312 (2010) 878.

[132] H.J. Möller, C. Funke, D. Kreßner-Kiel, S. Würzner, Energy Procedia 3 (2011) 2.

[133] S. Togawa, X. Huang, K. Izunome, K. Terashima, S. Kimura, J. Cryst. Growth 148 (1995) 70.

[134] M. Stavola, J.R. Patel, L.C. Kimering, P.E. Freeland, Appl. Phys. Lett. 42 (1983) 73.

[135] A.A. Istratov, H. Hieslmair, E.R. Weber, Appl. Phys. A 70 (2000) 489.

[136] H. Ono, T. Ishizuka, C. Kato, K. Arafune, Y. Ohshita, A. Ogura, Jpn. J. Appl. Phys. 49 (2010) 110202.

[137] 1990 Annual Book of ASTM Standards, 10. 05, F1 Proposal P 225, American Society for Testing and Materials, Philadelphia, 1990.

[138] X. Yu, J. Chen, X. Ma, D. Yang, Mater. Sci. Eng., A R 74 (2013) 1.

[139] T. Nozaki, J. Electrochem. Soc. 117 (1970) 1566.

[140] M. Trempa, C. Reimann, J. Friedrich, G. Müller, J. Cryst. Growth 312 (2010) 1517.

60 Crystal Growth of Si Ingots for Solar Cells Using Cast Furnaces

[141] M. Beaudhuin, G. Chichignoud, P. Bertho, T. Duffer, M. Lemiti, K. Zaidat, Mater. Chem. Phys. 133 (2012) 284.
[142] C. Reimann, M. Trempa, J. Friedrich, G. Müller, J. Cryst. Growth 312 (2010) 1510.
[143] X. Liu, H. Harada, Y. Miyamura, X. Han, S. Nakano, S. Nishizawa, K. Kakimoto, J. Cryst. Growth 499 (2018) 8.
[144] R.C. Newman, J. Wakefield, J. Phys. Chem. Solids 19 (1961) 230.
[145] J. Li, R.R. Prakash, K. Jiptner, J. Chen, Y. Miyamura, H. Harada, K. Kakimoto, A. Ogura, T. Sekiguchi, J. Cryst. Growth 377 (2013) 37.
[146] J.P. Goss, I. Hahn, R. Jones, Phys. Rev. B 67 (2003) 045296.
[147] M. Tajima, Y. Kamata, Jpn. J. Appl. Phys. 52 (2013) 086602.
[148] T. Itoh, T. Abe, Appl. Phys. Lett. 53 (1988) 39.
[149] J. Friedrich, C. Reimann, T. Jauss, A. Cröll, T. Sorgenfrei, Y. Tao, J.J. Derby, J. Cryst. Growth 475 (2017) 33.
[150] J. Champliaud, R. Voytovich, D. Rey, C. Dechamp, C. Huguet, B. Drevet, D. Camel, N. Eustathopoulos, The 28th european photovoltaic solar energy conference and exhibition, Paris, France, September 30–October 4, 2013.
[151] H. Matsuo, R.B. Ganesh, S. Nakano, L. Liu, K. Arafune, Y. Ohshita, M. Yamaguchi, K. Kakimoto, J. Cryst. Growth 310 (2008) 2204.
[152] H. Matsuo, R.B. Ganesh, S. Nakano, L. Liu, Y. Kangawa, K. Arafune, Y. Ohshita, M. Yamaguchi, K. Kakimoto, J. Cryst. Growth 310 (2008) 4666.
[153] H. Matsuo, R.B. Ganesh, S. Nakano, L. Liu, K. Arafune, Y. Ohshita, M. Yamaguchi, K. Kakimoto, J. Cryst. Growth 311 (2009) 1123.
[154] T. Tachibana, T. Sameshima, T. Kojima, K. Arafune, K. Kakimoto, Y. Miyamura, H. Harada, T. Sekiguchi, Y. Ohshita, A. Ogura, Jpn. J. Appl. Phys. 51 (2012) 02BP08.
[155] N. Inoue, K. Wada, Iron Steel 73 (1987) 31.
[156] J. Schön, A. Youssef, S. Park, L.E. Mundt, T. Niewelt, S. Mack, K. Nakajima, K. Morishita, R. Murai, M.A. Jensen, T. Buonassisi, M.C. Schubert, J. Appl. Phys. 120 (2016) 105703.
[157] H.J. Möller, L. Long, M. Werner, D. Yang, Phys. Status Solidi 171 (1999) 175.
[158] J.D. Murphy, K. Bothe, M. Olmo, V.V. Voronkov, R.J. Falster, J. Appl. Phys. 110 (2011) 053713.
[159] M.I. Bertoni, D.P. Fenning, M. Rinio, V. Rose, M. Holt, J. Maser, T. Buonassisi, Energy Environ. Sci. (2011), https://doi.org/10.1039/c1ee02083h (On line).
[160] T. Fukuda, Y. Horioka, N. Suzuki, M. Moriya, K. Tanahashi, S. Simayi, K. Shirasawa, H. Takato, J. Cryst. Growth 438 (2016) 76.
[161] M.G. Tsoutsouva, V.A. Oliveira, J. Baruchel, D. Camel, B. Marie, T.A. Lafford, J. Appl. Crystallogr. 48 (2015) 645.
[162] S. Kishino, M. Kanamori, N. Yoshihiro, M. Tajima, T. Iizuka, J. Appl. Phys. 50 (1979) 8240.
[163] Y. Nagai, S. Nakagawa, K. Kashima, J. Cryst. Growth 401 (2014) 737.
[164] I. Yonenaga, K. Sumino, K. Hoshi, J. Appl. Phys. 56 (1984) 2346.
[165] W. Fukushima, H. Harada, Y. Miyamura, M. Imai, S. Nakano, K. Kakimoto, J. Cryst. Growth 486 (2018) 45.
[166] M.C. Schubert, J. Schön, F. Schindler, W. Kwapil, A. Abdollahinia, B. Michl, S. Riepe, C. Schmid, M. Schumann, S. Meyer, W. Warta, IEEE J. Photovolt. 3 (2013) 1256.
[167] Y. Chettat, M. Boumaour, S. Senouci, M. Salhi, H. Slamene, L. Hamidatou, A. Benmounah, The 28th European Photovoltaic Solar Energy Conference and Exhibition, Paris, France, September 30–October 4, 2013.
[168] M.A. Martorano, J.B. Ferreira Neto, T.S. Oliveira, T.O. Tsubaki, Metall. Mater. Trans. A 42 (2011) 1870.

[169] F.A. Trumbore, Bull Syst. Tech. J. 39 (1960) 205.
[170] M.M. Mandurah, K.C. Saraswat, C.R. Heims, T.I. Kamins, J. Appl. Phys. 51 (1980) 5755.
[171] K. Tang, E.J. Øvrelid, G. Tranell, M. Tangstad, in: K. Nakajima, N. Usami (Eds.), Crystal Growth of Silicon in Solar Cells, Springer, Berlin Heidelberg, 2009 (Chapter 12).
[172] S. Gindner, P. Karzel, B. Herzog, G. Hahn, IEEE J. Photovolt. 4 (2014) 1063.
[173] O.V. Feklisova, X. Yu, D. Yang, E.V. Yakimov, Semiconductors 47 (2013) 232.
[174] X. Li, D. Yang, X. Yu, D. Que, Trans. Nonferrous Metals Soc. China 21 (2011) 691.
[175] Z. Xi, D. Yang, H.J. Möller, Infrared Phys. Technol. 47 (2006) 240.
[176] E.B. Yakimov, Metal impurities and gettering in crystalline silicon, in: D. Yang (Ed.), Crystalline Silicon Growth, Hand Book of "Photovoltaic Silicon Material, Springer, Berlin Heidelberg, 2018, pp. 1–46 (On line).
[177] A.A. Istratov, C. Flink, H. Hieslmair, E.R. Weber, Phys. Rev. Lett. 81 (1998) 1243.
[178] E.R. Weber, Appl. Phys. A 30 (1983) 1.
[179] J. Lindroos, D.P. Fenning, D.J. Backlund, E. Verlage, A. Gorgulla, S.K. Estreicher, H. Savin, T. Buonassisi, J. Appl. Phys. 113 (2013) 20906.
[180] Z. Xi, D. Yang, J. Chen, T. Sekiguchi, Mater. Sci. Semicond. Process. 9 (2006) 304.
[181] G. Coletti, P.C.P. Bronsveld, G. Hahn, W. Warta, D. Macdonald, B. Ceccaroli, K. Wambach, N.L. Quang, J.M. Fernandez, Adv. Funct. Mater. 21 (2011) 879.
[182] A.A. Istratov, H. hieslmair, E.R. Weber, Appl. Phys. A 69 (1999) 13.
[183] T.U. Nærland, S. Bernardini, H. Haug, S. Grini, L. Vines, N. Stoddard, M. Bertoni, J. Appl. Phys. 122 (2017) 085703.
[184] T. Isobe, H. Nakashima, K. Hashimoto, Jpn. J. Appl. Phys. 28 (1989) 1282.
[185] Y. Hayama, I. Takahashi, N. Usami, J. Cryst. Growth 468 (2017) 610.
[186] M. Hansen, K. Anderko, Constitution of Binary Alloys, McGraw-Hill, New York, 1958.
[187] L. Liu, S. Nakano, K. Kakimoto, J. Cryst. Growth 292 (2006) 515.
[188] J. Schön, A. Haarahiltunen, H. Savin, D.P. Fenning, T. Buonassisi, W. Warta, M.C. Schubert, IEEE J. Photovol. 3 (2013) 131.
[189] J. Chen, D. Yang, Z. Xi, T. Sekiguchi, Phys. Biol. 364 (2005) 162.
[190] M. Fukuzawa, M. Yamada, M.D.R. Islam, J. Chen, T. Sekiguchi, J. Electron. Mater. 39 (2010) 700.
[191] S. He, S. Danyluk, I. Tarasov, S. Ostapenko, Appl. Phys. Lett. 89 (2006) 111909.
[192] J. Schön, H. Habenicht, M.C. Schubert, W. Warta, Solid State Phenom. 156–158 (2010) 223.
[193] T. Narushima, A. Yamashita, C. Ouchi, Y. Iguchi, Mater. Trans. 43 (2002) 2120.
[194] R.H. Hopkins, R.G. Seidensticker, J.R. Davis, P.R. Choudhury, P.D. Blais, J.R. McCormick, J. Cryst. Growth 42 (1977) 493.
[195] B.C. Sim, K.H. Kim, H.W. Lee, J. Cryst. Growth 290 (2006) 665.
[196] Y. Endo, Y. Yatsurugi, Y. Terai, T. Nazaki, J. Electrochem. Soc. 126 (1979) 1422.
[197] R.I. Scace, G. Slack, J. Chem. Phys. 30 (1959) 1551.
[198] Wikipedia, https://ja.wikipedia.org/wiki/Silicon.
[199] Google, https://oshiete.goo.ne.jp/qa/257397.html.
[200] K. Nakamura, Study of the Diffusion of Point Defects and the Formation of Secondary Defects during Growth Processes of Si Single Crystals, The thesis for the degree of Doctor of Philosophy of Tohoku University, 2002.

CHAPTER 2

Basic characterization and electrical properties of Si crystals

2.1 Methods to measure electrical properties

Macdonald [1] studied the lifetime and recombination in Si multicrystals in detail. The minority carrier lifetime is important in controlling the recombination rate in the low-injection conditions, and it has no meaning in conditions other than low-injection. When the excess carrier concentration approaches the dopant density in the high-injection conditions, the electron and hole lifetimes become nearly equal. In this case, the recombination lifetime is generally used for measurement over large ranges of carrier densities, from low-injection to high-injection conditions. The effective recombination lifetime is used for the total recombination lifetime caused by some independent processes. The effective lifetime is used for cases where the measured quantity actually represents recombination. When the main recombination mechanisms limiting the minority carrier lifetime, τ are simply caused by bulk recombination centers such as point defects, grain boundaries and dislocations, τ is given by the following relation,

$$1/\tau = 1/\tau_b + 1/\tau_{gb} + 1/\tau_d, \tag{2.1}$$

where τ_b, τ_{gb} and τ_d correspond to the lifetime due to bulk recombination, recombination at grain boundaries and recombination at dislocations, respectively [2]. The bulk recombination lifetime characterizes the contamination level of a wafer [3]. In the simple case of a sample with a constant bulk lifetime, τ_b and a small constant surface recombination velocity, S that characterizes surface recombination and is the same on each surface, the effective lifetime, τ_{eff} can be written as follows,

$$1/\tau_{eff} = 1/\tau_b + 2S/W. \tag{2.2}$$

Crystal Growth of Si Ingots for Solar Cells Using Cast Furnaces
ISBN 978-0-12-819748-6
https://doi.org/10.1016/B978-0-12-819748-6.00002-5

Copyright © 2020 Elsevier Inc.
All rights reserved.

where W is the thickness of the bulk sample. On the condition of small S, the surface recombination rate is not limited by the diffusion of carriers to the surface [1]. When surface recombination is dominant over bulk recombination, $\tau_{eff} \ll \tau_b$ will be established. The minority carrier lifetime is directly related to the photoconductance. The effective lifetime, τ_{eff} can be also written as follows [4,5],

$$\tau_{eff} = \sigma_L / \{J_{ph}(\mu_n + \mu_p)\}, \tag{2.3}$$

where σ_L is the excess photoconductance, J_{ph} is the current density generated under illumination. μ_n and μ_p are electron and hole mobilities, respectively. The effective minority carrier diffusion length can be estimated by Eq. (2.7) using the minority carrier diffusion coefficient as D and the effective minority carrier lifetime as τ.

To evaluate the electrical quality of Si ingots, the effective minority carrier lifetime is measured by the microwave photoconductive decay (μ-PCD: Semilab WT-2000) method [1] and the quasi steady state photoconductivity (QSSPC: Sinton WCT-120) method [1,4,6]. PCD involves the generation of excess carriers by a sharp pulse of illumination, and the decay of carriers back to their equilibrium concentrations is monitored by the photoconductance to calculate the effective recombination lifetime [1]. Compared to the μ-PCD method, the QSSPC method allows the measurement of very low lifetime without short light pulses [7]. The QSSPC method can measure many parameters of solar cells such as open-circuit voltage and saturation current density without depositing metallic contacts or using patterning techniques [4]. The constant generation rate of 0.1 sun corresponds to average minority carrier densities ranging from 5×10^4 to 5×10^5 cm^{-3}, depending on the lifetime of samples. Generally, the QSSPC lifetime at an excess carrier density or an injection level of 1×10^{15} cm^{-3} is reported as the effective recombination lifetime or simply effective minority carrier lifetime of the samples. Usually, the effective minority carrier lifetime is simply used as the minority carrier lifetime which expresses characteristics of wafer quality. For p-type Si, when the concentration of excess carriers or the doping level is comparable with the deep defect density, the defects are filled with minority carriers, which results in the abrupt increase of carrier recombination rate and the abrupt decrease of the lifetime with injection level [8]. When the concentration of excess carriers exceeds the deep defect density, the lifetime values do not depend on injection level [8].

Before the measurement, the P diffusion gettering (PDG) or simply P gettering treatment is performed at a high temperature. Using the Sinton method, the effective minority carrier lifetime of ingots is measured with passivation of a 20 nm aluminum oxide (Al_2O_3) layer deposited at 200 °C by the thermal atomic layer deposition (ALD) following a 30 min anneal at 350 °C in N_2 ambient before and after PDG using a $POCl_3$ diffusion [9,10]. PDG is performed by P in–diffusion in a $POCl_3$ tube furnace using two processes. The first high-throughput process includes loading the sample at 700°C followed by a 25 min plateau at 845°C, and immediate unloading at 845°C. The second process includes a controlled cooling step after the 845°C plateau and a 2 h anneal at 650°C before unloading.

2.2 Electrical properties

2.2.1 Electrical properties of Si crystals with defects

Crystal defects such as impurities, grain boundaries, dislocations and precipitates reduce the efficiency of solar cells because the generated minority carriers recombine there before they reach the *pn*-junction. Especially, minority carrier lifetime varies with dislocation density, grain boundary misorientation and the CSL nature of the boundaries [11]. Recombination of charge carriers proceeds through localized states in the band gap which interact with carriers from both the valence and conduction bands [1]. The recombination activities of impurities, grain boundaries and dislocations are important to improve the quality of grown ingots. The normalized recombination strength, Γ is used to express the recombination activity. L can be calculated by the following equation,

$$\Gamma = \gamma/D_m, \tag{2.4}$$

where D_m is the minority carrier diffusion coefficient and γ is the recombination strength defined as the number of recombination per time, dislocation length and excess carrier density [12,13]. The areal density of recombination centers at a grain boundary is related to the density of dangling bonds and the density of metal impurities segregated to the grain boundary [14]. The EBSD contract is approximately 2−3 times higher at the random grain boundary than at $\Sigma 3$ and $\Sigma 9$ CSL grain boundaries in the whole range of temperatures (50−300 K) [15]. Impurities do not have as much effect on the electrical activity of $\Sigma 3$ and $\Sigma 9$ grain boundaries as that on random grain boundaries [15]. Thus, random grain boundaries contain many recombination centers, but $\Sigma 3$ and $\Sigma 9$ grain boundaries are almost electrically inactive even when the impurity levels are high [15].

The DLTS is used for measurement of signals from impurities or defects with energy levels located within the bandgap. For example, the signals from $E_v + 0.10$ eV and $E_v + 0.45$ eV are Fe–B complex and F_i, respectively, which can be used to determine the diffusion coefficient of F_i in Si [16] (see Section 1.4.4). The electroluminescence (EL) imaging technique enables fast and non–destructive determination of a variety of parameters over the entire area of solar cells. In Si crystal solar cells, dislocations give rise to localized states inside the energy bandgap, which can act as centers of radiative recombination in solar cells. Dislocations emit their own luminescent energies associated with deep localized states in the energy bandgap [17]. This radiation is known as sub–bandgap luminescence. The novel EL images can detect regions with high dislocation density in Si multicrystalline solar cells because of their high degree of polarization of the sub–bandgap EL [18]. The polarization of luminescence usually occurs for an anisotropic distribution of electric charges and corresponds with the orientation of dislocations in the crystal [17,18].

When one type of carrier has a small capture cross section, carriers are captured by the level for a certain short time before ejected back into the band from which they came. Such states are referred to as traps, and the states interact with one type of carrier only [1]. Both crystallographic and impurity-related levels can act as traps. Two types of minority carrier trapping centers occur in p-type multicrystalline Si, which are related to dislocations and B-impurity complexes [19]. Trap density increases as dislocation density increases even after gettering. The B-impurity traps can be removed by gettering, dissociated by thermal annealing and do not contribute to recombination [19]. The trap-affected lifetime is different from the recombination lifetime even at the carrier density higher than the trap density and it decreases as the injected carrier density increases using high suns [20].

2.2.2 Recombination centers (recombination activity of impurities)

p-Type Si crystals with majority carrier of holes are prepared by B doping and n-type Si crystals with majority carrier of electrons are prepared by P doping. The minority carriers of p- and n-type Si are electron and holes, respectively. The diffusivity of holes is roughly 2–3 times lower comparing to electrons. The capture cross section of holes is smaller than that of electrons for metal impurities. A p-type Si ingot has a homogeneous resistivity along the whole ingot height due to the large segregation coefficient of B, but an n-type Si ingot has a large variation in the resistivity along the ingot height due to the small segregation coefficient of P. A lower base resistivity caused by a higher doping concentration leads to a lower minority carrier lifetime and a lower charge carrier lifetime due to an increased Auger-recombination [21].

Macdonald and Geerloigs [22] studied recombination activity of transition metals. Interstitial transition metal impurities such as Fe, Cr, Cu, Mo and Co are most likely to produce relatively large concentrations of point defect recombination centers in Si crystals. Particularly, Cu, Ni and Fe are more likely to be effective [23]. Recombination through defect levels induced by Fe or Cr point defects depends on injection [24]. These transition metal impurities tend to occupy interstitial sites at room temperature and have a much larger capture cross section for electrons than holes. The ratio of electron to hole for capture cross sections of Fe is approximately 700. For Fe with deep levels, the low-injection carrier lifetime is limited by the capture of minority carriers in p-type Si and it is much lower than that in n-type Si. The carrier lifetime of p-type Si rapidly increases and that of n-type Si is almost constant with injection level. Their high-injection carrier lifetimes are almost equal for both types of Si because the smaller hole-capture cross section dominates the high injection lifetime. Unlike precipitates, these contaminants at interstitial or substitutional lattice sites have a possibility to evenly distribute throughout wafers. They tend to produce donor states and have much greater recombination strength in p-type Si than n-type Si.

Cu has a significantly lower impact on lifetime than Fe at the same concentrations [25]. Sachdeva et al. [26] studied the impact of Cu on the minority carrier diffusion length, which is determined by formation of Cu precipitates. In n-type Si, the effective density of recombination centers is initially high and increases almost linearly with the indiffused Cu concentration, whereas in p-type Si, the effective density of recombination centers initially remains very low and then sharply increases over a Cu concentration of 10^{16} cm^{-3} at which Cu precipitates start to be formed. Cu precipitates are positively charged in p-type Si and increase their capture cross section for minority charged carriers. The charge state of Cu precipitates is determined by the Fermi level position, which depends on the concentrations of B as shallow acceptors and Cu_i as shallow donors. When the Cu_i concentration becomes enough large for the Fermi level to exceed the electroneutrality level of the Cu precipitates, the charge state changes from positive to negative or neutral. In n-type Si, Cu precipitates are negatively charged or neutral and their nucleation occurs much easier because the Fermi level is very close to the electroneutrality level.

2.2.3 Recombination centers (recombination activity of grain boundaries)

The main factors to limit the solar-cell performance are recombination at crystal defects. Metallic impurities precipitated at crystal defects enhance the recombination activities of these defects. Grain boundaries originally have shallow levels [27]. Most of clean grain boundaries are electrically inactive at RT and are not active recombination centers, except small angle grain boundaries or sub-grain boundaries [28,29]. The recombination rate depends on the magnitude of misorientation, which does not change the electrical characterization of induced defects but rather changes the defect density [30]. Small angle grain boundaries are particularly recombination active and are dominating the number of active grain boundaries, but the recombination rate of small angle grain boundaries does not increase for boundaries with misorientation above 2° [11]. The EBIC contrast of grain boundaries is usually weaker at 300 K than at 100 K because these grain boundaries have only shallow level defects and are electrically inactive at 300 K. The EBIC contrast attributed to the density of defects related to the boundary plane [31]. On the view point of EBIC, the small angle grain boundaries can be categorized into two groups depending on the different tilt misorientation angle of 0—5° or 2—3°, which have the different dislocation densities and the different EBIC contrasts such as weaker for the 0—5° (maximum at 2°) general type and higher for the 2—3° special type at 300 K [32,33]. The general type appears in clean multicrystalline Si with only shallow energy levels and the special type originates from the deep and shallow energy levels due to high density of defects or the impurity contamination [33,34]. Fig. 2.1 shows the images measured by (A) scanning electron microscopy (SEM), (B) EBSD and (C) EBIC at 300 K and (D) EBIC at 100 K for grain boundaries in as-grown crystal contaminated-free multicrystalline Si [35]. As shown in this figure, the EBIC contrasts of all the grain boundaries are week at 300 K, and the EBIC contrasts of Σ 3, 9 and 11 grain boundaries and random grain boundaries (R) are still weak at 100 K. The EBIC contrasts of small angle grain boundaries (SA1—5) in as-grown multicrystalline Si are weak at 300 K and strong at 100 K because they have high density of shallow levels. Thus, the strong EBIC contrast of

Basic characterization and electrical properties of Si crystals 69

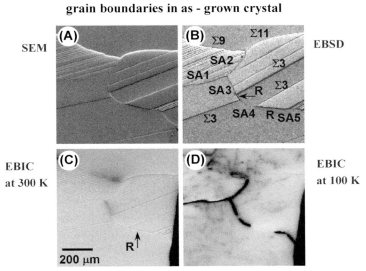

Fig. 2.1 Images measured by (A) SEM, (B) EBSD and (C) EBIC at 300 K and (D) EBIC at 100 K for grain boundaries in as-grown crystal contaminated-free multicrystalline Si. The EBIC contrasts of all the grain boundaries are week at 300 K, and Σ 3, 9 and 11 grain boundaries and random grain boundaries (R) still weak at 100 K. The EBIC contrasts of small angle grain boundaries (SA1–5) are weak at 300 K and strong at 100 K because they have high density of shallow levels. *(Referred from Fig. 1 in J. Chen et al. Scripta Mater. 52 (2005) 1211.)*

the small angle grain boundaries at 100 K suggests that they have higher density of shallow levels than the Σ and random grain boundaries [35].

In Fe contaminated multicrystalline Si, small angle grain boundaries act as strong recombination centers even at 300 K because they are easily contaminated with Fe compared to other grain boundaries, which are originated in their boundary dislocation structure [35]. Grain boundaries do not primarily degrade the lifetime, but intra-grain defects such as heavy-metal decorated dislocations degrade the lifetime extensively [29]. Grain boundaries are accompanied by deep levels, which are attributed to intrinsic structural defects such as dangling bonds or extrinsic impurity concentrations [27]. The deep levels act as recombination centers of minority carriers. Shallow level defects become EBIC active at lower temperatures whereas deep level defects become active at RT [27]. Impurity levels significantly impact the electrical properties of as-grown grain boundaries [36]. The recombination activity of grain boundaries varies by the grain boundary geometry or type, which has an impact on the ability to aggregate metal impurities [37] (see Sections 1.4.3 and 1.4.4).

The CSL grain boundary types and lifetime at grain boundaries have a good correlation, and the higher CSL indexes show stronger lifetime contrast of the spatially correlated lifetime-calibrated PL images [38]. Generally, grain boundaries with lower atomic coincidence such as random or $\Sigma 27$ grain boundaries tend to be more recombination active than grain boundaries with higher atomic coincidence such as $\Sigma 3$ or $\Sigma 9$ grain boundaries [36]. The recombination activity at small angle grain boundaries is also higher than that at $\Sigma 3$ grain boundaries [39]. Usually, the $\Sigma 3$ grain boundaries are originally electrically inactive and the boundary plane has no significant effect on their recombination activity [31]. Thus, grain boundaries with smaller CSL indexes are more eligible as an ingot. The main feature defining the electrical activity of large angle grain boundaries such as the CSL and random grain boundaries is whether the boundary plane is symmetric or asymmetric, since symmetric planes can be constructed without producing broken bonds [11]. Symmetric $\Sigma 3$ grain boundaries such as perfect twins exhibit a weak recombination activity or nearly inactive and become recombination active only when they are contaminated by impurities such as Fe [11,40]. On the other hand, asymmetric $\Sigma 3$ grain boundaries act as stronger minority carrier recombination centers [41]. The EBIC contrast decreases with increasing temperature, shows a minimum around 250 K, then increases again with increasing temperature [23]. The temperature dependence comes from a shallow level related to an inherent grain boundary structure and a deep level related to impurities segregated at grain boundaries [23]. In the as-grown contamination-free multicrystalline Si, the EBIC contrast is in the order of $\{111\}$ $\Sigma 3 < \{110\}$ $\Sigma 3 < \{112\}$ $\Sigma 3$ grain boundaries at both 300 and 100 K [31]. The order is caused by the difference in the defect density related to the boundary plane [31]. Thus, $\{112\}$ $\Sigma 3$ and zig-zag $\{110\}$ $\Sigma 3$ grain boundaries are much more recombination active than $\{111\}$ $\Sigma 3$ grain boundaries because $\{112\}$ $\Sigma 3$ and zig-zag $\{110\}$ $\Sigma 3$ grain boundaries have more defects than $\{111\}$ $\Sigma 3$ grain boundaries [27,36,40]. The $\{112\}$ $\Sigma 3$ grain boundary is estimated to be electrically non–active intrinsically and the observed deep states in the grain boundary is estimated to be extrinsic effects such as additional defects or impurities [42]. This means that electrical properties such as the recombination activity depend on the orientation of the boundary plane as well as the misorientation.

Transition-metal impurities gettered by grain boundaries act as recombination centers of carriers [40]. In lightly contaminated Si multicrystals, {111} $\Sigma 3$ grain boundaries show no obvious EBIC contrast at 300 K, and $\Sigma 9$ grain boundaries with symmetrical interface have a weak electrical activity, whereas {112} $\Sigma 3$, zig-zag {110} $\Sigma 3$, {221} $\Sigma 9$ symmetric tilt, $\Sigma 27$ and random grain boundaries become electrically active, increase the EBIC contrasts and become visible at 300 K, when contaminated with Fe [23,27,31,40]. Fe impurity is preferentially gettered as precipitates at the irregular parts of the {112} $\Sigma 3$ and {110} $\Sigma 3$ grain boundaries during slow cooling and exists in the form of clusters [31,40]. Especially, highly Fe contaminated Si, the EBIC contrasts of all the $\Sigma 3$ grain boundaries increase due to formation of Fe precipitates [31]. Generally, grain boundaries with symmetric interfaces show weaker contrast than those with asymmetric interfaces at any temperature [23]. The $\Sigma 9$ grain boundary with straight segments is recombination active because it forms triangles that contains {112} $\Sigma 3$ grain boundaries which become recombination active by impurity decorations [27,36,43]. As the boundary planes of large angle grain boundaries are smooth and small angle grain boundaries have array of edge dislocations, the large angle grain boundaries have a small amount of shallow levels incorporated in them, while the small angle grain boundaries have a large amount of shallow levels or electrically active impurities introduced by dislocations which act as effective recombination centers [28]. Thus, not only the grain boundary character but also the grain boundary plane affects the recombination activity of grain boundaries [27]. Active grain boundaries are significant in reducing the lifetime and electrically active dislocations are clustered around grain boundaries [11].

Clean grain boundaries are electrically inactive in nature, and their recombination activity is due to decoration by impurities such as Fe [27,36]. Thus, the recombination activity of grain boundaries depends heavily on the contamination levels in the materials [44]. The higher metal impurities level in p-type wafers can contribute to a higher recombination velocities of grain boundaries. Recombination active grain boundaries lead to a local reduction of lifetime and excess carrier concentration at the defects, resulting in a local reduction of both the open circuit voltage and the short circuit current of solar cells [45]. The small angle grain boundaries or sub-grain boundaries can be observed as dense dark lines by the EL imaging under forward bias because they act as electrically active sites to decrease the EL intensity [46].

The recombination activity of grain boundaries is principally determined by the getting ability of grain boundaries [27]. The recombination activity of a random grain boundary strongly increases after the PDG process [12]. Random grain boundaries have the strongest gettering ability, whereas {111} $\Sigma 3$ grain boundaries have the weakest [27]. The high-Σ grain boundaries in contaminated specimen have more gettering sites for impurities than the low-Σ grain boundaries [27]. The recombination activities of the random grain boundary and other crystal defects are lowered after an additional firing of hydrogenated SiN because of H bulk passivation [12]. H is released from the SiN:H layers and enters the bulk during firing where it passivates the defects (see Section 2.4.1).

Sio et al. [36] studied the relation between grain boundary effective surface recombination velocity, S_{GB} and the thermal processes. S_{GB} represents the intrinsic recombination properties of a grain boundary. A higher S_{GB} corresponds to greater recombination active and a much lower S_{GB} than 100 cm s^{-1} corresponds to recombination inactive. The average intragrain lifetime of wafers cut from the middle part of an ingot is the highest due to lower impurity concentrations in the wafers and S_{GB} is the lowest because of slightly decorated grain boundaries with impurities. The median S_{GB} values of grain boundaries from the middle part of the ingot are higher than those from the top and bottom after the P gettering or hydrogenation. Thus, the high temperature steps have more substantial influence on grain boundaries in cleaner wafers than grain boundaries in more contaminated wafers.

2.2.4 Recombination centers (recombination activity of dislocations)

Most of dislocations are already recombination active in the as–grown state before any processing [45]. Dislocations formed during growth exhibit a higher degree of disorder (kinks, jogs, tangles) because dislocation mobility, dislocation climbing and kink formation are favored at high temperature [13]. The degree of disorder of dislocation microstructure can be known by the shape of etch pits of dislocations. The ellipticity of the etch pits of dislocations corresponds to the degree of the disorder which exhibits the recombination activity of dislocation lines or clusters. The eccentricity of dislocation clusters estimated from the elliptical shape of the etch pits increases as the recombination strength increases [47]. Thus, highly recombination-active dislocation clusters exhibit a high degree of disorder such as curvilinear entanglement [47]. This relationship is similar to that for grain boundaries. In the same way as Eq. (2.4), the normalized recombination strength of dislocations can be written by

$$\Gamma_D = V_D/D_m, \tag{2.5}$$

where V_D is the dislocation recombination velocity. Γ_D differs for dislocations in a wafer or across an ingot because the recombination properties of dislocations largely vary even in neighboring areas [48].

The recombination behavior of dislocations increases with heavy O decoration and contamination with metallic impurities [48]. As the temperature drops, supersaturated metals precipitate at the most energetically favorable heterogeneous nucleation sites like dislocations [13]. The recombination strength relates to metal impurity decoration of dislocation clusters. In a high Γ_D region, clustering of Fe and Cu precipitates is shown around dislocation etch pits by the synchrotron-based nanoprobe X-ray fluorescence (XRF) mapping [13]. The recombination activity of Fe-contaminated dislocations significantly increases [3]. Thus, the recombination-active dislocations contain a high degree of nano-scale Fe and Cu decoration. The recombination-active dislocations lead to a local reduction of lifetime and excess carrier concentration at the defects [45]. A critical dislocation density is 10^6 cm^{-2}, which is the minimum dislocation density leading to a lowered lifetime compared with non dislocated surroundings [49].

The recombination behavior of dislocations can be improved by P gettering and H-passivation [48]. The Γ_D-values of dislocation clusters are increased during the PDG process because supersaturated impurity atoms at dislocations or random grain boundaries are emitted during PDG at high temperature [12]. The recombination activity of dislocations and grain boundaries is strongly enhanced by PDG at 900 °C because metallic impurities become mobile and are internally gettered by extended defects after PDG from their surrounding regions [50]. The Γ_D-values of dislocation clusters are effectively decreased by SiN:H firing or hydrogenation [12]. Thus, the recombination activity of dislocations is reduced by firing of hydrogenated Si_3N_4 which is a suitable method for H passivation of the extended defects after P diffusion. The extended defects are stronger gettering sites than the P layer at typical process temperature of 900 °C, and the external gettering is insufficient at most of these defects [50]. Metal impurities and precipitates trapped in dislocation cores cannot be readily removed during gettering. These impurities form deep-level recombination centers, which deteriorate the solar cell performance [51].

2.2.5 Diffusion length and minority carrier lifetime

Fig. 2.2 shows the diffusion length of minority carriers (μm) measured by the surface photovoltage (SPV) method with a prove size of 0.1 cm$^{\varphi}$ as a function of the sub-grain boundary density (mm^{-1}) in an ingot [52]. A spatially resolved X-ray rocking curves with angle resolution better than 0.01° can be applied for analysis of the sub-grain boundary density [52]. The number of sub-grain boundaries is estimated to be (n-1), where n is the number of peaks in an X-ray rocking curve, and the density of sub-grain boundaries is estimated to be 0.5 (n-1) mm^{-1}. These data were measured in the center and bottom of the ingot, which correspond to the position of 1.0—2.0 cm and 0—1.0 cm from the bottom of the ingot, respectively. The solid lines are drawn using Eq. (2.14) by employing S_g (cm s^{-1}) as a fitting parameter [52] (see Section 2.6.1). S_g is a constant recombination velocity at sub-grain boundaries, which does not depend on the angular difference. By fitting the data using S_g in Fig. 2.2, the values of S_g can be estimated in the center and bottom of the ingot as 7.4×10^3 and 2.2×10^4 cm s^{-1}, respectively. The bottom part contains a lot of metallic impurities. The diffusion length of minority carriers decreases as the sub-grain boundary density increases. The estimated effective diffusion

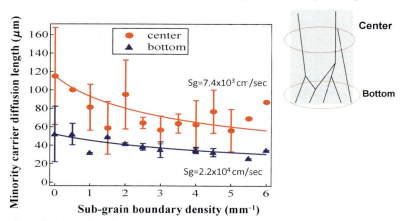

Reduction of dislocations is very important to obtain high-efficiency solar cells

Fig. 2.2 Minority carrier diffusion length (μm) as a function of the sub-grain boundary density (mm^{-1}) in an ingot. The sub-grain boundary comprises a dislocation array with many dislocations. *(Referred from Fig. 6 in K. Kutsukake et al. J. Appl. Phys. 105 (2009) 044909.)*

length also decreases as the dislocation density increases and it is close to the diffusion length in the dislocation-free semiconductor for low dislocation densities [53,54]. Therefore, the reduction of dislocations is very important to obtain high-efficiency solar cells.

The minority carrier lifetime can be determined from a measured difference in the conductivity by QSSPC. The conductivity is proportional to the injection level and the sum of the minority carrier mobility and the majority carrier mobility [55]. The mobility describes the drift and diffusion of charged carriers, and it is in inverse proportion to the resistivity. The minority carrier mobility directly corresponds to the diffusion coefficient or diffusion length. The mobility reduction due to the presence of crystal defects in multicrystalline Si ranges within 5% [55]. To identify and quantify lifetime-limiting defects, the injection-dependent lifetime spectroscopy can be used as a rapid and powerful method [56].

The dislocation related lifetime, τ_d is also in inverse proportion to the dislocation density and the minority carrier diffusion coefficient [53]. The effective lifetime, τ_e can be written by τ_d and the lifetime in the dislocation-free semiconductor, τ_0 as follows,

$$1/\tau_e = 1/\tau_0 + 1/\tau_d \tag{2.6}$$

High carrier lifetimes (>100 μs) are only measured in areas with low dislocation density ($<10^5$ cm^{-2}), and low carrier lifetimes (<20 μs) are observed in areas with high dislocation density ($>10^6$ cm^{-2}) [57]. After PDG is applied, the carrier lifetime increases by about a factor of three in lowly dislocated regions, whereas no gettering efficiency is observed in highly dislocated areas because impurities locked at dislocations remain and determine the local carrier recombination activity [57]. The recombination velocity for electrons in a p-type Si wafer is higher than that for holes in an n-type Si wafer because the electron capture cross section is larger than the hole capture cross section. For grain boundaries, the recombination velocity for electrons can be reduced by P gettering and hydrogenation [14,36]. The minority carrier lifetime of holes is longer than that of electrons at a same doping concentration because of such difference of the capture cross sections. Thus, the average minority carries lifetime of n-type wafers is higher than that of p-type wafers, and this difference becomes significant especially inside grains [49]. If the minority carrier lifetimes are same for n- and p-type wafers, the diffusion length of p-type wafers can be estimated as 1.8 times longer than that of n-type wafers using the same equation as Eq. (2.7), because the diffusion coefficient of electrons is about 3 times larger than that of holes.

The minority carrier lifetime of n-type CZ crystals does not significantly depend on the C concentration that is in the relatively high range between 1×10^{15} and $4 \times 10^{16} \, cm^{-3}$, but it depends on the dopant (P) concentration [58]. On the other hand, C impurities in Si crystals act as heterogeneous nucleation sites for O precipitates which reduce the minority carrier lifetime [59]. The low C concentration can reduce O content and the high C concentration forms O complex [59]. Thus, the level of the C concentration is very important for its effect on the minority carrier lifetime. N doped Si crystals are normally prepared by the float-zone (FZ) method to prevent agglomeration of intrinsic defects [60]. In this case, the lifetime is strongly reduced by vacancy related defects and completely recovered by annealing at higher than $1000°C$ in O because of annihilation of the recombination active defects [60].

The conversion efficiency of a solar cells is partly determined by the number of minority carriers that diffuse to the collecting junction before recombining, and hence it is affected directly by the minority carrier diffusion length, rather than the minority carrier lifetime which is a better comparability of the material quality [21,45]. In general, minority carrier diffusion lengths in solar cells should be equal to or exceed the wafer thickness [3]. The main factor to limit the diffusion length in p-type multicrystalline Si wafers is dissolved-impurities which homogeneously distribute as recombination centers [21]. The redistribution of precipitates and dissolution of impurities are large factors for carrier lifetime in n-type multicrystalline Si wafers [21]. In comparison to p-type Si, the diffusivity of the minority carriers in n-type Si is lower by about a factor of 2 or 3 depending on the doping concentration [61]. The minority carrier diffusion length, L can be calculated the following equation,

$$L = \sqrt{D\tau}, \tag{2.7}$$

where D is the diffusivity or diffusion coefficient. As the minority carrier diffusivity in n-type Si is typically a factor of 3 lower than that in p-type Si, the minority carrier lifetime of n-type Si must increase by a factor of 3 to achieve the same diffusion length [22]. As the minority carrier lifetime in n-type Si is usually much higher than that in p-type Si, such higher carrier lifetime can offset the lower diffusivity. Therefore, the diffusion length of minority carriers in n-type Si is higher by a factor of 2 or 3 than that in p-type Si after all processes such as the P gettering and hydrogenation [21]. The voltage of solar cells depends on the minority carrier lifetime, while the current depends on the diffusion length. For the comparison between p- and n-type solar cells, both the minority carrier lifetime and the diffusion length should be considered.

2.3 Optical measurement method and optical properties of Si crystals

Luminescence from a semiconductor is related to its energy bandgap and radiative recombination involving defects with energy levels located within the bandgap [17]. PL spectroscopy is a useful technique to detect defects and impurities in semiconductors [29]. PL imaging usually takes advantage of the band-to-band carrier recombination in Si wafers [24]. PL spectroscopy and topography provide the spectroscopic and structural information of the lifetime-limiting defects with the advantages of non-destructiveness, noncontact and rapidity [62]. Using these methods, the typical broadband appeared at 0.8 eV at RT in multicrystalline Si crystals is known to consist of both O precipitation-related and dislocation-related components, which are defects, impurities and O precipitations around or on dislocations [62]. The thermal donors are produced by O aggregation which appears by annealing at 450°C, and the deep-level PL emissions of such thermal donors-related defects are observed at 0.767, 0.926 and 0.965 eV [63]. PL imaging is beneficial for detecting a variety of defects in Si crystal solar cells. C is another unwanted impurity in Si and it acts as a nucleation site for O precipitates [64]. Typical C related PL emissions are the C-line at 0.789 eV and the G-line at 0.969 eV. They correspond to the C_i-O_i complex and the C_i-C_s complex, respectively [64]. The C08-band appears at RT near the C-line and has a very similar origin to that of the C-line [64]. Luminescence from random grain boundaries show strong PL imaging contrast while the contrast at $\Sigma 3$ and $\Sigma 9$ CSL grain boundaries can barely be seen [15]. The intensity of radiative band-to-band recombination decreases at defective random grain boundaries, due to non-radiative recombination centers at the defects. Grain boundaries in multicrystalline Si show radiative recombination in the near infrared region (1.0−0.7 eV) at RT and these luminescence centers are not attributed to dislocations [65]. Carrier recombination at these grain-boundary luminescence centers in the near infrared region is negligible and these centers are not lifetime limiting [65].

Sub-band-gap PL spectroscopy is a powerful technique to study defects and impurities in Si solar cells [66]. In an as-grown wafers, the PL intensity of D_{a1} band decreases with increasing the misorientation angle of sub-grain boundaries [67]. The average dislocation distance decreases or the dislocation density increases as the misorientation angle increases [68]. The D_{a1} peaks are attributed to secondary defects and/or impurities trapped by the strain field around dislocations and exist at approximately 0.78 eV [67]. Sub-grain boundaries emit four distinct sub-band-gap peaks of D-lines called D1. D2, D3 and D4. The doublet D1/D2 is thought to be emitted from decorating defects and impurities trapped around dislocations and the emission is a better indicator of high recombination activity at sub-grain boundaries [66]. The doublet D3/D4 reflects intrinsic properties of dislocations. High densities of metal impurities are present at locations with strong D1/D2 emission. but not present at locations with low D3/D4 emission [66]. The intensity of the doublet D3/D4 is reduced more quickly as the temperature rises than that of the doublet D1/D2 [66]. The carrier recombination properties at sub-grain boundaries are strongly related to the PL emission in the high-energy region between 0.83 eV and 0.87 eV (D_b peak), and the D_b peaks are considered to be related to O precipitates [67].

The combination of the EBSD measurement and laser confocal microscopy images is effective for a fast analysis of rough and textured surfaces such as as-cut and artificially modified surfaces [69]. The method can detect the orientation of small surface elements which is depicted in an orientation distribution function of surface normal. The typical resolution of the microscope is $10-20$ nm in height and 100 nm in lateral direction [69].

As the absorption coefficient, α is 2×10^3 cm^{-1} at wavelength, λ of 0.7 μm for Si [70] and the thickness of 20 μm is necessary to absorb 98% of solar light, a thick layer is required for a Si solar cell. So, it is very important to reduce crystal defects such as dislocations in the Si single and multicrystalline ingots because long diffusion length is necessary for such thick layers. The internal quantum efficiency (IQE) decreases at higher dislocation density than 10^5 cm^{-2} [71].

2.4 Processes to control electrical and optical properties of Si crystals

2.4.1 Gettering and hydrogenation

Some types of impurities can be removed from a wafer by gettering which concentrates impurities in a specific region of the wafer such as surface.

This gettering is referred as extrinsic or external gettering. In the case that the specific regions are internal features such as dislocations, grain boundaries and O precipitates, the gettering is referred as intrinsic or internal gettering. All gettering processes are caused by the very large segregation coefficients achieved between the gettering region and the bulk [1]. During formation of pn junctions and firing of metal contacts, the gettering and hydrogenation are in common solar cell fabrication steps. The electrical properties of Si crystals can be significantly changed after these processes [49].

Using the external gettering, impurity concentrations can be reduced in the device region of a wafer by localizing them in separate pre-defined regions of the wafer where they cannot affect the performance [3]. To drive impurities out of a crystal wafer into a P-rich layer, reduce metallic contamination and improve the minority carrier lifetime, the external P gettering is usually performed through subjecting wafers by a $POCl_3$ diffusion at a high temperature (about $825°C$) and annealing in an N_2 ambient for several hours at a lower temperature. $POCl_3$ is used as carrier gas and a phosphor-silicate glass forms on the Si surface, which acts as the doping source for the P in–diffusion [3]. Such P diffusion is effective in gettering grown-in impurities in low quality Si wafers [44]. The P diffusion gettering transports impurities to P-rich layers or surfaces to form substitutional impurity-P pairs [72]. It enhances the diffusion of substitutional metals toward the gettering layer through the injection of Si self-interstitials [3]. The gettering based on segregation is driven by a gradient or a discontinuity of the impurity solubility, and the higher solubility region acts as a sink for impurities from the lower solubility region [3]. It decreases the metal concentration below its equilibrium solubility at the annealing temperature [3]. The solubility of metals increases as the diffusivity or diffusion length of the metals becomes higher, so the supersaturation or the efficient gettering is achieved at much higher temperature, corresponding to higher diffusivity [3]. The P gettering is very effective to improve the intra-grain lifetimes especially on wafers with higher levels of impurities and its efficiency is anti-corelated with the number of lattice distortions [36]. The P gettering depends on the concentration of structural defects and the O and C concentrations [3]. Even though the P gettering significantly improves the lifetime in intragrain regions, it activates grain boundaries, resulting in a reduction of the average minority carrier lifetime. The external gettering is much stronger than the internal gettering [12].

The internal gettering transports dissolved impurities to precipitates such as SiO_2 or defects. The internal-gettering process requires an impurity supersaturation which occurs during cooling, and saturated impurities precipitate in gettering sites. The internal gettering is related to the impurity gettering effect by grain boundaries which purifies the region just near the grain boundary [15]. The impurity gettering abilities of grain boundaries are as following, low-Σ grain boundaries < high-Σ grain boundaries < random grain boundaries < small angle grain boundaries [73]. Thus, the small angle grain boundaries show stronger impurity-gettering ability than the large angle grain boundaries, and they have the strongest impurity gettering ability and tend to be the most detrimental grain boundaries because of their boundary dislocation structure [33,73]. The P diffusion increases the recombination strength of most of grain boundaries because the high-temperature process such as the P diffusion or the rapid quenching process leads to a re-distribution of metal impurities around the grain boundaries which causes a change in their recombination behaviors [6].

The decoration of dislocations with impurities cannot be dissolved during the gettering process, whereas impurities from non-distorted sites are successfully removed from wafers [57]. In Si multicrystals, the best response for P gettering or the significant enhancement of the recombination lifetime by P gettering is obtained mainly in areas of wafers where the initial lifetime is relatively high, the dislocation density is relatively low and the concentration of mobile impurities is high [1,74]. The high lifetime regions experience the largest lifetime enhancements after P gettering [74]. High dislocation density reduces the gettering effect [75]. The transition metal contamination is most significant in regions with a high dislocation density [74]. The lifetime in these regions is limited by the recombination activity of dislocations or nongetterable impurities decorated or precipitated around the dislocations [74].

If the lifetime increases during PDG, the lifetime in the as-grown wafers is dominated by getterable and fast-diffusing metal impurities such as Cu, Ni, Co, Fe, Cr and Mn at concentrations that are low enough to be effectively removed during processes [76]. The diffusivity of Co in Si is estimated to be larger than that of Fe [77]. PDG at temperature as low as 820°C fully removes Cu-, Ni-, and Co-rich particles to below detection limits of synchrotron-based micro-X-ray fluorescence (μ-XRF), and PDG at 920°C can remove all precipitated transition metals including Fe to below detection limits [78]. Minority carrier increases with increasing precipitate dissolution by PDG at 820−920°C [78]. In the case that PDG does not lead to sufficiently high bulk lifetime and the defect density is low, the material performance is limited by either non-getterable impurities, impurity complexes such as B-O, or by slowly-diffusing metal impurities such as Ti, vanadium (V), Zn, platinum (Pt) and Au [76]. The diffusion length in n-type wafers is more significantly increased by the P gettering comparing with that in p-type wafers [21].

On the other hand, a B diffusion leads to a thermal degradation of the sample lifetime because it is performed at higher temperature compared to the P diffusion [61]. The material degradation after the B diffusion is attributed to the higher temperature (890°C) compared with the P diffusion (800 °C). Thus, the standard B diffusion is not effective at gettering metal impurities like the P diffusion. The high temperature leads to a stronger dissolution of precipitates. The B diffusion degrades the material quality of both n- and p-type wafers compared with the initial state. However, the diffusion length in n-type wafers is still on a high level (≥ 300 μm) and much larger than in p-type wafers [21] (see Sections 2.4.2 and 2.5.2).

The hydrogenation or H-passivation is performed through firing H-rich Si_3N_4 coated wafers at a high temperature in an N_2 ambient to produce bulk hydrogenation. H comes from the H-rich Si_3N_4 layers deposited on both sides of the wafers for use as an antireflection coating. The Si_3N_4 layers are deposited by plasma-enhanced chemical vapor deposition (PECVD). Such layers are frequently used as an antireflection coating to increase the absorption of incident sunlight. A greater amount of H is introduced into the layers grown by the higher frequency reactor [49]. The high temperature annealing/firing steps drive H from the dielectric films into the bulk with a concentration of $\sim 10^{15}$ cm^{-3}, passivating the defects [79]. H monoatoms bond with the detrimental dangling bonds of crystal defects or wafer surface. The H-passivation takes place almost within the whole volume and not only in a thin layer below the surface [50]. On the other hand, the surface passivation can be done by H-rich amorphous Si (a-Si:H) deposited in a PECVD system. The hydrogen passivation can be done by a forming gas (90% N_2, 10% H_2) annealing to reduce the recombination–active defects [80]. H forms neutral complexes with B and P, which contain one H atom and the H concentration can be monitored by measuring the electrically active B or P profile [72].

The hydrogenation benefits both the intra-grain regions and grain boundaries. Performing both the P gettering and then the hydrogenation results in the highest intra-grain lifetimes. Moreover, the hydrogenation is very effective in reducing the recombination strength of all gettered grain boundaries because it is more effective in passivating impurities such as interstitial impurities or metal precipitates re-distributed after the P gettering [6]. Thus, the hydrogenation is much more effective on gettered grain boundaries compared with as-grown grain boundaries [36]. Chen et al. studied the effect of the hydrogenation in detail by EBIC [81]. In the low contaminated multicrystalline Si, the effect of the hydrogenation depends on both the grain boundary character and impurity contamination level. Σ grain boundaries are more effectively passivated than the random and small angle grain boundaries. In the heavy contaminated multi-crystalline Si, the grain boundary character has no significant impact on the effect of the hydrogenation. Thus, the hydrogenation works less efficiently in the heavy contaminated multicrystalline Si than in the low contaminated multicrystalline Si. In Si multicrystalline wafers with grain boundaries as recombination centers, a hydrogenation step is applied after PDG to effectively passivate crystal defects [76]. The hydrogenation can significantly improve the overall lifetimes due to the effective passivation of grain boundaries and the neutralization of detrimental influence of the P high-temperature diffusion such as the re-distribution of metal impurities around the grain boundaries [6,44,49].

The major effect of hydrogenation occurs at grain boundaries and dislocations [82]. The reason is that the highest density of recombination centers at the grain boundaries and more defects can be passivated. Dislocations are also passivated almost all over the volume [82]. However, the recombination lifetime is only significantly enhanced in regions where the dislocation density is low due to the lower transition metal concentrations [74]. Thus, the lowest dislocation density causes the largest lifetime enhancements after hydrogenation. In the low lifetime regions, high impurity concentrations cause a dislocation decoration, resulting in poor gettering and passivation effects of both P gettering and hydrogenation [74]. Thus, the hydrogenation is not efficient on dislocation clusters which remain strongly recombination active after both processes [44]. The diffusion length in p-type wafers is more significantly increased by the hydrogenation comparing with that in n-type wafers [83].

2.4.2 Gettering and low-temperature annealing for Fe impurities

Fe impurities are very detrimental elements for Si and have a strong negative influence on minority carrier lifetime in the crystal [84]. Fe_i and Fe-B pairs influence the recombination lifetime in Si multicrystalline crystals though the proportions in the total Fe concentration are small [85]. The Fe_i capture cross section for electron is nearly three orders of magnitude higher than for holes [86,87]. The Fe_i point defect has a capture cross section ratio for electron and holes of $k = 186$ [76] or 260 [87], and it exhibits a large injection dependence. Fe_i forms an energy level at $E_v + 0.38$ eV and is not stable in the Si lattice due to its very low solubility at RT [3]. The capture cross sections for electron and holes of the energy level of $E_v + 0.38$ eV for Fe_i are 5×10^{-14} cm^{-2} and 7×10^{-17} cm^{-2}, respectively [1]. From a least-squares fit of many reported data, the hole capture cross section of Fe_i can be expressed as $\sigma(Fe_i) = (3.9 \pm 0.5) \times 10^{-16} \times \exp(-(0.045 \pm 0.005 \text{ eV})/k_B T)$ cm^2 in the temperature range from 150 K to 300 K [88].

The carrier lifetime is often limited by Fe_i and precipitated Fe is less recombination active than Fe_i, even though the Fe_i concentration is smaller than the total Fe concentration [83]. The higher total Fe concentration leads to higher Fe_i concentration and most of Fe in low dislocated grains is in the interstitial state [85]. The Fe_i concentration is shown by Coletti et at [86]. The as-grown Fe_i concentration in a p-type Fe doped ingot is about 10^{13} cm^{-3} which represent about 10% of the total Fe concentration. During PDG, Fe_i is highly mobile and diffuses into the P-doped regions. The Fe_i concentration in the ingot decreases to about 10^{11} cm^{-3} after PDG and it is further reduced by a factor 2 after subsequent hydrogenation [86]. The density of Fe precipitates between 3×10^4 and 5×10^5 is reduced by 70%−90% due to PDG [83]. The slow cooling rate during the crystallization process grows much amount of large precipitates by giving Fe enough time to diffuse. Therefore, for the slow cooling rate of 0.35 K min^{-1}, the density of large precipitates, which typically include most of the precipitated Fe, is reduced only slightly during PDG [83]. For the fast cooling rate of 35 K min^{-1}, the density of large precipitates is largely reduced and most of Fe is already dissolved during PDG [83]. Thus, the PDG is more effective after the fast cooling rate, due to the high fraction of getterable Fe_i. However, a complete dissolution of precipitates results in high ungetterable Fe_i concentration after PDG due to the limited gettering capacity [83]. Fig. 2.3 shows the clear lifetime maps of neighboring wafers from the middle of the p-type Fe doped ingot after each solar cell process such as (A) as grown, (B) after P diffusion, (C) after P diffusion + hydrogenation and (D) after B/P co-diffusion + hydrogenation [86]. The lifetime increases extremely after P diffusion especially in the intragrain areas and the darker areas improve after hydrogenation [86]. After B/P co-diffusion + hydrogenation, the

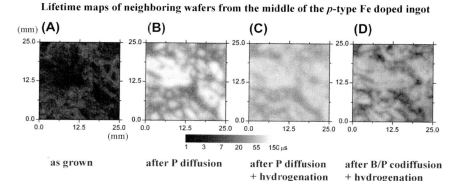

Fig. 2.3 Lifetime maps of neighboring wafers from the middle of the p-type Fe doped ingot after each solar cell process such as (A) as grown, (B) after P diffusion, (C) after P diffusion + hydrogenation and (D) after B/P codiffusion + hydrogenation. *(Referred from Fig. 11 in G. Coletti et al. J. Appl. Phys. 104, 104913 (2008).)*

darker areas shift toward low lifetime because the B diffusion is a high temperature process and the thermal degradation occurs during the process [61]. The Fe_i concentration in the ingot shown in Fig. 2.3 is about 10^{12} cm^{-3} after B/P co-diffusion + hydrogenation [86].

Precipitate sizes become smaller after PDG relative to the as-grown state, and the higher-temperature gettering processes can reduce Fe-rich precipitate sizes [89]. However, even the highest temperature P diffusion (920°C) does not fully dissolve the largest precipitates [89]. The PDG process incompletely dissolves these precipitated metals, but the higher-temperature PDG process can largely reduce the precipitated metal content and this reduction can significantly improve lifetime in Fe-dominated crystals [89].

The Fe_i concentrations in areas of high dislocation density do not significantly depend on the concentration level because grain boundaries and grains with high dislocation density have a remarkable lower Fe_i concentration than good grains [85]. The reduction of the precipitate density inside a grain is three times higher than at grain boundary during PDG [83]. The Fe_i concentration further decreases by a factor for 2 after hydrogenation. The lifetime of a p-type Si ingot with Fe impurities increases extremely after PDG especially in intragrain areas and the average lifetime more increases after both PDG and hydrogenation. On the other hand, the most harmful impurities in n-type wafers are mainly substitutional ones because they are ungetterable defects. Only the surface atoms of the precipitates serve as recombination centers [12]. Dissolved precipitates result in high point-defect concentrations of impurities which can be removed during the extended thermal annealing [76].

All Fe and Cu precipitates seen in Si crystals with Fe + Cu are co-located, and they appear generally smaller and have fewer atoms than those seen in Fe-only crystals [89]. In comparison the Fe, Fe + Cu, and Cu contaminated wafers with each other, the as-grown lifetimes in these wafers are dominated by Fe, since the as-grown lifetimes in the Cu only wafer are much higher [89]. Fe and Cu contaminations lead to an increase of the EBIC contrast of electrically active grain boundaries [90]. On the other hand, Fe and Cu strongly decorate dislocations after growth and form complexes and/or precipitates with low recombination activity, thus dislocations in multicrystalline Si are not effective getters of Fe and Cu [90]. The P gettering is not sufficiently effective to remove the Fe and Cu precipitates [91].

Low-temperature processing is effective to reduce the Fe_i concentration under some conditions [25]. Rinio et al. [12,92] studied the effect of the low-temperature anneal which is performed at temperature from 300 to 800°C directly after P diffusion. Grain boundaries and dislocations are weakly recombination active after crystallization with slow cooling from 1400°C to RT because some impurities have enough time to reprecipitate. However, they show increased recombination activity after the P emitter diffusion for 90 min at 900°C followed by rapid cooling because the precipitates can emit dissolved atoms again during the P diffusion. The low-temperature anneal effectively improves the deteriorated area with a high amount of Fe precipitates near the crucible wall. Real effect is not seen in the middle of an ingot with low Fe concentration by the annealing. The best improvement can be obtained near 600°C. The required annealing time is about 1 h. The reason for the positive effect of the low-temperature anneal is mainly due to the external P gettering into emitters or surfaces. The diffusion length of Fe impurities during anneal, L_{Fe} can be calculated by the following equation,

$$L_{Fe} = \sqrt{D_{Fe}t}, \tag{2.8}$$

where D_{Fe} is the diffusion coefficient for Fe and t is the annealing time.

Fe exists as Fe_i and precipitated iron (Fe_p) in a Si crystal. The effect of the low-temperature annealing was also confirmed by Fenning et al. [25]. The low-temperature annealing does not change Fe precipitate size distribution and affects the density of precipitates minimally. However, the low-temperature annealing with the 900 °C P-diffusion process increases the medium precipitate size and decreases the precipitate density. The P-diffusion followed by the low-temperature annealing decreases the Fe_i concentration over three orders of magnitude and increases the bulk lifetime. The low-temperature annealing acts on Fe_i and has little effect on Fe_p. After the low-temperature annealing, the effective lifetime is dominated by Fe that remains as recombination-active precipitates.

The capture cross sections for electron and holes of the acceptor level of E_c - 0.23 eV for Fe-B pairs are 5×10^{-15}–3×10^{-14} cm^{-2} and 2×10^{-15}–3×10^{-15} cm^{-2}, respectively [1,87]. In B doped p-type Si, Fe_i has a particularly strong recombination activity [22,87]. Fe_i is mobile and can diffuse short distance in wafers at RT, and quickly forms immobile Fe-B pairs which are stable or low recombination active at RT and can be easily dissociated by annealing at 200 °C for a few minutes or illuminating the wafer with high intensity light [3,93]. Fe-B pairs have shallow levels with low recombination activity at RT even though Fe_i and Fe precipitates have deep levels with strong recombination activity at RT [31]. Fe-B pairs form a donor level at E_v + 0.1 eV and an acceptor level at $E_c - (0.26 \pm 0.03)$ eV, where E_c is the conduction level in the band structure [3]. The level of the orthorhombic configuration (E_v + 0.1 eV) of the Fe-B pair is closer to the valence band edge than that of the trigonal configuration [88]. The acceptor level of the $\langle 100 \rangle$ orthorhombic configuration of Fe-B pairs is E_c − 0.43 eV [94]. Fe_i is transformed from Fe-B pairs dissociated by strong illumination or minority carrier injection [88]. Fe_i and Fe-B pairs affect the recombination lifetime though the proportion in total Fe concentration is generally small [85]. The lifetime limited by Fe-B pairs is inversely proportional to the concentration of Fe-B pairs and becomes higher as the B concentration becomes lower [3,93]. Incidentally, Cr-B pairs can be also thermally dissociated as well as Fe-B pairs and B-O defect complex [24].

2.4.3 Tabula Rasa process

Oxide precipitates are usually SiO_x ($x = 1$–2) and their nucleation is facilitated by O_2V complexes that are formed during cooling process after high temperature rapid thermal processing [95]. In order to mitigate O-related defects, a high-temperature homogenization anneal call Tabula Rasa (TR) has been used in the electronics industry [96]. The TR process can be used for solar-grade wafers to dissolve oxide precipitates and to homogenize the lifetime across Si wafers with swirl defects caused by oxide precipitates. The activation energy governing O precipitate dissolution is 2.6 ± 0.5 eV, which is in the same range as the O interstitial migration enthalpy of ~ 2.6 eV, suggesting the TR process to be kinetically limited by O diffusion and meaning the long-time and high-temperature process [96]. Harmless free O atoms can be obtained from such O-related defects by the TR process with the high-temperature annealing. Jensen and Youssef et al. [97,98] reported the effect of the TR process on O-related defects. Even though ingots contain some amount of O impurities, the relative recombination strength of O-related defect regions becomes weaker and the minority carries lifetime is largely improved using the TR process following by the PDG process. The conversion efficiency of solar cells also largely increases after TR prior to PDG. However, the minority carries lifetime is uniformly degraded after TR only. The TR process consists of a timed anneal at 1090 °C in a clean tube furnace with a rapid cooling rate above 100 °C min^{-1}. Such TR process can remove O-related defects such as swirl patterns throughout an ingot and obtain higher lifetime. Swirl patterns are associated with point-defect clusters–voids, stacking faults and dislocation loops, which are favorable nucleation sites for metal-impurity precipitation [10].

Thus, the TR process dissolves or reduces the size of O precipitates, but precipitates are again formed by PDG when the precipitates are not completely dissolved by TR. The effect of the TR process is enhanced by the extended (EXT) process at low temperature [10]. The EXT process consists of a 25 min plateau at 845 °C followed by a 2 h anneal at 650 °C before unloading. Recombination-active precipitates of Cr, Cu, Ni, and Fe respond well to the EXT gettering process [10] (see Section 7.5.1).

2.4.4 Thermal donors

The thermal donor is related to O impurity, which affects the stability of the resistivity of Si crystal and devices [99]. The thermal donors are usually generated in CZ Si crystal by heat treatments in the temperature range of 350–500°C, and they are eliminated by heat treatments in the temperature above 500°C [100]. The O_i concentration is the important factor for the formation of thermal donors, and it is related to the concentration of thermal donors which are O-related centers acted as donor [100–102]. The thermal donors are the small O clusters formed through the aggregation of fast-diffusing di-O atoms such as $O_2 + O_2 = O_4$ (thermal donor) [100].

88 Crystal Growth of Si Ingots for Solar Cells Using Cast Furnaces

C exerts an inhibited effect on the formation of thermal donors due to the formation of C–O complexes [100]. N impurity in Si annealed at 450°C can suppress the formation of the O related thermal donors because N interacts with O atoms to form N–O complexes, and the shallow thermal donor is related to the N–O complex [99,100].

The generation of thermal donors decreases the resistivity of n-type CZ Si samples and increases that of p-type CZ samples [103]. A high concentration of thermal donors affects the electrical performance of solar cells [103].

2.5 Si crystal solar cells

2.5.1 Dopants

B, Al, Ga and In can be used as p-type dopant and N, P, As and Sb can be used as n-type dopant. Most multicrystalline solar cells are based on B doped p-type Si wafers with minority carrier of electrons. Solar cells prepared by P doped n-type Si wafers have minority carrier of holes. The segregation coefficients of B and P in Si (B or P in Si crystal/B or P in Si melt) are 0.7–0.8 [86,104,105] and 0.35 [86], respectively. The segregation coefficients of As in Si is 0.3–0.8 [103,106]. The segregation coefficients of Al, Ga, In and N, Sb are 0.002 [107,108], 0.008 [104], 0.000–0.0005 [107–110], 0.0007 [71,111] and 0.023 [103], respectively and they are two and three orders of magnitude lower than B in Si.

Al-, Ga- and In-doped solar cells show no light induced efficiency degradation [112]. Al, Ga and In concentrations increase in the ingots with the growth height due to their quite small segregation coefficients. The homogeneous distribution of dopants is more difficult to be obtained for the Al, Ga and In elements. Al is totally ionized and occupy the substitutional sites in Si [112]. A high content of Al causes extensive micro-cracking in the surroundings of Al phases in the Si ingot due to the thermal mismatch between Si and Al inclusions [113]. The average conversion efficiency of Al-doped solar cells is 0.34%–0.62% lower than that of Ga-doped ones because of Al-related deep level recombination centers [112,114]. The growth of Ga-doped Si multicrystalline ingots is explained in Section 8.4. In is a volatile dopant and the evaporation rate of In from a Si melt is estimated to be 1.6×10^{-4} cm s^{-1} [109]. The unstable cellular growth easily occurs with nucleation of micro-precipitates of In in a highly In-doped CZ Si ingot because of the constitutional supercooling caused by the low solubility and quite low segregation coefficient of In

Ref. [110] (see Section 1.2.1.4). The electrical activity of In dopants in Si decreases from 100% to 20% as the doping level increases from 10^{14} cm^{-3} to 10^{17} cm^{-3} because of formation of micro-precipitates [110]. The double feeding of Al and In leads to uniform impurity profiles for each element over the entire length of a crystal [108].

The evaporation rates of P and Sb from a Si melt are very similar in the order of 5.5×10^{-5} cm s^{-1}, which are controlled by the impurity transport through the Si melt and strongly depend on the melt convection [115]. P has a strong tendency to segregate to grain boundaries [15]. For As in Si, As and Si in a tetrahedral configuration are nearly the same size [103,104]. P and As decrease the melting point of Si, but B increases the melting point of Si [106].

2.5.2 Performance of p- and n-type Si solar cells

For preparation of Si solar cells, alkali solution is used to form inverted pyramid texture to reduce the surface reflection of sun light from the solar cell surface. A SiN$_x$ layer is used to form a uniform antireflection film by PECVD. After them, front and back screen metallization and metallization firing are performed.

The conversion efficiency, η (%) of a solar cell is determined by the following relation,

$$\eta = (J_{SC} V_{OC} FF) \times 100 / P_{in}, \tag{2.9}$$

where J_{sc}, V_{oc} and FF are the short circuit photo-current, open–circuit voltage and fill factor of the solar cell, respectively. P_{in} is the power incident on the solar cell per unit area ($P_{in} = 100$ mW cm^{-2} under standard AM1 conditions). AM 1 is the solar spectrum through the direct optical pass on the earth.

Generally, a lower base resistivity increases V_{oc}, and potentially leads to higher conversion efficiency [21,116]. However, the low resistivity also increases Auger and radiative recombination. For industrial multicrystalline solar cells, a commonly used base resistivity of $1-3$ Ω cm provides a good balance between these parameters [116]. As the electron capture cross-section is much larger than the hole capture cross-section, for p-type solar cells with electron as minority carriers, the optimum base resistivity strongly depends on defect concentration. For p-type solar cells with Fe$_i$ or B–O defects, such defects drive the optimum resistivity to be higher [116]. The lifetime roughly increases with the resistivity because the lifetime increases as the base doping concentration decreases. Such behavior for Fe_i is caused by the fact that the hole capture cross-section is much smaller than the electron capture cross-section. When holes are the minority carrier, a small hole capture cross-section will result in a low recombination rate and high lifetime [116]. As a high-resistivity material is more tolerant to contamination, the carrier lifetime in p-type Si is enhanced with increasing base resistivity [57]. For n-type solar cells, recombination due to Fe_i is much lower and nearly independent of resistivity [116].

The dissolved impurities lead to efficiency loss for p-type solar cells [21]. For p-type solar cells, processes of the P gettering and hydrogenation are normally incorporated in the solar cell fabrication (see Section 2.4.1). The carrier lifetime after illumination increases using the P gettering because the metastable B–O defects decrease by the thermal process [101]. Using H–passivation triggered by forward bias, the illumination stability is improved and the light–induced degradation becomes much smaller for both Al back–surface-field (Al–BSF) and Passivated emitter and rear contact (PERC) solar cells because the recombination-active B–O defects are removed [117]. The n-type Si has a smaller impact of certain metal impurities such as Fe on electrical properties [21,22]. Solar cells prepared from n-type Si crystals have higher conversion efficiencies than those of p-type solar cells [21]. For n-type solar cells, the B diffusion is typically required for the formation of pn junction. The V_{oc} can be significantly improved without the B diffusion, particularly for single wafers such as mono–like wafers [61]. The decorated crystal defects after the B-diffusion lead to efficiency loss for n-type solar cells [21]. The degradation during the B diffusion is evident and this process should be improved (see Sections 2.4.1 and 2.4.2).

The performance (V_{oc} and fill factor) of high-efficiency Passivated emitter rear locally-diffused (PERL) and PERC solar cells prepared by gettered Si multicryslalline wafers is limited almost wholly by the bulk recombination lifetime [1]. The presence of metallic impurities such as Fe_i, Cu and Cr reduces the recombination lifetime. Dimitriadis [2] theoretically studied the relation between the conversion efficiency and the dislocation density, ρ_d. For small dislocation density ($<10^3$–10^4 cm^{-2}), the conversion efficiency does not depend on the dislocation density and significantly decreases as the effective diffusion length decreases. The effective diffusion length is due to the bulk and grain boundary recombination. For high dislocation density ($>10^3$–10^4 cm^{-2}), the conversion efficiency, η initially decreases slowly with the dislocation density (η is proportional approximately to $\rho_d^{-0.5}$), then rapidly decreases with the dislocation density (η is proportional approximately to ρ_d^{-1}). The conversion efficiency is independent of the effective diffusion length at higher dislocation density than 5×10^5 cm^{-2}.

Needleman et al. [14,118] simulated the relation between the conversion efficiency and the dislocation density. The conversion efficiency of Al–BSF solar cells strongly depends on the dislocation density and starts to decrease above the dislocation density of 1×10^3 cm^{-3}. When the average dislocation density is same, a higher conversion efficiency can be generally reached if the dislocations are clustered than if they are homogeneously distributed. The low recombination strength of the dislocation clusters or the relatively clean (undecorated) and/or well-passivated dislocations can obtain the high conversion efficiency. The simulated conversion efficiency, short-circuit current density and open-circuit voltage of solar cells decrease as the density of metallic impurities (recombination centers) decorating a grain boundary increases [119]. The simulated conversion efficiency is also very sensitive to lifetime and steeply increases with increasing lifetime, especially in the low lifetime range [76]. The simulated Voc and conversion efficiency increase as the grain size increases and the areal density of recombination centers at a grain boundary decreases. To obtain a high efficiency solar cells, larger grains and lower grain boundary defect concentrations are required. n-Type solar cells that operate at higher injection levels can reduce the impacts of dislocations and grain boundaries [120].

2.5.3 B-O complexes

B has a strong affinity for pairing with O in p-type Si like interstitial boron (B_i)- Fe$_s$ complex [121]. The activation energy to form B-O complexes or defects is $0.2-0.4$ eV [103] or $0.42-0.56$ eV [122]. The activation energy to dissociate B-O complexes is ~ 1.37 eV and the binding energy of B-O complexes is 0.93 eV [122]. B doped p-type Si solar cells suffer from B-O-related defects because the minority carrier lifetime is limited by the metastable B-O-related defects, which are formed due to illumination or minority-carrier injection [101]. The bulk lifetime of B-doped Si wafers decreases during illumination because the B-O-related defect concentration increases by illumination, especially for wafers with a low resistivity (<5 Ω cm) due to a high B doping concentration [123]. As the O$_i$ concentration of Si multicrystalline wafers is considerably lower than that of Si CZ wafers, B_i-O_i concentration is largely reduce and the light-degraded lifetime is greatly improved [124]. The O_i concentration in a Si multicrystalline ingot decreases from the bottom to the top. The carrier lifetime after illumination increases by a factor of 2−3.5 using the P gettering because the metastable defects decrease by the thermal process and the fast cooling rate [101]. The B-O defect concentration does not necessarily reach an equilibrium after processing, but an equilibrium can be established if the samples are annealed in darkness such as annealing at 200°C for 50 h in darkness, and higher annealing temperatures lead to lower equilibrium defect concentrations [125]. The defect pair of B_i and O_i is identified as a deep-level defect [86].

Schmidt et al. [114,124] and Bothe et al. [101] proved the defect pair as a B_s-O_{2i} complex by B_s and O_{2i}. B_s is largely immobile in Si and B_i exists in negligible amounts in Si. O_{2i} which is made up of two O_i atoms can diffuse fast in Si. The O_{2i} concentration increases with the O_i concentration. The diffusivity of O_{2i} in Si is several orders of magnitude higher than that of O_i. The diffusion activation energy of the neutral O_{2i} dimers is 1.3 eV and the diffusivity can be expressed as $D_{O2i} = 1.0 \times \exp(- 1.3 \ eV/k_B T) \ cm^2 \ s^{-1}$ [122]. The fast-diffusing O_{2i} dimers are captured by immobile B_s to form the lifetime-limiting B_s-O_{2i} metastable complex, which acts as a highly effective recombination center. Thus, the B_s-O_{2i} defect formation process is governed by the diffusion of O_{2i} and it is a thermally activated process. The B_s-O_{2i} defect concentration increases with increasing illumination time and increasing B_s, and the defect generation rate increases with the illumination intensity at low intensities [114]. The resistivity increases as the O_i concentration increases [114]. Thus,the carrier lifetime after illumination decreases with increasing the O_i and B_s concentrations [101]. The B_s-O_{2i} defects with deep level completely dissociate at temperature above 200°C, thus the degradation of the carrier lifetime can be completely recovered by a low-temperature anneal (350−500°C) [124]. The B_s-O_{2i} defect concentration also decreases by annealing for formation of thermal donors or small O clusters at 450 °C because the O_{2i} concentration decreases by the thermal process [114]. Moreover, from the detailed analysis by Schmidt and Glunz et al. [126,127], the fundamental recombination center created during illumination has an energy level between $E_v + 0.35$ and $E_c = 0.45$ eV. The defect concentration is approximately proportional to the O_i concentration and the complex may be like BO_{i5}, which have five O atoms in the complex.

Re-formation of the B_i-O_i pairs (B_iO_i) is prevented at RT by a competing reaction of B_i with C_s, leading to the formation of B_i-C_s pairs

(B_iC_s) which are much less effective recombination centers than B_i–O_i pairs [124]. The B_i–C_s pairs can be dissolved to B_i and C_s by illumination and the B_i–O_i pairs can be dissolved to B_i and O_i by annealing, which are proposed as a model for the bulk lifetime degradation and recovery mechanism observed in B-doped CZ Si [124]. In this case, B_iO_i has a chance to be formed again by the reaction of O_i with dissolved B_i instead of C_s and B_iC_s has a chance to be formed again by the reaction of C_s with dissolved B_i instead of O_i. This competing reaction is also associated with the capture of O_{2i} by C_s [114]. The formation is in direct competition with the formation of the B_s–O_{2i} complex (see Section 2.4.1).

Lifetime degradation is suppressed for B doped p-type Si single wafers with low O content because the light degradation increases with the O concentration [128]. No degradation is observed even for solar cells prepared by B doped p-type Si multicrystalline wafers with low O concentration [57]. Comparing with p-type Si, the bulk lifetime of P doped n-type Si wafers shows no change during illumination [124]. P doped n-type Si has advantages like the absence of the B–O-related degradation and the lifetime degradation effect does not occur for P-doped Si wafers [114]. Even for p-type Si single wafers, the lifetime degradation is suppressed for Ga-doped Si single wafers even with high O content [128]. The degradation does not occur for the bulk lifetime of Ga- and In-doped p-type Si wafers except for Al-doped Si wafers with Al-related defects [114]. Ga dopants strongly suppress the recombination center such as radiation induced defects of C_i–O_i complex, which is one of the most stable complexes of all the interstitials related deep levels in Si [121]. According to Schmidt's mode [114], the B atom is 25% smaller than the Si atom and the O_{2i} dimer is preferentially accommodated near a B_s atom. On the other hand, Al, Ga and In atoms are 8%–23% larger than the Si atom. The tetrahedral covalent radius of B is 0.8 Å and that of Ga is 1.26 Å [121]. The larger size of a Ga atom produces larger stress comparing to a B atom. Therefore, there is not enough space available near these atoms to accommodate the O_{2i} dimer. The bonding energy between B and Si is relatively weak, and B_s unstably exists in a Si crystal as compared with substitutional Ga (Ga_s) [121]. Thus, B is more active to make a complex with impurities as compared with Ga. The Ga and In doped Si is suitable material for high-efficiency solar cells. A high-temperature oxidation near 800 °C is effective to obtain no light degradation of Si solar cells using Ga–doped wafers [128]. However, the segregation coefficients of Ga and In are much lower than B in Si, which cause a high variation of resistivity along the growth direction (see Section 8.4).

Recently, Sun et al. [129] measured the activation rate of the slowly forming recombination center of the B-O defect in n- and p-type Si under high-injection conditions by micro PL measurements using the excitation wavelength of 532 nm. The B-O defects are activated by diffusing excess carriers as well as injection excess carriers. The degradation of the integrated band-to-band PL signal is caused by the activation of the B-O defects. The effective saturated concentration of slowly forming recombination centers has the same activation rate constant when fixing the high-injection level in both n- and p-type Si wafers. The activation rate constant depends only on the difference between the B and P concentrations (net doping concentration), but not on the O_i and B concentrations. This observation is against the above mentioned B_s-O_{2i} model.

The Ge doping retards the reaction of B-O defects because Ge atoms capture O_i atoms and generate a large energy barrier for diffusion of O_i to form O_{2i} [130] (see Section 8.5.2). The C co-doping with B can effectively suppress the formation of B-O complexes, and the saturated B-O complexes concentration decreases the C concentration increases [131]. The C co-doping offers the low light-induced degradation effect.

2.6 Theoretical estimation for the characterization of crystals

2.6.1 Estimation of the diffusion length of minority carriers

Spatial distribution of minority carrier diffusion length can be measured by SPV. The recombination velocity at sub-grain boundaries is estimated using the following simple analysis [52]. The minority carrier diffusion length in grains, L_0 (cm) is assumed to be constant and not to be affected by sub-grain boundaries. The sub-grain boundaries are assumed to be aligned perpendicular to the sample surface and have a longer distance than L_0 between adjacent them. The recombination velocity at sub-grain boundaries, S_g (cm s^{-1}) is assumed to be constant. Lower S_g indicates the lower recombination activity and week recombination strength of sub-grain boundaries. S_g is assumed not to depend on the angular difference and the surface recombination is neglected. The number of carriers recombined in a grain in unit time and volume, G_0 (cm^{-3} s^{-1}) is expressed as

$$G_0 = \frac{\Delta n}{\tau_0} = \frac{\Delta n D}{L_0^2}, \tag{2.10}$$

where Δn τ_0, and D are density (cm^{-3}), lifetime (s), and diffusion constant ($\mathrm{cm}^2\ \mathrm{s}^{-1}$) of excess minority carriers, respectively. The number of carriers recombined at sub-grain boundaries, G_g ($\mathrm{cm}^{-3}\ \mathrm{s}^{-1}$) is expressed as

$$G_g = 2\Delta n S_g N, \tag{2.11}$$

where N (cm^{-1}) is the density of sub-grain boundaries. Therefore, the number of recombined total carriers, G is

$$G = \frac{\Delta n D}{L_0^2} + 2\Delta n S_g N, \tag{2.12}$$

Using the measured diffusion length, L (cm), G ($\mathrm{cm}^{-3}\ \mathrm{s}^{-1}$) can be express as follows,

$$G = \frac{\Delta n D}{L^2}, \tag{2.13}$$

From Eqs. (2.12) and (2.13), L can be expressed as

$$L = \sqrt{\frac{1}{\dfrac{1}{L_0^2} + \dfrac{2 S_g N}{D}}}, \tag{2.14}$$

The solid lines shown in Fig. 2.1 are calculated using Eq. (2.14) and S_g as a fitting parameter.

References

[1] D.H. Macdonald, Recombination and Trapping in Multicrystalline Silicon Solar Cells, The thesis for the degree of Doctor of Philosophy of the Australian National University, May 2001.
[2] C.A. Dimitriadis, J. Phys. D Appl. Phys. 18 (1985) 2489.
[3] A.A. Istratov, H. Hieslmair, E.R. Weber, Appl. Phys. A 70 (2000) 489.
[4] R.A. Sinton, A. Cuevas, Appl. Phys. Lett. 69 (1996) 2510.
[5] A. Cuevas, R.A. Sinton, Prog. Photovoltaics Res. Appl. 5 (1997) 79.
[6] R.A. Sinton, A. Cuevas, M. Stuckings, Proc. 25th IEEE Photovoltaic Specialists Conf. (1996) 457.
[7] R.A. Sinton, A. Cuevas, M. Stuckings, 25th PVSC, May 13-17, 1996. Washington D. C.
[8] S.Z. Karazhanov, J. Appl. Phys. 89 (2001) 332.
[9] M. Kivambe, D.M. Powell, M.A. Jensen, A.E. Morishige, K. Nakajima, R. Murai, K. Morishita, T. Buonassisi, in: 40th IEEE Photovoltaic Specialists Conf., 2014.
[10] M. Kivambe, D.M. Powell, S. Castellanos, M.A. Jensen, A.E. Morishige, K. Nakajima, K. Morishita, R. Murai, T. Buonassisi, J. Cryst. Growth 407 (2014) 31.
[11] G. Stokkan, S. Riepe, O. Lohne, W. Warta, J. Appl. Phys. 101 (2007) 053515.

[12] M. Rinio, A. Yodyunyong, S. Keipert-Colberg, D. Borchert, A. Montesdeoca-Santana, Phys. Status Solidi 208 (2011) 760.

[13] M.I. Bertoni, D.P. Fenning, M. Rinio, V. Rose, M. Holt, J. Maser, T. Buonassisi, Energy Environ. Sci. (2011), https://doi.org/10.1039/c1ee02083h (On line).

[14] D.B. Needleman, H. Wagner, P.P. Altermatt, T. Buonassisi, IEEE J. Photovol. 6 (2016) 817.

[15] F. Liu, C.-S. Jiang, H. Guthrey, S. Johnston, M.J. Romero, B.P. Gorman, M.M. Al-Jassim, Sol. Energy Mater. Sol. Cells 95 (2011) 2497.

[16] T. Isobe, H. Nakashima, K. Hashimoto, Jpn. J. Appl. Phys. 28 (1989) 1282.

[17] M.P. Peloso, J.S. Lew, P. Chaturvedi, B. Hoex, A. Aberle, Prog. Photovoltaics Res. Appl. 20 (2012), 661.

[18] M.P. Peloso, B. Hoex, A.G. Aberle, Appl. Phys. Lett. 98 (2011) 171914.

[19] D. Macdonald, A. Cuevas, Solar Energy Mater.Solar Cells 65 (2001) 509.

[20] D. Macdonald, R.A. Sinton, A. Cuevas, J. Appl. Phys. 89 (2001) 2772.

[21] F. Schindler, B. Michl, A. Kleiber, H. Steinkemper, J. Schön, W. Kwapil, P. Krenckel, S. Riepe, W. Warta, M. Schubert, IEEE J. Photovolt. 5 (2015) 499.

[22] D. Macdonald, L.J. Geerligs, Appl. Phys. Lett. 85 (2004) 4061.

[23] Z.-J. Wang, S. Tsurekawa, K. Ikeda, T. Sekiguchi, T. Watanabe, Interface Sci. 7 (1999) 197.

[24] M.C. Schubert, H. Habenicht, W. Warta, IEEE J. Photovol. 1 (2011) 168.

[25] D.P. Fenning, J. Hofstetter, M.I. Bertoni, S. Hudelson, M. Rinio, J.F. Lelièvre, B. Lai, C. del Cañizo, T. Buonassisi, Appl. Phys. Lett. 98 (2011) 162103.

[26] R. Sachdeva, A.A. Istratov, E.R. Weber, Appl. Phys. Lett. 79 (2001) 2937.

[27] J. Chen, T. Sekiguchi, D. Yang, F. Yin, K. Kido, S. Tsurekawa, J. Appl. Phys. 96 (2004) 5490.

[28] J. Chen, B. Chen, T. Sekiguchi, M. Fukuzawa, M. Yamada, Appl. Phys. Lett. 93 (2008) 112105.

[29] H. Sugimoto, M. Inoue, M. Tajima, A. Ogura, Y. Ohshita, Jpn. J. Appl. Phys. 45 (2006) L641.

[30] T. Kojima, T. Tachibana, Y. Ohshita, R.R. Prakash, J. Appl. Phys. 119 (2016) 065302.

[31] J. Chen, D. Yang, Z. Xi, J. Appl. Phys. 97 (2005) 033701.

[32] T. Sekiguchi, J. Chen, W. Lee, H. Onodera, Phys. Status Solidi C 8 (2011) 1347.

[33] J. Chen, T. Sekiguchi, Jpn. J. Appl. Phys. 46 (2007) 6489.

[34] W. Lee, J. Chen, B. Chen, J. Chang, T. Sekiguchi, Appl. Phys. Lett. 94 (2009) 112103.

[35] J. Chen, T. Sekiguchi, R. Xie, P. Ahmet, T. Chikyo, D. Yang, S. Ito, F. Yin, Scripta Mater. 52 (2005) 1211.

[36] H.C. Sio, S.P. Phang, T. Trupke, D. Macdonald. IEEE J. Photovol. 5 (2015) 1357.

[37] T. Buonassisi, A.A. Istratov, M.D. Pickett, M.A. Marcus, T.F. Ciszek, E.R. Weber, Appl. Phys. Lett. 89 (2006) 042102.

[38] P. Karzel, M. Ackermann, L. Gröner, C. Reimann, M. Zschorsch, S. Meyer, F. kiessling, S. Riepe, G. Hahn, J. Appl. Phys. 114 (2013) 244902.

[39] K. Arafune, E. Ohishi, H. Sai, Y. Terada, Y. Ohshita, M. Yamaguchi, Jpn. J. Appl. Phys. 45 (2006) 6153.

[40] B. Chen, J. Chen, T. Sekiguchi, M. Saito, K. Kimoto, J. Appl. Phys. 105 (2009) 113502.

[41] T. Tachibana, T. Sameshima, T. Kojima, K. Arafune, K. Kakimoto, Y. Miyamura, H. Harada, T. Sekiguchi, Y. Ohshita, A. Ogura, J. Appl. Phys. 111 (2012) 074505.

[42] M. Kohyama, R. Yamamoto, Y. Watanabe, Y. Ebara, M. Kinoshita, J. Phys. C Solid State Phys. 21 (1988) L695.

[43] G. Stokkan, J. Cryst. Growth 384 (2013) 107.

[44] H.C. Sio, S.P. Phang, P. Zheng, Q. Wang, W. Chen, H. Jin, D. Macdonald, Jpn. J. Appl. Phys. 56 (2017) 08MB16.
[45] H.C. Sio, D. Macdonald, Sol. Energy Mater. Sol. Cells 144 (2016) 339.
[46] N. Usami, K. Kutsukake, K. Fujiwara, I. Yonenaga, K. Nakajima, Appl. Phys. Express 1 (2008) 075001.
[47] S. Castellanos, M. Kivambe, J. Hofstetter, M. Rinio, B. Lai, T. Buonassisi, J. Appl. Phys. 115 (2014) 183511.
[48] H.J. Möller, C. Funke, M. Rinio, S. Scholz, Thin Solid Fims 487 (2005) 179.
[49] S. Gindner, P. Karzel, B. Herzog, G. Hahn, IEEE J. Photovolt. 4 (2014) 1063.
[50] M. Rinio, E. Zippel, D. Borchert, The 20th European Photovoltaic Solar Energy Conf. & Exhibition, Barcelona, Spain, June 6-10, 2005.
[51] S. Woo, M. Bertoni, K. Choi, S. Nam, S. Castellanos, D.M. Powell, T. Buonassisi, H. Choi, Sol. Energy Mater. Sol. Cells 155 (2016) 88.
[52] K. Kutsukake, N. Usami, T. Ohtaniuchi, K. Fujiwara, K. Nakajima, J. Appl. Phys. 105 (2009) 044909.
[53] C. Donolato, J. Appl. Phys. 84 (1998) 2656.
[54] S. Pizzini, A. Sandrinelli, M. Beghi, D. Narducci, F. Allegretti, S. torchio, J. Electrochem. Soc. 135 (1988) 155.
[55] F. Schindler, J. Geilker, W. Kwapil, W. Warta, M.C. Schubert, J. Appl. Phys. 110 (2011) 043722.
[56] A.E. Morishige, M.A. Jensen, D.B. Needleman, K. Nakayashiki, J. Hofstetter, T.A. Li, T. Buonassisi, IEEE J. Photovol. 6 (2016) 1466.
[57] O. Schultz, S.W. Glunz, S. Riepe, G.P. Willeke, Prog. Photovoltaics Res. Appl. 14 (2006) 711.
[58] Y. Miyamura, H. Harada, S. Nakano, S. Nishizawa, K. Kakimoto, J. Cryst. Growth 486 (2018) 56.
[59] S. Kishino, M. Kanamori, N. Yoshihiro, M. Tajima, T. Iizuka, J. Appl. Phys. 50 (1979) 8240.
[60] N.E. Grant, V.P. Markevich, J. Mullins, A.R. Peaker, F. Rougieux, D. Macdonald, Phys. Status Solidi RRL, Wiley Online Library, 2016.
[61] S.P. Phang, H.C. Sio, C.F. Yang, C.W. Lan, Y.M. Yang, A.W.H. Yu, B.S.L. Hsu, C.W.C. Hsu, D. Macdonald, Jpn. J. Appl. Phys. 56 (2017) 08MB10.
[62] M. Tajima, IEEE J. Photovol. 4 (2014) 1452.
[63] F. Higuchi, M. Tajima, A. Ogura, Jpn. J. Appl. Phys. 56 (2017) 070308.
[64] M. Tajima, Y. Ishikawa, H. Kiuchi, A. Ogura, Appl. Phys. Express 11 (2018) 041301.
[65] F. Dreckschmidt, H.J. Möller, Phys. Status Solidi C 8 (2011) 1356.
[66] H.T. Nguyen, M.A. Jensen, L. Li, C. Samundsett, H.C. Sio, B. Lai, T. Buonassisi, D. MacDonald, IEEE J. Photovol. 7 (2017) 772.
[67] M. Funakoshi, N. Ikeno, T. Tachibana, Y. Ohshita, K. Arafune, A. Ogura, Jpn. J. Appl. Phys. 53 (2014) 112401.
[68] T. Sameshima, N. Miyazaki, Y. Tsuchiya, H. Hashiguchi, T. Tachibana, T. Kojima, Y. Ohshita, K. Arafune, A. Ogura, Appl. Phys. Express 5 (2012) 042301.
[69] T. Behm, C. Funke, H.J. Möller, Surf. Interface Anal. 45 (2013) 781.
[70] S.M. Sze, Physics of Semiconductor Devices, John Wiley & Sons, New York, 1969.
[71] H.J. Möller, C. Funke, D. Kreßner-Kiel, S. Würzner, Energy Procedia 3 (2011) 2.
[72] E.B. Yakimov, Metal impurities and gettering in crystalline silicon, in: D. Yang (Ed.), Crystalline Silicon Growth, Hand Book of "Photovoltaic Silicon Material", Springer, Berlin Heidelberg, 2018, pp. 1−46 (On line).
[73] J. Chen, T. Sekiguchi, D. Yang, Phys. Status Solidi 4 (2007) 2908.
[74] A. Bentzen, A. Holt, R. Kopecek, G. Stokkan, J.S. Christensen, B.G. Svensson, J. Appl. Phys. 99 (2006) 0935509.

98 Crystal Growth of Si Ingots for Solar Cells Using Cast Furnaces

[75] D. You, J. Du, T. Zhang, Y. Wan, W. Shans, L. Wang, D. Yang, in: Proceedings of 35th IEEE Photovol. Specialists Conf., 2010, pp. 2258–2261.
[76] J. Hofstetter, C. del Cañizo, H. Wagner, S. Castellanos, T. Buonassisi, Prog. Photovoltaics Res. Appl. 24 (2016) 122.
[77] M.C. Schubert, J. Schön, F. Schindler, W. Kwapil, A. Abdollahinia, B. Michl, S. Riepe, C. Schmid, M. Schumann, S. Meyer, W. Warta, IEEE J. Photovolt. 3 (2013) 1250.
[78] A.E. Morishuge, M.A. Jensen, J. Hofstetter, P.X. Yen, C. Wang, B. Lai, D.P. Fenning, T. Buonassisi, Appl. Phys. Lett. 108 (2016) 202104.
[79] S. Kleekajai, F. Jiang, M. Stavola, V. Yelundur, K. Nakayashiki, A. Rohatgi, G. Hahn, S. Seren, J. Kalejs, J. Appl. Phys. 100 (2006) 093517.
[80] K. Arafune, M. Nohara, Y. Ohshita, M. Yamaguchi, Sol. Energy Mater. Sol. Cells 93 (2009) 1047.
[81] J. Chen, D. Yang, Z. Xi, T. Sekiguchi, Phys. Biol. 364 (2005) 162.
[82] M. Rinio, A. Hauser, H.J. Möller, in: WCPEC-3 Osaka, May 11-18, 2003.
[83] J. Schön, A. Haarahiltunen, H. Savin, D.P. Fenning, T. Buonassisi, W. Warta, M.C. Schubert, IEEE J. Photovol. 3 (2013) 131.
[84] M.J. Alam, M.Z. Rahman, Appl. Mech. Mater. 440 (2014) 82.
[85] J. Schön, H. Habenicht, M.C. Schubert, W. Warta, Solid State Phenom. 156–158 (2010) 223.
[86] G. Coletti, R. Kvande, V.D. Mihailetchi, L.J. Geerligs, L. Arnberg, E.J. Øvrelid, J. Appl. Phys. 104 (2008) 104913.
[87] D. Macdonald, T. Roth, P.N. Deenapanray, Appl. Phys. Lett. 89 (2006) 142107.
[88] A.A. Istratov, H. hieslmair, E.R. Weber, Appl. Phys. A 69 (1999) 13.
[89] D.P. Fenning, A.S. Zuschlag, M.I. Bertoni, B. Lai, G. Hahn, T. Buonassisi, J. Appl. Phys. 113 (2013) 214504.
[90] O.V. Feklisova, X. Yu, D. Yang, E.B. Yakimov, Phys. Status Solidi C 9 (2012) 1942.
[91] X. Li, D. Yang, X. Yu, D. Que, Trans. Nonferrous Metals Soc. China 21 (2011) 691.
[92] M. Rinio, A. Yodyunyong, S. Keipert-Colberg, Y.P.B. Mouafi, D. Borchert, A. Montesdeoca-Santana, Prog. Photovoltaics Res. Appl. 19 (2011) 165.
[93] M. Yli-Koski, M. Palokangas, V. Sokolov, J. Storgårds, H. Väinölä, H. Holmberg, J. Sinkkonen, Phys. Scripta T 101 (2002) 86.
[94] H. Nakashima, T. Sadoh, T. Tsurushima, Phys. Rev. B 49 (1994) 16983.
[95] B. Wang, X. Zhang, X. Ma, D. Yang, J. Cryst. Growth 318 (2011) 183.
[96] E.E. Looney, H.S. Laine, A. Youssef, M.A. Jensen, V. LaSalvia, P. Stradins, T. Buonassisi, Appl. Phys. Lett. 111 (2017) 132102.
[97] M.A. Jensen, V. LaSalvia, A.E. Morishige, K. Nakajima, Y. Veschetti, F. Jay, A. Jouini, A. Youssef, P. Stradins, T. Buonassisi, Energy Procedia 92 (2016) 815.
[98] A. Youssef, J. Schön, S. Mack, T. Niewelt, K. Nakajima, K. Morishita, R. Murai, T. Buonassisi, M.C. Schubert, in: Proceedings of the 43th IEEE Photovoltaic Specialists Conference (43th PVSC), 2016, pp. 68–72.
[99] D. Yang, D. Que, K. Sumino, J. Appl. Phys. 77 (1995) 943.
[100] X. Yu, J. Chen, X. Ma, D. Yang, Mater. Sci. Eng., A R 74 (2013) 1.
[101] K. Bothe, R. Sinton, J. Schmidt, Prog. Photovoltaics Res. Appl. 13 (2005) 287.
[102] S. Singh, R. Singh, B.C. Yadav, Physica B 404 (2009) 1070.
[103] X. Yu, D. Yang, Growth of crystalline silicon for solar cells: czochralski Si, in: D. Yang (Ed.), Chapter 13, Section Two, Crystalline Silicon Growth, Hand Book of "Photovoltaic Silicon Material", Springer, Berlin Heidelberg, 2018, pp. 1–45 (On line).
[104] F.A. Trumbore, Bull Syst. Tech. J. 39 (1960) 212.
[105] M. Sanati, S.K. Estreicher, Phys. Rev. B 72 (2005) 165206.
[106] M.M. Mandurah, K.C. Saraswat, C.R. Heims, T.I. Kamins, J. Appl. Phys. 51 (1980) 5755.

[107] W. Dietze, W. Keller, A. Mühlbauer, J. Grabmaier, in: Crystals—Growth, Properties and Applications, vol. 5, Springer-Verlarg, 1982.

[108] Z.A. Agamaliev, Z.M. Zakhrabekova, V.K. Kyazimova, G.K. Azhdarov, Inorg. Mater. 52 (2016) 244.

[109] S. Haringer, A. Giannattasio, H.C. Alt, R. Scala, Jpn. J. Appl. Phys. 55 (2016) 031305.

[110] X. Yu, X. Zheng, K. Hoshikawa, D. Yang, Jpn. J. Appl. Phys. 51 (2012) 105501.

[111] J. Li, R.R. Prakash, K. Jiptner, J. Chen, Y. Miyamura, H. Harada, K. Kakimoto, A. Ogura, T. Sekiguchi, J. Cryst. Growth 377 (2013) 37.

[112] S. Yuan, X. Yu, X. Gu, Y. Feng, J. Lu, D. Yang, Superlattice. Microst. 99 (2016) 158.

[113] T. Orellana, E.M. Tejado, C. Funke, S. Riepe, J.Y. Pastor, H.J. Möller, J. Mater. Sci. 49 (2014) 4905.

[114] J. Schmidt, K. Bothe, Phys. Rev. B 69 (2004) 024107.

[115] M. Porrini, R. Scala, V.V. Voronkov, J. Cryst. Growth 460 (2017) 13.

[116] L.J. Geerligs, D. Macdonald, Prog. Photovoltaics Res. Appl. 12 (2004) 309.

[117] M. Xie, C. Ren, L. Fu, X. Qiu, X. Yu, D. Yang, Front. Energy 11 (2017) 67.

[118] D.B. Needleman, H. Wagner, P.P. Altermatt, Z. Xiong, P.J. Verlinden, T. Buonassisi, Sol. Energy Mater. Sol. Cells 158 (2016) 29.

[119] D.B. Needleman, H. Wagner, P.P. Altermatt, T. Buonassisi, Energy Procedia 77 (2015) 8.

[120] D.B. Needleman, A Private Communication, 2017.

[121] A. Khan, M. Yamaguchi, Y. Ohshita, N. Dharmarasu, K. Araki, T. Abe, H. Itoh, T. Ohshima, M. Imaizumi, S. Matsuda, J. Appl. Phys. 90 (2001) 1170.

[122] M. Xie, X. Yu, Y. Wu, D. Yang, J. Electron. Mater. 47 (2018) 5092.

[123] W.W. Mullins, R.F. Sekarka, J. Appl. Phys. 35 (1964) 444.

[124] J. Schmidt, A. G. Aberie, R. Hezel, 26th PVSC, Anaheim, CA, USA, September 30-October 3, 1997, pp. 13-18.

[125] D.C. Walter, R. Falster, V.V. Voronkov, J. Schmidt, Sol. Energy Mater. Sol. Cell. 173 (2017) 33.

[126] S.W. Glunz, S. Rein, J.Y. Lee, W. Warta, J. Appl. Phys. 90 (2001) 2397.

[127] J. Schmidt, A. Cuevas, J. Appl. Phys. 86 (1999) 3175.

[128] T. Saitoh, X. Wang, H. Hashigami, T. Abe, T. Igarashi, S. Glunz, S. Rein, W. Wettling, I. Yamasaki, H. Sawai, H. Ohtuka, T. Warabisako, Sol. Energy Mater. Sol. Cell. 65 (2001) 277.

[129] C. Sun, H.T. Nguyen, H.C. Sio, F.E. Rougieux, D. MacDonald, IEEE J. Photovol. 7 (2017) 988.

[130] X. Zhu, X. Yu, D. Yang, J. Cryst. Growth 401 (2014) 141.

[131] Y. Wu, S. Yuan, X. Yu, X. Qiu, H. Zhu, J. Qian, D. Yang, Sol. Energy Mater. Sol. Cell. 154 (2016) 94.

CHAPTER 3

Growth of Si multicrystalline ingots using the conventional cast method

3.1 Unidirectional growth of Si multicrystalline ingots

3.1.1 Design of the furnace with a hot zone

For the growth of a Si multicrystalline ingot, the temperature field or distribution, the temperature gradient and the stress distribution strongly affect the microstructure and crystal defects in the ingot. The temperature field and the temperature gradient strongly affect the growth rate, the interface shape and the thermal stress. The shape of the solid/liquid or growing interface determined by the temperature field affects the thermal stress, crystal defects and the impurity distribution in the ingot. The smaller convexity at the interface leads to the smaller thermal stress in the ingot. Therefore, it is very important to design a cast furnace with a suitable hot zone which can control the temperature field and the stress distribution to obtain a high-quality ingot. As an example of the cast furnace, a schematic illustration of an industrial cast furnace (GT DSS 450) is shown in Fig. 3.1A [1,2]. The maximum ingot size of 84 cm × 84 cm × 36 cm (generation five: G5) can be grown by the furnace. Generally, inside a stainless chamber, the furnace has many contents such as graphite resistance heaters, graphite insulators, a graphite case, a graphite susceptor, a graphite partition block, a heat-exchanger, a gas shield, a protecting graphite cover-plate against contamination, a graphite block for a crucible, a quartz crucible and thermocouples. To obtain a Si ingot with low O concentration, pure inert Ar gas continuously flows inside the chamber to remove O which evaporates from a Si melt, prevent oxidation of graphite materials and purify the growth environment. The furnace design for the gas flow is important to obstruct or make the upward Ar gas flow near the heaters by adding an insulation layer on the side insulation or a hole on the top

Crystal Growth of Si Ingots for Solar Cells Using Cast Furnaces
ISBN 978-0-12-819748-6
https://doi.org/10.1016/B978-0-12-819748-6.00003-7

Copyright © 2020 Elsevier Inc.
All rights reserved.

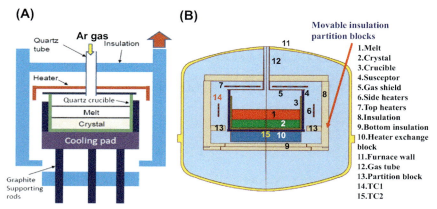

Fig. 3.1 (A) Schematic illustration of an industrial cast furnace (GT DSS 450). The maximum ingot size of 84 cm × 84 cm 36 cm (generation five: G5) can be grown by the furnace. (B) Cast furnace with movable insulation partition blocks. ((A) Referred from Fig. 1 (B) in C. W. Lan et al. J. Cryst. Growth 360 (2012) 68. (C) Referred from Fig. 1 in Z. Wu et al. J. Cryst. Growth 426 (2015) 110.)

insulation, and it can significantly influence the transport of CO and reduce the C concentration in the Si melt [3]. The nucleation and ingot growth are mainly controlled by lifting the insulation made by graphite to decrease the melt temperature by increasing the upward heat flow. The cooling pad below the crucible is a graphite heat exchanger of a water-cooling system or a graphite chamber through Ar gas flow. The major heat loss during ingot growth is radiation heat loss from the insulations and the heat-exchanger block, and the thermal conductivity of the insulations plays an important role in determining the heat loss rate due to the heat conduction from the insulation layers to the surroundings [2,4]. The quality of the crucible strongly affects the quality of grown ingot because Fe is found in large quantities within industrial standard ceramic crucible materials [5].

To optimize the design of the hot zone with a suitable thermal field, the effects of the heater position and partition block on the temperature distribution, the shape of growing interface and the thermal stress should be clearly known. The heater position has a significant impact on the temperature distribution and the interface shape. The side and top heaters should be used together to control the temperature distribution and the interface shape [6]. The partition block increases the vertical temperature gradient in a Si melt because the heat flow in the vertical direction is partially blocked by the partition block [7]. The design of the partition block is very sensitive for the growth of high-quality ingots. Numerical simulation is a useful tool to visualize the temperature distribution in the furnace, and even an effect of minor geometry modifications such as adding small blocks on energy saving can be known [8]. The temperature gradient is mainly determined by the heat loss between insulators, heat power and latent heat caused by crystallization, and the radial temperature gradient is strongly affected by the side heater [9].

The growing interface is controlled using the optimized temperature profile in the hot zone. A flat or convex shape of the interface in the growth direction is effective to suppress the grain growth from the crucible wall and to distribute impurities and defects outward. The direction of heat flux at the growing interface determines the shape of growing interface [7]. To obtain a flat or convex shape, the melt temperature keeps slightly higher near the crucible wall than near the center of the Si melt. The growth rate, the melt flow and the thermal conductivity of crucible affect the shape of growing interface. The shape changes from concave to convex in the growth direction as the growth rate increases, and the growth rate can be higher using an insulation partition block [7,10,11]. Therefore, a large vertical temperature gradient in an ingot and a small temperature difference in the radial direction near the growing interface is effective to obtain a flat shape. By measuring temperatures at the upper and lower points of Si melt, the thickness of grown ingot can be estimated and the growth rate at the interface can be known [3]. A large thermal conductivity of crucible wall increases the outgoing heat flux through the crucible wall resulting in a larger melt flow rate and a larger temperature gradient in the radial direction near the growing interface, and the growing interface becomes more concave in the growth direction [12].

Especially, for a seed-assisted cast furnace, a large vertical temperature gradient in a melt is necessary to prevent melting seed plates set on the bottom of a crucible. An extruding insulation partition block made by graphite is used below the heater to obtain the suitable temperature gradient and melt flow in the hot-zone of a furnace as shown in Fig. 3.1B. Radiation heat loss in a furnace without an insulation partition block is higher than that in a furnace with an insulation partition block due to higher temperature beneath the heat exchange block [10]. The vertical temperature gradient in an ingot is significantly increases using the insu-lation partition block because the downward heat flow in the vertical direction is prevented by the partition block, while the radial temperature gradient decreases [7]. The growth rate increases using the insulation partition block because of larger temperature gradients in both an ingot and a melt, resulting in reduction of the heating power consumption [10,11]. As the melt flow is determined by the thermal field in the hot zone, the partition block can influence the melt flow and impurity transport in a Si ingot [7]. The position and thickness of the partition block can influence the temperature gradient and the interface shape [7,11]. The thinner partition block can obtain the smaller temperature gradient and reduce the deflection of the growing interface, but the energy consumption increases [7]. Thus, adding a partition block can increase the thermal stress in an ingot because of increasing the vertical temperature gradient in the ingot [10].

The optimized temperature gradient is also required to obtain a flat growing interface. To increase the effect of the partition block, sometimes movable insulation partition blocks along the side insulation wall are used to obtain a favorable seed-melt interface and a higher temperature gradient (or a higher growth rate) by controlling the position of the partition block as shown in Fig. 3.1B, which is JJL500 by Jing Gong Science & Technology [6,13]. The right furnace in Fig. 3.1B has the movable partition block in the vertical direction. The position of the partition block strongly affects the radial heat flux reaching the peripheral area of the melt near the growing interface. The lower position of the partition block can enhance the heat flux reaching the peripheral melt area and lower the growth rate in the periphery resulting in a flat growing interface [13]. The raise velocity of the partition block has also a great effect on the growing interface, and initially slowly and then rapidly raising the partition block is more favorable to obtain a slightly convex interface and a low thermal stress even though the growth rate is fast [14]. Thus, the growth rate in the periphery and the total shape of the growing interface are sensitively controlled using the movable partition blocks, resulting in growth of a low stressed ingot. The movable partition blocks can obtain a favorable melt convection distribution for the ingot growth and affect the distribution of temperature gradient and the thermal stress in the ingot by changing its radiated area [13]. The consumption of the total heat power for the ingot growth can be also decreased using the movable partition blocks by optimizing the radiation hear loss from the gap between side and bottom insulators [11]. Especially, this method is effective for the growth of mono-like ingots which are grown under a suitable temperature gradient and a flat seed-melt interface [7,10,11,13].

3.1.2 Unidirectional growth using quartz crucibles

Si ingots are traditionally manufactured using electronic grade Si feedstock, which is obtained from metallurgical-grade Si through the Siemens process or complex purification processes. Sometimes, upgraded metallurgical-grade Si is used to aim at a cost-effective and energy-efficient approach for production of Si ingots, which is three orders of magnitude less pure than traditional Si feedstock [15]. For growth of a Si ingot using metallurgical-grade Si, the bottom and middle are purer than the initial metallurgical Si, but impurities accumulate at the top [16]. A Si multicrystalline ingot is usually grown using a Si_3N_4 coated quartz crucible as shown in Fig. 3.2. The Si feedstock or source material is put in the crucible and heated up above the melting point of Si (1414 °C) to completely melt the source material. Nucleation of Si crystals can occur in the growth melt on the Si_3N_4 coated crucible wall [17]. To unidirectionally grow the ingot from the bottom to the top, the temperature gradient is made as the melt temperature near its top is set to be higher than that near its bottom. The solidification is controlled by movement of the temperature gradient which is made by moving the insulation basket upward (cooling) or downward (heating) as shown in Fig. 3.1A or moving the crucible upward (heating) or downward (cooling) inside the heaters. Thus, the unidirectional growth can be performed by moving the temperature gradient from the bottom to the top or cooling the furnace temperature. During growth, the ingot has columnar structure with parallel grain boundaries to the growth direction. Generally, the quality of the ingot deteriorates with the ingot height due to

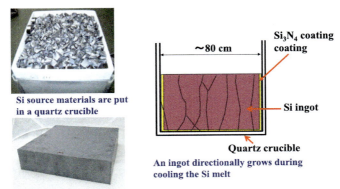

Fig. 3.2 Process to grow a Si multicrystalline ingot.

accumulation of impurities and generation of crystal defects such as dislocations. To reduce the scattering effect of grain boundaries on diffusion of minority carriers, the ingot with the columnar structure is cut in parallel to the growth direction to small bricks with a square of 15.6 cm.

The bricks are horizontally sliced to wafers with a thickness of 180 μm by multi-wire sawing. The wire has a speed between 5 and 20 m s^{-1} in production saws [18]. The wire material is usually stainless steel. Two types of sawing technique are used such as the diamond wire sawing and the slurry sawing. The former one uses diamond coated wires which allows a higher cutting-speed, but is more sensitive to hard inclusions in crystals [19]. A scratch which occurs in sawing with diamond coated wires consists of a strongly damaged region and microcracks [20]. The latter one has a slurry, a suspension of abrasive powders, mostly SiC, and a carrier fluid such as polyethylene glycol [19]. SiC particles with a broader size distribution and a more spherical shape yield a deeper saw damage [18]. The low-cost diamond wire sawing is mainly used for sawing a strong Si single ingot or mono-like ingot and the latter one is mainly used for sawing a brittle Si multicrystalline ingot. The wafer thickness is going to be thinner aiming to 100−150 μm. The mechanical properties for ultra-thin wafer can be strengthened by introduction of artificially manipulate Σ3 and Σ9 grain boundaries in the ingot [21].

In the cast method, solidification starts from the crucible bottom, where high thermal and mechanical stresses cause plastic deformation. The resulting crystal contains a variety of defects, including grain boundaries, stacking faults and dislocations, which can be in average densities as high as 10^5 cm^{-3} [22]. The concentration of in-diffused impurities is high ($\sim 10^{15}$ cm^{-3} for Fe), as the Si_3N_4 coated crucible is in contact with solidified Si from the onset of crystallization [22]. The structure control for the growth of a Si multicrystalline ingot is done to reduce crystal defects such as grain boundaries and dislocations which are introduced to reduce stress, sometimes to arrange the grain orientation. The main origin of stress is the non-uniform thermal distribution in the ingot and the expansion force during solidification. To reduce such stress, many growth techniques have been tried for the cast methods. The stress, purity, structure and crystal defect are basic factors to be controlled to obtain a high-quality Si multicrystalline ingot using a crucible. In these factors, the purity control is simplest but very difficult because the key point is usage of high-purity materials for every parts and sources, and perfect elimination of contaminations. The growth rate and solid/liquid interface shape influence thermal stress and impurity distribution in a grown ingot. Stirring a Si melt during the directional growth of an ingot affects the melt convection, the segregation of impurities, the interface stability and the shape of growing interface [23,24]. The shape of growing interface changes from slightly convex toward the melt to largely convex as the stirrer rotation increases due to the stronger convection in the vertical section [24].

The curvature of the growing interface can be easily controlled using the traveling magnetic fields (TMFs) due to melt stirring [25]. The TMFs assisted cast method is also effective to grow inclusion-free ingots with large grains (several cm) [25]. The dopant concentration in ingots is related to its evaporation from melt caused by the melt stirring. The evaporation of P depends on the intensity of melt stirring and enlarges with increasing of TMF strength because applied TMF significantly enhances the melt stirring [26]. The initial dopant (P) concentration in the melt decreases using TMF, but the dopant distribution in the ingot varies in accordance with the Scheil's model [26,27].

As a reusable crucible, a Si_3N_4 crucible can be used in several crystallization runs without sticking of a grown ingot to the crucible, but comparing with a quartz crucible, the melting time is longer because of its higher thermal conductivity even though crystallization rate is reduced [28]. The ingot is contaminated with B, Al and P from the crucible, especially the C concentration is higher because the Si_3N_4 crucible contains small amounts of SiC [28]. B-doped HP ingots grown using Si_3N_4 crucible have the higher C concentration, but it does not largely influence the carrier lifetime because substitutional dissolved C in Si is not electrically active [28,29]. Ba doped silica crucible is used to prevent the dislocation formation in a Si ingot due to attachment of brownish small particles of O-deficient cristobalite on the growing interface [30]. Cristobalite is a silica polymorph that is stable at temperatures above $1470\,^{\circ}C$ up to the melting point at $1705\,^{\circ}C$. The generation of brownish cristobalite can be suppressed completely by Ba doping in the silica crucible when the Ba concentration is higher than 30 ppm [30]. A carbon crucible is tried to use for Si ingot growth to reduce the difference between the thermal expansion coefficients of the ingot and the crucible over the temperature range $20-700\,^{\circ}C$ [31]. The acceptable limits for the average thermal expansion coefficient are estimated to be $3-4.5 \times 10^{-6}\ K^{-1}$ [31].

Among the crystal defects, dislocations strongly affect the conversion efficiency of solar cells because the diffusion length and minority carrier lifetime strongly depend on the dislocation density. The reduction of dislocations is very important to obtain high-efficiency solar cells. The stress in an ingot is naturally reduced by introduction of dislocations or arrangement of grains during crystal growth. Therefore, the structure control should be done to obtain grains with proper size (small or large) to reduce stress and dislocations or to obtain a large single grain with small stress and few dislocations.

3.1.3 Ar and N gas flow

Ar is the third-most abundant gas in the Earth's atmosphere (0.934%). Ar is chemically inert under most conditions and forms no confirmed stable compounds at RT. Ar is mostly used as an inert shielding gas in high-temperature processes such as the cast method and an Ar atmosphere is used in the furnace with graphite heaters to protect the graphite materials. The Ar gas usually flows through flow tubes equipped inside the furnace chamber. The cast furnace is filled with high-purity Ar gas, which continuously flows inside the chamber to remove O from a Si melt as evaporated SiO gas and prevent oxidation of graphite materials. The O impurity in the melt is carried toward the free surface where the Ar gas takes the evaporated SiO gas away. The O concentration in the melt depends on the evaporated SiO gas concentration and the flow rate of Ar gas. Increasing the flow rate and pressure of the Ar gas can reduce the O impurity in an ingot [32]. To enhance the Ar gas flow rate near the free surface and carry a large amount of the SiO gas outside the furnace, a gas flow guidance tube is installed in the furnace chamber [33]. The O concentration distribution in the melt and the SiO concentration distribution in the Ar gas are strongly affected by the enhanced flow pattern by the gas tube, resulting in the lower O concentration in the melt [33]. The C_s, O_i and CO concentrations in an ingot also strongly depend on Ar flow rate. They increase as the Ar flow rate decreases because their transport away from the melt surface increases by the Ar flushing [34]. Thus, the higher Ar flow rate can reduce or avoid formation of SiC precipitates. The Ar gas flow can change the temperature distribution and enhance removement of impurity from the melt surface [6]. Moreover, the convection is caused by the Ar gas, which takes the impurity away from the melt surface. To strengthen the convection flow, an extended cover made by tungsten (W) or Mo is attached around the Ar gas flow tube, and the distance between the cover and the melt surface is controlled to increase the convection flow and reduce C flux into the melt [35]. However, such gas shield design affects the heat transfer in the furnace. As the Ar gas flow becomes stronger near the melt surface, the deflection of the growing interface also increases [36].

The Ar gas has tendency to recirculate in the furnace owing to the existing of the partition block and the effect of buoyancy of Ar gas and causes the back transport of CO to the melt surface, resulting increase of C impurity in a Si ingot [7,35]. In the case, the partition block should be removed or largely moved to downwards. The Ar gas has a property to easily discharge under a reduced pressure. A little N_2 gas is added in the Ar flow to prevent the discharge even for a furnace with high–voltage carbon heaters.

Yuan et al. [37] studied the N gas flow as cheaper ambient gas substituting for Ar. The N gas was also studied as a N doping source and the doping amount could be controlled by the partial pressure and flow rate of N_2 during melting stage [37]. The mechanical strength of wafers grown under N is significantly higher than that of wafers grown under Ar. The doped N increases the flexure strength of Si single crystals because N atoms form the bonds with Si and O atoms and N helps the formation of O precipitates which pin dislocations [38]. The larger strength of wafers is important for the cost performance because breakage or warpage is a serious problem for Si multicrystalline wafers with a high defect density. The maximum value of the N concentration of 9.4×10^{15} cm^{-3} can be attained by controlling the N pressure [37]. However, even though N_2 does not react with Si melt, N_2 can react with the solid surface of Si feedstocks and generate Si_3N_4 particles. High–N partial pressure causes many cracks in the crystal, and these cracks are filled with such Si_3N_4 particles formed during a long melting time of feedstocks [37].

N–O complexes formed in N-doped Si multicrystals act as shallow donors, compensate the B acceptors and increase the resistivity of the ingots [37,39]. In a Si multicrystalline ingot, the concentration of the N–O complexes is as high as 3.4×10^{15} cm^{-3}, which is 44% of the total N content, for $[O] = 3.2 \times 10^{17}$ cm^{-3}, and the N–O complexes are barely detectable for $[O] = 0.4 \times 10^{17}$ cm^{-3} [40]. The N–O complexes are higher at the ingot bottom than at the ingot top and N dimers are higher at the ingot top than at the ingot bottom [41]. The doped N in CZ Si wafers significantly enhances the nucleation of O precipitates together with the rapid thermal processing in the temperature range of 800–1000 °C because of co-reaction with vacancies induced by the thermal processes [42].

3.2 Growth behavior of Si multicrystalline ingots

3.2.1 Development of the in-situ observation system and basic observation of Si melt growth

The precise observation of crystal growth in a Si melt is difficult because of the high melting temperature of Si and the evaporation of Si at such high temperature. The in-situ observation system is quite useful to directly characterize the growing interface of a Si crystal at a higher temperature than 1414 °C, which was developed by Fujiwara et al. [43–46] as shown in Fig. 3.3. The main parts of the system consist of a microscope (KEYENCE VH-5000) and a furnace (MATELS *inc.* MAT-20CG) which has a window at the top of it. To clearly observe the growing interface, a lens used has a deep depth of field of 0.03–0.83 cm at 35–245 magnifications and a long focal distance. To obtain the unidirectional growth, the furnace has a temperature gradient formed by controlling the power of two zone heaters. A water-cooled tube is inserted from one side of the sample to obtain a temperature gradient of around 9 K mm^{-1}. The furnace is filled with ultrahigh-purity Ar gas which is continuously flowed during heating the furnace. Si wafer chips are placed in a fused silica crucible with the size of 1.0 × 2.0 × 1.0 cm. The silica crucible is set inside the small furnace and heated up above the melting point of Si to completely melt the wafer chips. After melting, the furnace is cooled normally at the rate of 10–30 K min^{-1}. The growth rate is controlled by changing the cooling rate while keeping the temperature gradient. The growing interface can be directly observed through the quartz window by the microscope and the image of

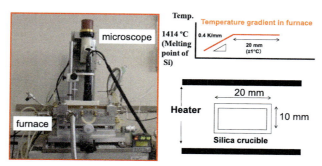

Fig. 3.3 In-situ observation system developed to directly characterize the growing interface of Si crystals at a higher temperature than 1400 °C. *(Referred from Fig. 1 in K. Fujiwara et al. J. Cryst. Growth 262 (2004) 124.)*

crystallization is recorded on videotapes at 250 frames s^{-1} [47]. The temperature under the quartz window is controlled to be about 2 K lower than its surrounding [48].

To more precisely observe the growing interface, some wafers with several orientations are horizontally set between flat quartz plates inside the furnace to keep the surface of melted part of the wafers flat during crystal growth [49]. Two heaters can be used to make more abrupt temperature gradient of about 8 K mm^{-1} by setting the heaters at different temperatures [49,50]. Thermocouples are set below the heaters. Melting is started from the end of a wafer and the temperature of the heaters is cooled at a constant rate before completely melting the wafers. Directional growth is initiated from an unmelted part of the wafers used as a seed. In this case, a high-speed camera with a light source is used to obtain a reflected image of the sample surface and to clearly observe the growing interface with several kinds of grain boundary [47].

Fig. 3.4 shows two kinds of arrangement of Si growth melt to make clear the effect of supercooling or undercooling in a Si melt on the dominant grain orientation [45]. In the case 1, Si {111} and {100} seed crystals are set side by side to directly contact with a Si melt. Crystals are grown on both seeds at the same time by cooling the melt. In the case 2, a very thin Si melt directly contacts with the crucible wall and bottom.

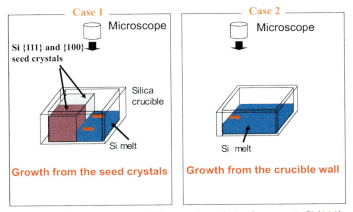

Fig. 3.4 Two kinds of arrangement of Si growth melt. In the case 1, Si (111) and (100) seed crystals are set side by side to contact with a melt. In the case 2, a melt directly contacts with the crucible wall. *(Referred from Fig. 2 in K. Fujiwara et al. J. Cryst. Growth 266 (2004) 441.)*

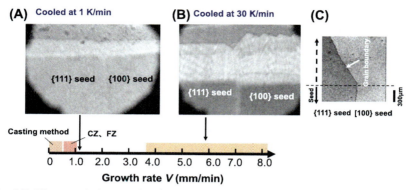

Fig. 3.5 Microscopic images for the Si melt growth in the case 1, using Si (111) and (100) seed crystals contacting with a Si melt. (A) Flat growing interfaces on the Si (111) and (100) seeds when the cooling rate is as small as 1 K min^{-1} and it is enough small not to make a supercooling. As the (111) face is most stable, the grain grown on the (111) seed gradually extends to lateral direction to shrink the grain grown on the (100) seed and finally becomes dominant in this case with a small supercooling. (B) Flat growing interface on the Si (111) seed and irregular one on the (100) seed when the cooling rate is as large as 30 K min^{-1} and it is enough large to make a supercooling. (C) Extended grain grown on the (100) seed to lateral direction to shrink the grain grown on the (111) seed because the growth rate on the (100) face is faster. *(Referred from Figs. 5 and 6 in K. Fujiwara et al. J. Cryst. Growth 266 (2004) 441.)*

Fig. 3.5 shows microscopic images for the Si melt growth in the case 1, using the Si {111} and {100} seed crystals contacting with the Si melt [45]. When the cooling rate is as small as 1 K min^{-1} and it is enough small not to make a supercooling, both growing interfaces on the Si {111} and {100} seed crystals are flat as shown in Fig. 3.5A. As the {111} face is most energetically stable, the grain grown on the {111} seed gradually extends to lateral direction to shrink the grain grown on the {100} seed. Finally, the {111} face becomes dominant in this case with a small supercooling. When the cooling rate is as large as 30 K min^{-1} and it is enough large to make a supercooling, the growing interface on the Si {111} seed is still flat, but that on the {100} seed is irregular as shown in Fig. 3.5B. The crystal grown on the {111} seed has an {111} flat facet and the crystal grown on the {100} face has many steps with {111} facets. Such interface with several {111} facets is formed after vanishing many small steps by so called bunching effect. The growth rate of the crystal grown on the {100} or {110} face is faster than that of the crystal grown on the {111} face because there are many kinks on the {100} face and many steps on the {110} face and the surface energy of the closed packed {111} face is smallest. As the growth rate on the {100} face is faster, the grain grown on the {100} seed gradually extends to lateral direction to shrink the grain grown on the {111} seed as shown in Fig. 3.5C. Thus, the overgrowth of grains strongly depends on growth conditions such as the cooling rate, the grain orientation and the energetic stability or stress distribution of each grain. The growing interface varies from flat to facet with the growth rate as known from Fig. 3.5A and B. On the {100} seeds, the growth rate of the facet interface is much faster than that of the flat interface. The growth rate on the {111} facet plane was determined to be $V = 6.9 \times 10^{-4}$ m s^{-1} or $V = 690$ μm s^{-1} by the direct observation [51]. The degree of supercooling for the growth on the {111} facet plane can be estimated as $\Delta T_s = 10$ K [52] (see Section 1.2.1.2). At the initial stage of this growth, the melt has a positive temperature gradient in the growth direction before crystal growth. During the crystal growth, the melt closed to the growing interface locally has a negative temperature gradient with a supercooling due to the latent heat. Required ΔT_s for the growth becomes smaller as the positive temperature gradient becomes smaller. In the case 2, when the cooling rate is 1K min^{-1}, many grains grow from the crucible wall at an almost same rate and the growing interface is flat. The growth rate is as small as 0.3 mm min^{-1}. As shown in the left figure of Fig. 4.7, grains with the most stable {111} face in the growth direction also extend to lateral direction to reduce the surface energy of the interface and become dominant in the case with a small supercooling. When the cooling rate is 30 K min^{-1}, grains with a faster growing plane such as {100} or {110} extended to lateral direction to shrink grains with slower growing planes such as {111} and become dominant in the case with a large supercooling as shown in Fig. 3.5C [44,45]. The critical ΔT_s changing from the slower growth rate to the faster growth rate is estimated as 3.8−6.3 K on the {111} face (see Section 1.2.1.3).

3.2.2 Shape of a growing interface during growth

For the unidirectional growth such as the cast method, grains grow outward for a convex growing interface in the growth direction between a crystal and a melt, and grains grow inward from the side wall of a crucible for a concave growing interface. The grain size generally becomes larger for the convex growing interface and many small grains appear for the concave growing interface because the crucible wall strongly acts as nucleation sites. Each part of the growing interface is determined by the radial temperature gradient or by perpendicularly to the heat flow at the interface. The flat or slightly convex growing interface is selected for the cast method. However, a convex growing interface with a large radius gives rise to severe thermal stress and consequent dislocation generation and cracking [53]. The convex growing interface is obtained by reducing the heat loss or maintaining the heat near the crucible wall during growth.

The shape of the growing interface strongly depends on the temperature gradient in a Si melt. When the temperature gradient in the melt at the interface is positive or negative, the crystal grows with a flat or zigzag interface, respectively [46]. Even though the melt has a positive temperature gradient all over the melt for the cast growth, the local negative temperature gradient is easily formed closed to the growing interface owing to a local supercooling and a high growth rate near the interface because the melt temperature near the interface becomes higher approaching to the melting point of Si due to the latent heat occurred during growth [54]. For the unidirectional growth, the {100} growing interface generally has periodically facetted structure surrounded by the stable {111} planes [43].

The precise growing interface can be directly observed using the in-situ observation system explained in Section 3.2.1. As shown in Fig. 3.6, the wavelike fluctuation appears on the growing interface at the initial stage of the growth and becomes larger as the crystal grows [48,54]. The initial growing interface varies from the flat one to the fluctuated and zigzag ones. The final growing interface has the zigzag facetted structure consisting of {111} facets like saw-teeth as shown in Fig. 3.6. The wavelike fluctuation is

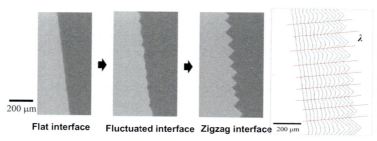

Flat interface **Fluctuated interface** **Zigzag interface**

Growing interface varies from the flat one to the fluctuated and zigzag ones.

Fluctuation on the growing interface becomes larger as it grows ahead, and finally the zigzag interface with {111} facets appears.

Fig. 3.6 Growing interface varied from the flat, fluctuated and zigzag ones. The final growing interface has the zigzag facetted structure consisting of {111} facets like saw-teeth. *(Referred from Fig. 3 in M. Tokairin et al. Phys. Rev. B 80 (2009) 174108.)*

amplified by the negative temperature gradient. The growth velocity or growth rate of the facet, v is mainly determined by the degree of supercooling in the Si melt close to the facet surface, which is caused by the negative temperature gradient in the melt near the facet. The peak height of the zigzag teeth, h increases, and the wavelength or length between troughs of the zigzag teeth, λ increases with unification of facets as the interface grows ahead [54]. The unification of facets occurs by difference for the growth rate of each facet as a facet with a higher growth rate catches up with another facet with a lower growth rate. Finally, λ takes a stable value of λs at the unification stage where the degree of supercooling becomes equivalent all over the $\{111\}$ facet area on the saw-teeth, and the shape of the growing interface becomes stable and immutable. λs is determined by area of the negative temperature gradient or area of the supercooling spread into the melt from the growing interface, and λs increases as v increases because the area increase as v increases at a constant temperature gradient [54]. λs also increases as the temperature gradient becomes smaller at a constant growth rate (see Section 1.2.1.3).

The protrusions rapidly extend into the melt at first, then the tip of them becomes flat because of lowering the melting temperature at the pointed tip [46]. The irregular interface finally becomes flat as time passes, simultaneously grains with the protrusions are coarsening to lateral direction over neighboring grains [44]. When the cooling rate is as slow as 1 K min^{-1}, the growing interface is flat from the beginning as shown in Fig. 3.5. If the melt contains a large amount of dopant, a larger temperature gradient in the melt and a lower growth rate are better to reduce constitutional supercooling, but the larger temperature gradient increases the thermal stress [53] (see Section 1.2.1.3).

3.2.3 Nucleation for Si multicrystalline ingots

The cooling rate is a dominant factor for the growth rate and the higher cooling rate gives a higher nucleation rate. The tips of dendrite-like crystals have the highest growth rate resulting in a large supercooling calculated by Eqs. (1.8 and 1.12), while lateral growth and the bulk growth are slower. This behavior is dominant for the samples with the highest cooling rate [17]. There is no correlation between the growth rate or grain size and the Si_3N_4 coating variation on a crucible. The wetting properties between the Si melt and the coating are not significant for nucleation mechanisms of Si, and the roughness of coated crucible does not largely affect nucleation conditions [17].

Nucleation undercooling strongly depends on the C and N concentrations in a Si melt [17,55,56]. The nucleation undercooling decreases with increasing N concentration in Si crystals because Si_3N_4 precipitates in the N-rich melt act as heterogeneous nucleation sites [55]. Similarly, the nucleation undercooling decreases with increasing C concentration in Si crystals because of SiC precipitates in the C-rich melt act as heterogeneous nucleation sites. The C concentration is controlled by methane (CH_4) in the atmosphere which determines C concentration in the melt [56].

3.2.4 Formation of grains in Si multicrystalline ingots

When the initial supercooling is much larger than 10 K, the area of grains with the {110} and {112} faces is dominant at the bottom of the ingot because dendrite crystals with the {110} and {112} upper surfaces are easy to appear as shown in Fig. 4.4 [57]. The grain structure approaches to a stable state with low energy as the ingot height increases and finally large grains with preferential orientations remain during grain growth as shown in Fig. 4.7, which was measured by EBSD [51,58]. The grains with preferential orientations are selected in such a way as to reduce the surface energy at the growing interface [45]. The area of grains with the {111} or {112} face which has the smallest or second smallest surface energy increases as the ingot height increases. The grain orientation is also changed by a twin formation at {111} facets. The large area of grains with the {115} face is related to formation of twins in the upper part of the ingot [59]. Near the bottom of the ingot, the stable distribution of grain orientation appears as shown in Fig. 1.6, in which {111} face contained with {112} face is dominant. Generally, the mean grain size largely increases with increasing the ingot height at the lower part of the ingot as shown in Fig. 5.2 [59]. The mean grain size is almost constant regardless of the ingot height at the middle and top parts of the ingot because the crystal growth is an essential process to go toward an energetical stable state [60]. Grain boundaries also nucleate on the crucible wall and they have dominant character of {111}/{111} Σ3 which means a Σ3 grain boundary with {111} grain boundary planes on both sides [61].

3.2.5 Formation of grain boundaries in Si multicrystalline ingots

A high level of stress and strain exist in grains of an ingot grown by the cast method because of an elastic bending of the crystal lattices. The bending energy is relaxed by introduction of dislocations. These dislocations aggregate to produce a small angle grain boundary or a sub-grain boundary which has a misorientation below 2° and a high density of dislocations. The residual stress and strain around dislocations can preferentially trap other defects and impurities, and become very recombination active [62].

Generally, grain boundaries can act as a recombination center by combination with point defects as shown in (Section 2.2.3). The $\Sigma3$ grain boundary is one of twin boundaries with the lowest order of the coincidence parameter and the highest coherency. More than 50% of total grain boundaries in the typical multicrystalline ingot is $\Sigma3$ grain boundaries as shown in (Section 1.3.2). The fraction between $\Sigma3$ grain boundary and random grain boundary strongly depends on growth conditions such as growth rate. Fig. 3.7 shows each fraction of $\Sigma3$ grain boundary and random grain boundary in a typical Si multicrystalline ingot as a function of the growth rate. $\Sigma3$ grain boundaries increase and random grain boundaries decrease as the growth rate decreases. When the growth rate is 0.4 mm min^{-1}, more than 50% of total grain boundaries are $\Sigma3$ grain boundaries. For the cast method, the growth rate normally is between 0.07 mm min^{-1} at the early stage and 0.2 mm min^{-1} at later stage [63]. Therefore, for such growth rate, the $\Sigma3$ grain boundaries easily appear in the typical multicrystalline ingot and each fraction of $\Sigma3$ grain boundary and random grain boundary becomes about 50% and 10%, respectively. As shown in Fig. 5.6, Trempa et al. [59] reported that at the bottom of an ingot, the fractions of $\Sigma3$ grain boundary and random grain boundary were

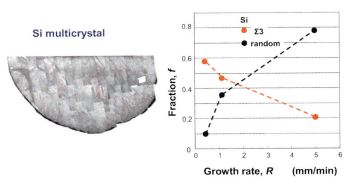

Fig. 3.7 Fraction of $\Sigma3$ grain boundary and random grain boundary in a typical Si multicrystalline ingot.

55% and 15%, respectively, and at the middle and upper parts of the ingot, the fractions were kept constant as 30% and 50% for $\Sigma 3$ grain boundary and random grain boundary, respectively. For their growth, the growth rate was 0.16 mm min^{-1}. This tendency is consistent with Fig. 3.7.

The high growth rate generates large thermal stress in the ingot. However, the $\Sigma 3$ or twin grain boundaries with a high coherency cannot largely reduce the stress unlike the random grain boundaries because the energy of twin boundaries is lower than 1/5 times of that of other grain boundaries [64]. A heterogeneous surface such as a crucible wall acts as nucleation sites of twin boundaries. About 42% of grain boundaries are every types of twin boundary [65]. The $\Sigma 3$ grain boundaries can appear under less thermal stress and more equilibrium growth condition with small supercooling than the random grain boundaries can. For a high growth rate of > 1.0 mm min^{-1} like an ingot grown by the floating zone (FZ) method, random grain boundaries disappear during growth with the ingot height and are replaced by low energy twin boundaries such as $\Sigma 3$ grain boundaries which are electrically inactive [66]. An appropriate amount of driving force is required to newly produce $\Sigma 3$ grain boundaries. This tendency is very different with Fig. 3.7. At a low growth rate of 0.16—0.2 mm min^{-1}, random grain boundaries keep their singular structure over the ingot height [66,67]. This tendency is similar with Fig. 5.6. On the other hand, even though it was reported that the fraction of twin boundaries was low at a smaller cooling rate than 10 K min^{-1} [68], the relation between the fraction of $\Sigma 3$ grain or twin boundary and the cooling rate is not still clear because there are few reports about it.

3.3 Crystal defects and impurities in Si multicrystalline ingots

3.3.1 Grain boundary character of Si multicrystalline ingots

There are many kinds of grain boundaries in Si multicrystalline ingots such as the small and large angle grain boundaries and the coherent and random grain boundaries. Laue scanner based on X-ray diffraction can measure the grain orientation and grain boundary type over the entire wafer area of Si multicrystals [69] (see Section 1.3.2). Oriwol et al. [70] studied the small angle grain boundaries in detail. The tilt angle of the small angle grain boundaries is several arc seconds ($= 1°/3600$) which is far below the angle resolution in the order of $1°$ for the EBSD measurement. A sub-grain is surrounded by small angle grain boundaries or sub-grain boundaries as

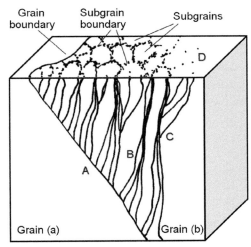

Fig. 3.8 Schematic illustration of sub-grains and sub-grain boundaries, which are formed by dislocations generated at grain boundaries near a solid-liquid interface. *(Referred from Fig. 10 in D. Oriwol et al. Acta Mater. 61 (2013) 6903.)*

shown in Fig. 3.8 [70]. The alignment and agglomerate of dislocations result in a rotation forming a sub-grain boundary which has a mixed character with tilt and twist components. Thus, sub-grain boundaries are formed due to rearrangement of dislocations during crystal growth and have a slightly tilt with a rotation of $< 1°$ around an axis in the growth direction owing to a high number of dislocations [70]. Sub-grains tend to be directed nearly parallel to the growth direction, and the strained sub-grains and dislocation density increase as growth proceeds. The sub-grains have an increasingly rotated orientation with the ingot height. The space between dislocations in sub-grain boundaries is between 300 and 30 nm and decreases with the misorientation angle.

Generally, grain boundaries with a high energy tend to minimize their energy during growth by the following two cases [67]. In the first case, the energy of angle deviation, θ of a grain boundary can be reduced by local incorporation of dislocations into the grain boundary. In the second case, a grain boundary with a high energy can be split into two or more grain boundaries with lower energy. The new grain boundaries can be either of the same grain boundary type with lower angle deviation [71] or of other grain boundary type with lower energy. For example, a $\Sigma 3$ grain boundary ($\theta = 3.5°$) can split into a $\Sigma 3$ grain boundary ($\theta < 1°$) and a $\Sigma 9$ grain boundary such as $\Sigma 3 = \Sigma 3 + \Sigma 9$. A $\Sigma 9$ grain boundary ($\theta = 2.5°$)

can split into two $\Sigma 3$ grain boundaries ($\theta = 0.7°$ and $1°$) such as $\Sigma 9 = \Sigma 3 + \Sigma 3$ [67].

During growth of a Si multicrystalline ingot, most of grain boundaries tend to follow the growth direction. Generally, $\Sigma 3$ and $\Sigma 9$ grain boundaries have inclined or tilt nature from the growth direction due to their crystallographic preference. Random grain boundaries also have tilt nature from the growth direction. The tilt angles of $\Sigma 3$, $\Sigma 9$ and random grain boundaries are estimated as $15-40°$, $10-14°$ and $0-35°$, respectively [72]. The twin nucleation is a key mechanism for the formation of such a tilted $\Sigma 3$ grain boundaries and the twin formation is more frequent at a higher growth rate [72].

Grain boundaries act as carrier recombination centers due to deep levels which are attributed to dangling bonds and impurity decoration. Grain boundaries originally have shallow levels and they have deep levels after impurity contamination [73]. The coherency of grain boundaries strongly affects the electric property and such grain boundary structure affects the recombination activity of grain boundaries. Thus, the electric property such as minority carries lifetime strongly depend on the character and impurity contamination of grain boundaries. Chen et al. [73,74] investigated the electrical property of several types of grain boundaries using the EBIC method. Both small and large angle grain boundaries without any impurity contamination show a weak EBIC contrast at 300 K. They are electrically inactive in nature and their intrinsic recombination activity is weak. All the grain boundaries after a metal decoration show a strong EBIC contrast at 300 K. The recombination activity of the grain boundaries becomes stronger as the contamination level increases and it is principally determined by the gettering ability of the grain boundaries. The recombination activity of $\Sigma 3$ {111} is weaker than that of $\Sigma 3$ {112} because the $\Sigma 3$ {112} has more defects than the $\Sigma 3$ {111} and the grain boundary plane affects the recombination activity. Large angle grain boundaries with Fe contamination as low as 10^{13} cm^{-3} show significant recombination activity at 300 K. The larger gettering ability causes stronger recombination activity or minority carrier recombination in the grain boundaries. In the case of a same contamination degree, the random and high-Σ grain boundaries have the stronger EBIC contrast and the stronger gettering ability than those of the highly coherent or low-Σ grain boundaries [73,74]. The small angle grain boundaries show stronger EBIC contrast or stronger gettering ability than the large angle grain boundaries because they consist of many dislocations and posse a high density of recombination centers.

The recombination velocity of minority carrier at sub-grain boundaries is comparable with that of random grain boundary [75]. Coherent grain boundaries such as CSL grain boundaries or twins are less detrimental than sub-grain boundaries and other grain boundaries with high planar mismatch because twins are not decorated by dislocations and even multiple twins do not reduce minority carrier lifetime [76]. The small angle grain boundaries with a high dislocation density or high impurity contamination are the most detrimental defects in Si multicrytalline ingots [74].

3.3.2 Dislocations and dislocation clusters in Si multicrystalline ingots

There are many kinds of crystal defects in Si multicrystalline ingots grown by the cast method. Among them, dislocations and dislocation clusters are the most influential on the efficiency of solar cells because the high dislocation density reduce the diffusion length and lifetime of minority carriers. Dislocations are usually positively charged by holes in p-type Si [77]. The effective diffusion length of minority carriers decreases with the average dislocation density higher than $10^4 \, cm^{-2}$ [78]. The effective bulk lifetime decreases with the dislocation density higher than $10^5 - 10^6 \, cm^{-2}$ [77]. The reduction of dislocation densities from $10^6 - 10^8 \, cm^{-2}$ to as low as $10^3 - 10^4 \, cm^{-2}$ leads to an improvement of the solar cell performance from 13 to 14% to more than 20% [77]. The dislocation density in Si multicrystalline ingots grown by the conventional cast method are between 10^4 and $10^6 \, cm^{-2}$. Regions with high and low dislocation densities are coexisted in an ingot. The dislocation density increases from the bottom to the top in an ingot and the dislocation density at the top part of the ingot is at least three times higher than that of at the bottom [79]. The dislocation density of a center brick is generally higher than that of a corner brick at the same height position of the ingot due to the impurity distribution [79]. The number of active slip planes such as $\{111\}$ during solidification influences the dislocation density in the ingot, and the high dislocation density causes in a growth direction close to the $<111>$ grain orientation [80]. The dislocation density can be estimated by the EPD of dislocations in a cross section of an ingot. To evaluate EPD, the surface of the cross section is mechanically polished to a mirror finish and then chemically etched with the Secco or Sopori solution [81,82]. For highly dense dislocations such as arrays in sub-grain boundary which have distances in order of tens of nm, the EPD tends to largely underestimate the real dislocation density [83]. The cell structure of dislocations also appears in the bottom of an ingot directly contacted on a Si_3N_4 coating [84].

The dislocation EBIC contrast in multicrystalline Si does not noticeably increase after metal contamination because they are already decorated with some impurities after growth [85]. Dislocation clusters contaminated with impurities are much more detrimental for solar cells than grain boundaries in multicrystalline ingots. During growth, many dislocations generate at grain boundaries near a solid–liquid interface due to thermo-mechanical stresses and dislocation pile-ups occur at the grain boundaries [70,86]. Thus, most dislocation clusters are generated in the lower part of an ingot, and most dislocation clusters (more than 90%) have their origin at grain boundaries and connect to the grain boundaries [77]. Individual dislocations minimize their energy through shortening their length by taking the shortest possible line direction to the liquid–solid interface, and the microstructure in dislocation clusters is in configurations that are energetically most favorable [78] (see Section 7.4.4). The dislocations line up in patterns creating dislocation line and sub-grain boundaries to minimize their energy. In such ordered dislocation clusters, polygonised structures can be observed in defect-etched directionally solidified multicrystalline Si wafers. These structures consist of parallel, similarly straight and well-ordered dislocations, with minimum contact-interaction [80]. In a polygon wall, dislocations are mostly straight and parallel, lining-up in parallel glide planes normal to a polygon wall, and no decorated precipitates are found in these ordered dislocation structures [80]. The structure of ordered dislocation clusters is formed near the melting point, at which the mobility of dislocations is very high. Therefore, dislocations can move fast to glide or climb to the outside of crystal, to grain boundaries and to annihilate with dislocations of opposite Burgers vectors, and remaining dislocations re-arrange in polygon walls [87]. Disordered dislocation structures consist of different-interacting types of dislocations. For the {111} plane, dislocations are expected to be lying in three planes such as (-11-1), (1-1-1) and (-1-11), depending on the number of activated slip systems [78]. Lineup of dislocations along {110} planes is commonly explained by slip on {110} planes [86]. From the recent three-dimensional (3-D) visualization of dislocation clusters in a HP ingot by processing PL images, sub-grain boundaries with angular deviation of less than $10°$ cause generation of dislocation clusters [88] (see Sections 1.3.4 and 5.3.2).

Ryningen et al. [86] studied dislocation clusters in a Si ingot in detail. Dislocation clusters emanate from grain boundaries such as $\Sigma 27$ and have their origin at the grain boundaries. Dislocations in dislocation clusters are elongated approximately parallel the growth direction during growth as shown in Fig. 3.8. The size of dislocation clusters increases and the number dislocations in the clusters also increases with the ingot height. The dislocation clusters are of square cm size at the top of the ingot. The dislocation etch-pits in the clusters emitted from the grain boundary line up in regular patterns of loops and lines. The dislocation cluster growth with slip on the $\{110\}$ <110> system is common in directionally solidified Si multi-crystalline ingots.

Arafune et al. reported that the minority carrier lifetime was inversely proportional to the EPD when the grain size was sufficiently large comparing with the minority carrier diffusion length [89]. The dislocations act as recombination centers of the minority carriers. From their results, the minority carrier lifetime largely decreases for the EPD larger than 1×10^5 cm^{-2}. A dislocation density of less than 10^3 cm^{-2} is very important for obtaining a large minority carrier diffusion length [73]. The dislocation density should be reduced to lower than 10^5 cm^{-2} in Si multicrystalline ingots. The dislocation density in an ingot is normally on the order between 10^4–10^6 cm^{-2} [74].

The stress in an ingot as the driving force to generate dislocations is also strongly relates to the grain distribution, grain boundary structure and grain orientation as explained in Section 5.3. Generally, small and large grains ununiformly distribute in a multicrystalline ingot. This distribution causes local stress and high density of dislocations. The generation of dislocations can be suppressed by reducing the thermal stresses [77]. On the other hand, small stress (<5 MPa) leads to a significant local reduction (>99%) of dislocation density [90]. The distribution of dislocations is inhomogeneous and regions with high dislocation density are scattered in regions with low dislocation density. Dislocations are mainly generated in the vicinity of the growing interface by inheriting them from solidified crystals during crystal growth because the dislocation velocity is much higher than the forward speed of the interface [70,91]. Dislocations or sub-grain boundaries generated at the initial stage of the growth tend to largely propagate in the growth direction toward the top of the ingot because dislocations perpendicularly extend to the growing interface owing to reduce their length or energy. The defect area with a higher EPD than 1×10^6 cm^{-2} rapidly increases with the ingot height.

The grain size generally increases as the ingot height increases and finally large grains with preferential orientations remain because the grain structure approaches to a stable state with low energy by grain growth [51,58]. The dislocation density in an ingot grown with active grain growth from initial small grains becomes significantly low and the distribution of the minority carrier lifetime is uniform [1]. This behavior is strongly related to the density and effect of random grain boundary (see Section 5.4). The grain growth can be controlled by the thermal gradient, cooling rate and supercooling during ingot growth. Dislocations or sub-grain boundaries at the top of the ingot are largely reduced by control of initial nucleation and growth.

The stress as driving force to generate dislocations is also strongly relates to the growth behavior of a Si multicrystalline ingot. As a supercooling in a melt increases during growth of an ingot, the growing interface becomes rough and sometimes a small melt region surrounded by several grains with random grain boundaries remains. This small melt region finally disappears by solidification to grow a small crystal. The finally solidified region has local stress owing to confinement of expansion of the solidified small melt. From observation of dislocations in a Si multicrystal during two-dimensional growth using the in-situ observation technique as explained in Section 3.2.1, Kutsukake et al. [92] shows that a small crystal or grain with a twin boundary is formed in such a small melt region during solidification. The quality of the small crystal can be confirmed by measuring the PL, EPD and EBSD images to obtain each image mapping as shown in Fig. 3.9. These images correspond to the same small crystal solidified at the small melt region. Many dislocations exist within this region and a twin boundary appears as a grain boundary because locally concentrated stress remains near the small crystal. The stress in a grain depends on the orientation of the surrounding grains and unfavorable combinations of the orientations of neighboring grains lead to very high stresses on several glide systems resulting in high local dislocation density [80]. Si multicrystalline ingots should be grown so as not to occur such small melt regions by keeping the supercooling near the growing interface enough small. Formation of dislocations is not observed in any small grains where twin boundaries are not observed. The similar formation mechanism of twin boundaries is also explained by Stokkan [93] (see Sections 1.3.3.4 and 1.3.3.5).

Grain boundaries that generate dislocation clusters are in general CSL grain boundaries and the $\Sigma 27$ grain boundary is the most common [91].

Growth of Si multicrystalline ingots using the conventional cast method   125

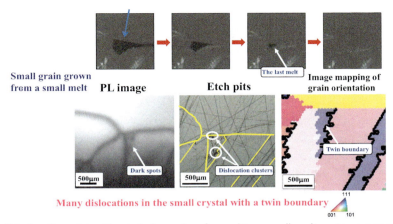

Fig. 3.9 Small crystal with a twin boundary formed in a small melt region surrounded by grains during solidification. Quality of the small crystal solidified at the small melt region is confirmed by the PL, EPD and EBSD images. Dislocations are generated near the small crystal. *(Referred from Figs. 4 and 5 in K. Kutsukake et al. J. Appl. Physics 110 (2011) 083530.)*

Oriwol et al. [83] studied behavior of dislocation clusters during ingot growth in detail. As dislocations largely propagate in the growth direction toward the top of an ingot, they tend to simultaneously build dislocation clusters with high density. The dislocation clusters are made of a network of sub-grain boundaries with linear arrays of dislocations arranged parallel to each other as shown in Fig. 3.8 [70,83]. They are mainly made of 60° dislocations. The formation of arrays can decrease elastic energy. A sub-grain boundary has a slightly different grain orientation with a minute tilt angle. In some dense clusters with condensed dislocations, the distance between dislocation lines is between 30 and 40 nm and the dislocation density is higher than 10^7 cm^{-2} [83]. Grains with dense dislocation clusters tend to have the <110> orientation and their area decreases as the ingot height increases because a high dislocation density restrains the grain growth. Grains with dense dislocation clusters have two slip planes lying nearly parallel to growth direction.

In general, grown-in dislocations cannot be easily removed by post thermal treatment regardless of applied stress, or rather, some dislocations are introduced during the annealing process [94]. However, the dislocation density decreases with increasing annealing temperature from 1100 °C to 1400 °C and increasing time from 10 min to 360 min [77,95]. High temperature annealing can effectively remove dislocations from Si multicrystalline wafers by pairwise annihilation, and the dislocation density reduction of $> 95\%$ are achieved in the Si multicrystalline wafers by this annealing at 1366 °C for 360 min [95]. A combination process of the thermal cycling and isothermal annealing provides a maximum reduction of the dislocation density, with 12 h of static annealing followed by 12 h of cyclic annealing because the cyclic annealing activates dislocations faster than the static annealing due to an oscillating stress [96]. A reduction of $> 60\%$ in dislocations is also observed for Si multicrystalline wafers after P gettering at 820 °C because the diffusion flux of impurities toward the gettering layer may drive dislocations to sink sites [97]. Moreover, the dislocation density in Si multicrystalline ingots depends on the O and C concentrations. The dislocation density decreases as the O concentration increases and the C concentration decreases because the dislocation generation rate is governed by the O concentration and the local stress is caused by the C agglomeration and precipitation [98].

3.3.3 Metallic impurities and precipitates in Si multicrystalline ingots

Main origins of metallic impurities are Si feedstock, fused silica crucible, Si_3N_4 coating, carbon heater and carbon-related material in a furnace. Among them, the Fe concentration in Si feedstock is below 2.5×10^{11} cm^{-3} [99]. As the purity of Si feedstock is typically high enough, the quartz crucible and the coating materials remain the main source for impurities [100]. Impurities diffuse into a Si melt and are incorporated into an ingot during growth. These impurities also solid–state-diffuse from the crucible or the coating into the ingot. Especially, harmful impurity Fe can be easily diffused after solidification due to a small value of activation energy of Fe diffusion in Si crystal [101–103]. As the main sources of Fe are the crucible or the coating, using a purified crucible with a coating powder of very high purity results in a notable reduction of the red zone in an ingot [104]. The red zone is a part with higher impurities and defects in the ingot, which often appears at the bottom and side of the ingot. The main source of impurities for contamination of the red zone is the crucible, and the coating has only a minor effect [104,105]. The Fe concentration is also estimated to be higher near the crucible wall and at the top of the ingot due to the diffusion of Fe from the crucible/coating and the segregation in the melt, respectively [102,103,106] (see Section 1.4.4).

Fe has more time to diffuse in the bottom than in the top of the ingot because the ingot bottom keeps longer in contact with the crucible wall at high temperature. Usually, a Fe concentration is high near the bottom of a multicrystalline ingot where Fe diffuses in from a Si_3N_4 coated crucible. In the middle and upper part of the ingot, the total Fe concentration becomes higher toward the top of the ingot due to the segregation of Fe [107]. The number and mean radius of Fe precipitates increase with the ingot height [107]. Typical concentration of total Fe in commercial Si multicrystalline ingots in in the range of 10^{13}–10^{15} cm^{-3} [108]. More than 99% of Fe is precipitated in a commercial standard ingot [109]. Especially, in the border region (2 cm wide) with lower lifetime of a p-type multicrystalline Si ingot close to the crucible wall, the F_i and Fe-B concentration is significant higher even though contents of O, C and other metals are almost at the same level as that in the bulk, but these F_i and Fe-B pairs can be effectively removed by P gettering resulting in higher lifetime [110,111]. Fe is a fast diffusing impurity $(D_{Fe} = 1.3 \times 10^{-3}\exp\left(-0.68 \text{ eV}/k_B T\right) cm^2\ s^{-1})$ and can be gettered [108,112] (see Section 1.4.4). Most of Fe in low dislocated grains is F_i, but the F_i concentration in areas of high dislocation density do not significantly depends on the contamination level [107]. Grain boundaries and grains with high dislocation density have a lower F_i concentration than good grains, but the higher total Fe concentration leads to higher F_i concentration [107]. The F_i concentration is strongly increases near the crucible wall and deceases within a few centimeters in a G1 Si ingot [5].

In most multicrystalline Si ingots, transition metal impurities such as Fe, Cr, Ni, and Cu are usually found in concentrations as high as 10^{14}–10^{16} cm^{-3} and they play a large role in decreasing the performance of solar cells. Only Fe, Cr, Ni and Cu can be detected in higher concentrations than 1×10^{14} cm^{-3} in multicrystalline Si solar cells [85,113]. In some case, Fe, Cr, Cu, Mo, Co, Ni and Ga are detected in higher concentrations than 1×10^{13} cm^{-3} in multicrystalline Si ingots [114,115]. Using high-purity quartz crucibles or Si_3N_4 crucibles, only Fe, Cr, Ni and Zn can be detected in higher concentrations than 1×10^{12} cm^{-3} and Co can be detected in lower concentration than 1×10^{11} cm^{-3} in multicrystalline Si ingots [29]. Fe, Cu and Ni precipitate at dislocations, grain boundaries and other crystal defects and enhance the recombination activities of these defects [116]. The influence of Fe, Cu and Ni atoms themselves is qualitatively similar [117]. Fe decorates extended defects, making them extremely recombination active [99]. Cu and Ni can more easily form silicide precipitations than Fe because of their faster diffusivity [99]. Cu preferably precipitates on grain boundaries as hetero-nucleation sites and increases the recombination activity of grain boundaries resulting in degradation of carrier lifetime [118,119]. In a Si ingot, Cu does not significantly vary along the ingot height because of Cu evaporation from the melt and the high Cu diffusivity in the ingot [120]. Cu can diffuse on few centimeters during ingot cooling, which contributes to the homogenization of Cu [120]. Ni can create highly recombination-active precipitates and precipitate at the O precipitates and extended defects [121]. Dissolved Ni_i easily precipitate during rapid cooling due to the small lattice mismatch between Si and $NiSi_2$ [121].

A high flux, non-destructive X-ray synchrotron-based technique, the μ-XRF microscopy can detect metal precipitates as small as a few tens of nanometers in diameter [122]. X-ray beam-induced current (XBIC) technique can be used together with μ-XRF to obtain more information about the elemental nature of precipitates and recombination activity [122,123]. Homogeneously distributed, recombination-active nano-precipitates separated by only a few microns have a much greater impact on the minority-carrier diffusion length than micron-sized clusters separated by several hundreds of microns, even if the total amount of Fe contained in these two distinct defect types can be comparable [124]. The precipitate density of iron-silicide is higher at a grain boundary than inside a grain, and it decreases with the precipitate size and increases as the cooling rate after solidification becomes faster because of increase of the F_i concentration [125]. The precipitate density largely decreases after P diffusion, especially for the smaller precipitates because of faster dissolution [125,126]. A complete removal of Fe from wafers is potentially possible if the gettering time is long enough [126]. The rapid thermal processing (RTP) at short high-temperature completely dissolves Cu and Ni precipitates, but only partly dissolves iron-silicide nanoprecipitates, and dissolved metals form smaller and more distributed recombination centers [127]. Submicron-sized Cu_3Si and $NiSi_3$ precipitates can significantly reduce the minority carrier diffusion length in multicrystalline Si which have a strong correlation for reduction of diffusion length [128]. The formation mechanism of co-precipitates of Cu, Ni and Fe is elucidated by μ-XRF [129] (see Section 1.4.3).

Metals in the highest concentrations are usually found in grain boundaries. Buonassisi et al. [130] studied metal contents in several types of grain boundaries. Grain boundaries with low CSL indices or low Σ values ($\Sigma3$, $\Sigma9$ and $\Sigma27$) generally contain lower concentration of precipitated metals as silicide than grain boundaries with high CSL indices or non-CSL boundaries. This tendency is consistent with that of the CSL boundary electrical activity [73]. The low-Σ grain boundaries have the high degree of atomic coincidence and low level of bond distortion which offer a lower density of energetically favorable sites for impurity atoms to segregate and precipitate. The low-Σ grain boundaries basically contain no dislocations which provide efficient diffusion pathways for impurities. On the other hand, larger concentrations of precipitated metals are found at random grain boundaries or non-CSL boundaries. In the case of dopant atoms, the fraction of As or P atoms segregating to grain boundaries is proportional to the volume density of grain boundary segregation sites and inversely proportional to the size of grains, but B atoms do not segregate to the grain boundaries [131]. The segregation which occurs at grain boundaries also occurs for the free surface [131].

3.3.4 SiO_2, Si_3N_4 and SiC precipitates

Only C and O can be detected in higher concentrations than 1×10^{16} cm^{-3} in multicrystalline Si solar cells as impurities [85,113]. C and O strongly affects the quality of Si crystals and the performance of solar cells in several forms such as precipitates. SiO_2, Si_3N_4 and SiC precipitates are formed in a melt and incorporated in an ingot during growth. Si_3N_4 inclusions have a lower thermal expansion coefficient, but a higher elastic modulus than Si [132]. SiC inclusions, which are the least deleterious ones, have a higher thermal expansion coefficient and a higher elastic modulus than Si [132]. Main origins of impurities in Si are Si feedstock, silica crucible, Si_3N_4 coating, carbon insulating materials, heaters, other contents in the furnace and furnace atmosphere. The dislocation density strongly affects the existence of impurities near them because the impurities tend to combine with dislocations. The dislocations act as efficient trapped centers for contaminated Fe atoms, which also act as recombination centers [89]. The main origin of contaminated Fe is a Si_3N_4 coating on the crucible inner wall and Fe elements move into an ingot by solid–sate diffusion from the coating material. Fe is relatively stable at O precipitates in Si with low C content because of the reduced strain by C stabilized Fe at the O precipitates [99].

Main origin of impurity O is silica crucible. The O solubility in a Si melt equilibrated with amorphous silica is larger than in the Si melt equilibrated with crystalline cristobalite which can be used on the inner surface of a silica crucible by coating of Ba oxides [30]. The O concentration in an ingot grown using crucibles with a Si_3N_4 coating is significantly lower than that in an ingot grown using crucibles without a Si_3N_4 coating [133]. The Si_3N_4

coating is effective for preventing a reaction between silica crucible and Si melt. In a typical ingot grown by the cast method, the O concentration is $1-5 \times 10^{17}$ cm^{-3}. The O concentration is normally higher in the ingot bottom and lower in the ingot top because the contact area between the melt and the crucible decreases with solidification and the segregation coefficient is larger than 1.0. The O distribution in the ingot bottom is mainly determined by evaporation of O and that in the ingot top is dominated by segregation of O [134]. The O concentration becomes lower as the growth rate becomes higher because O elements in the melt have not enough time to transport to the solid–liquid interface [135]. On the other hand, the lower growth rate is better to buy time for the O evaporation in the form of SiO from the melt surface and decrease the O concentration in the melt. Anyway, it is much important to use a technique to forcibly remove SiO from the melt surface. O atoms exist in the ingot as O_i which can be measured by the FTIR analysis.

C contamination is one of the serious problems for the growth of a Si ingot by the cast method. Main origins of C contamination are carbon heaters, carbon insulators, internal structure made by carbon, crucibles and Si feedstocks. C can transfer into a Si melt in the form of CO from these carbon–based materials according to the following reactions [35]. At first, volatile silicon monoxide SiO (g) is formed by the reaction between the Si melt and the silica crucible (SiO$_2$(s)) as shown in Eq. (3.1). In other words, a dissolved O atom from SiO$_2$(s) in the Si melt combines with a Si atom to form SiO (g) near the surface of Si melt [32]. SiO (g) is carried to graphite materials (C) in the furnace chamber by the Ar gas and CO (g) is formed by the reaction between SiO (g) and C as shown in Eq. (3.2).

$$SiO_2 \ (s) + Si(l) = 2SiO \ (g), \tag{3.1}$$

$$SiO \ (g) + 2C \ (s) = SiC \ (s) + CO \ (g), \tag{3.2}$$

where subscript s, l and g stand for solid, liquid and gaseous states. The reaction as shown in Eq. (3.1) occurs at any point of the solid/liquid interface, but it is intense at the solid/liquid/vapor triple line where the evacuation of SiO species is much easier [136]. The CO (g) diffuses or flows back to the Si melt surface and dissolves as C and O atoms in the melt. The Si melt is saturated with C at the beginning of solidification and the C concentration is $3-4 \times 10^{17}$ cm^{-3} at the beginning of the ingot [135]. The melt becomes supersaturated by C at the solid/liquid interface during growth due to the segregation coefficient of 0.07. Thus, the C concentration normally increases with the ingot height during growth. A smaller growth rate leads to a lower C concentration with more uniform distribution [135]. The C concentration tends to radially have a low and uniform distribution by a large convective flow which suppresses supersaturation caused by the segregation. The C concentration is normally higher than that in a CZ ingot.

The C concentration is high in high–EPD regions in an ingot [89]. C contains at the origin of dislocations [137]. C is easily precipitated as SiC from a C-condensed Si melt during growth when the C concentration is larger than 1 at%. The formation of SiC precipitates is also suppressed by the large convective flow. SiC has a shape like monocrystalline filaments and it is β-SiC which is stable below 2000 K [138]. Such SiC particles grow on other SiC precipitates, Si_3N_4 particles, grain boundaries, dislocations and a crucible wall and extend parallel to the growth direction, along grain boundaries and dislocations [135]. Agglomerates of only SiC particles are also observed [138]. C_s can be measured by the FTIR analysis.

Main origin of impurity N is Si_3N_4 coating materials. While a Si melt is maintained at a high temperature, N dissolves from the surface of a Si_3N_4 coating into the Si melt by the following reaction and the solubility increases with temperature [138],

$$Si_3N_4 \ (s) = 3Si \ (l) + 4N, \tag{3.3}$$

The N concentration in a cast ingot is $2 \times 10^{15} - 1 \times 10^{16} \ cm^{-3}$ and it tends to radially and axially have a low and uniform distribution by increasing a convective flow [135]. N is saturated and precipitated in a Si melt as Si_3N_4 as explain in Section 3.4. More Si_3N_4 nucleate at the former grown Si_3N_4 particles and at the Si_3N_4 coating during cooling. The shape of Si_3N_4 is like hexagonal rods of single crystal as shown in Fig. 3.11C. The melt convection plays an important role for the distribution of C and N, the formation of SiC and Si_3N_4 particles in the melt and the realization of free of these particles in the ingot [35]. The ratio of melt volume to wetted coating surface is suitable to be as large as possible to reduce the transfer rate of N into the melt.

Si_3N_4 and SiC precipitates mainly appear as large clusters (75−385 µm) at the top of an ingot because of their low segregation coefficients [138]. Big butterfly-shaped precipitates appear at the middle part of the ingot, which nuclear on the Si_3N_4, SiN_x and SiC particles [139]. They consist of filament-like and fiber-like precipitates with some SiN_x composition and the distribution of these two precipitates is affected by the convection flow in a melt. Si_3N_4 and SiC precipitates and inclusions are locally distributed in a Si multicrystalline ingot, and they strongly affect the quality of the ingot, the wafer sawing process as obstacles and the performance of solar cells. The existence of SiC inclusions are main cause for wire sawing-related defects on the wafer surface, while the effect of Si_3N_4 is not notable [140]. Si_3N_4 and SiC precipitates cause local stress around them and crystal defects such as dislocations, and they also decrease the mechanical strength. Large tensile stresses are observed surrounding the β-Si_3N_4 and β-SiC inclusions because of large mismatch between their thermal expansion coefficients to Si matrix, and the average stresses are 12 MPa at β-Si_3N_4/Si and 24 MPa at β-SiC/Si interfaces, respectively, when thermally-induced residual stress in the Si matrix is less than 5 MPa [141]. Si_3N_4 and SiC precipitates also act as heterogeneous nucleation centers for grain growth.

Trempa et al. [142] studied impurities of C and N during growth in detail. The C concentration become higher in the center of an ingot as the growth rate becomes higher because the growing interface becomes more concave due to the increased latent heat. For the low growth rate $(0.03 \text{ mm min}^{-1})$, the C concentration $(3-4 \times 10^{17} \text{ cm}^{-3})$ does not largely vary in the growth direction. When Si_3N_4 precipitates are formed in a melt or at the surface of the melt, the N concentration in the ingot largely decreases and almost constant $(2-4 \times 10^{15} \text{ cm}^{-3})$ in the center of the ingot even though the calculated concentration by segregation increases in the growth direction. The N concentration is affected by the growth rate because Si_3N_4 precipitates are more easily formed at the higher growth rate $(0.36 \text{ mm min}^{-1})$. The convection flow in the Si melt also strongly affects the distribution of C atoms in the ingot. The convection flow is axisymmetric and a large torus with down-stream at the crucible wall and upstream in the center appear in the melt. The number and amount of SiC and Si_3N_4 precipitates increase with growth rate because the high growth rate causes the piling-up of C and N at the growing interface. Thus, a precipitate-free ingot can be obtain using the low growth rate or the high convection flow velocity and improved convection pattern.

SiC particles are captured and engulfed into crystal as inclusions near the growing interface of a multicrystalline Si ingot. The capture of SiC particles during directional solidification of Si is simulated using a phase-field model [143]. The particle shape and size largely affect the critical growth velocity that allows the particles to move with the same speed and direction as the growing interface, and they affect the strength of the repulsive force working against gravity and drag force into the crystal. At the equilibrium force condition, the repulsive force balances with the gravity and drag forces. Moreover, melt flow and buoyance cause lift forces which pull the particles away from the interface. Sedimentation of the particles due to gravity plays a role only for very large particles [144]. The mechanism of particle engulfment under fluctuating solidification velocities during directional solidification of Si is also simulated using a finite-element model [145]. Higher-amplitude growth oscillations at the same frequency result in particle engulfment.

To obtain a homogeneous O, C and N distribution without any precipitate formation, high purity materials, a small growth rare, a flat solid/liquid interface and a well mixed system are generally required [135]. As C transfers into a Si melt in the form of CO from the gas atmosphere, special contrivances are necessary to effectively prevent the CO flow (see Sections 1.4.2 and 1.4.4).

3.4 Si_3N_4 coating materials

3.4.1 Growth of Si multicrystals from Si melt with floating Si_3N_4 particles on the surface

For the growth of a Si ingot using the cast furnace, the inner wall of a silica crucible is generally coated with α-Si_3N_4 particles to prevent the sticking due to the reaction between crucible wall and a Si melt during growth. The thickness of the coating is a few hundred μm-thick, which is made by spraying a slurry consisting of a submicronic powder of α-Si_3N_4 and a polymer binder (PVA) dissolved in deionized water [146]. Silica (SiO_2) and Si_3N_4 are separated by a thin nanometric silicon oxinitride layer [147]. The minimum O content corresponds to the initial Si_3N_4 power, usually lying in the range 1—2 wt% [147]. The coating exhibits a microscopic porosity between individual Si_3N_4 grains and a macroscopic porosity formed by bubbles of some tens of microns is produced by the spraying process [147]. The pore radius is estimated to be 0.115 μm [147]. The coated crucible is oxidized in air at a temperature in the range 900—1000 °C in order to burn off the binder and to oxidize the Si_3N_4 particles [131]. During oxidation, oxide bridges are formed at the areas of contact between particles [147]. The coating is fully and deeply deoxidized on its outer surface when the heating temperature reaches at higher than 1300 °C [147]. The coating is porous and consists of oxidized Si_3N_4 particles surrounded by a thin (nm thick) SiO_2 skin layer which ensures coating cohesion and slows down Si infiltration to avoid the sticking [133]. Such porous Si_3N_4 coating layer is well wetted by the Si melt and it also acts as an interface releasing agent between Si and the crucible [136]. As the coating is very porous, the infiltration of the coating by the Si melt is possible and the reaction of Eq. (3.1) is occurred between the silica crucible and a Si melt. The reaction is enhanced by removal of SiO species [146].

The detachment of a Si ingot from the coating at high temperature is related to the segregation of Si_3N_4 small crystals at the ingot surface, which forms a so-called self-crucible around the ingot from the beginning of the growth [146]. The self-crucible is effective to detach the grown ingot from the coated crucible. The Si_3N_4 small crystals are formed on the surface of the coating from a N-saturated Si melt, which prevent the direct contact between the Si melt and the coating. The key point in the self-crucible formation is the coating decohesion due to deoxidation [146]. To avoid the sticking of the ingot on the crucible caused by direct contact of the Si melt to the crucible wall, the thickness of the coating should be locally increased. To reduce the pollution by O, the deoxidation flux should be controlled by choice of the suitable value of Ar gas flux. The low Ar gas flux is preferable to lead a slow deoxidation and to keep down the infiltration rate of the Si melt to the crucible [146]. Deoxidation is limited by the diffusion of gaseous species inside the pores except at the vicinity of the outer surface [147]. Before introduction of Ar flow in the furnace, the coating can be safely heated under high vacuum up to 1150 °C, which can favor furnace degassing [147]. A nearly continuous layer of large and clearly faceted β-Si_3N_4 particles exists between the α-Si_3N_4 coating and the solidified Si crystal, and the initial nucleation of Si crystals occurs on the large facets of β-Si_3N_4 particles at low supercooling [148]. These β-Si_3N_4 particles are formed by dissolution of the α-Si_3N_4 coating followed by re-precipitation [133,148] (see Section 3.4.3).

Some Si_3N_4 particles detach from the Si_3N_4 coating material and float on the surface of the Si melt before ingot growth occurs. Such floating Si_3N_4 particles usually appear at the top of the grown ingot. These particles do not strongly affect to the casting growth because the crystal growth starts from the bottom of the crucible, and the floating Si_3N_4 particles do not significantly influence the structure of the ingot. To study behavior of the floating Si_3N_4 particles in a Si melt in detail, the inner wall of a crucible was coated with a mixture of 97%—98% α-Si_3N_4 particles and 2%—3% β-Si_3N_4 particles. There are two cases in which floating Si_3N_4 particles were formed on the surface of a Si melt [133]. In the first case (case I), Si_3N_4 particles float due to their removal from the Si_3N_4 coating material on the crucible wall, and the removed Si_3N_4 particles are transformed into other floating

Si$_3$N$_4$ particles [149–152]. In the second case (case II), Si$_3$N$_4$ particles float due to precipitating from the N-saturated Si melt during cooling when N is dissolved into the Si melt from the surface of the Si$_3$N$_4$ coating material [135,139,153]. In this case, the number of floating Si$_3$N$_4$ particles is limited by the saturated concentration of N. In both cases I and II, the floating Si$_3$N$_4$ particles become attached to the ingot as small particles during crystal growth, and some of them act as nucleation sites.

p-Type ingots are prepared using crucibles coated with a mixture of α- and β-Si$_3$N$_4$ particles using the NOC method [149–152]. In both cases I and II, floating Si$_3$N$_4$ particles are formed on the surface of the Si melt. The removal of Si$_3$N$_4$ particles from the Si$_3$N$_4$ coating layer is the key point for both cases. For case I, a p-type ingot with small particles on its top surface is grown. Before the growth, a large number of Si$_3$N$_4$ particles detached from the Si$_3$N$_4$ coating material are floating on the surface of the Si melt, and some of them act as nucleation sites during crystal growth. The ingot has many small grains distributed from a large island of small Si$_x$N$_y$ particles on the top surface. The Si$_x$N$_y$ particles have several morphologies, such as needle-like, columnar, and leaf-like structures [133]. For case II, Fig. 3.10A and B show a p-type ingot with a diameter of 23 cm and small particles on its top surface, respectively. Before the growth of the ingot, small particles precipitated from the N-saturated Si melt are floating on the surface of the Si melt. The ingot has many small grains distributed radially from the small particles on the top surface. Small particles observed on the surface of the ingots are analyzed by scanning electron microscopy (SEM) and energy-dispersive X-ray spectroscopy (EDX). Fig. 3.10C shows a SEM image of the same small particles on the top surface of the ingot, which can be observed at 5000x magnification. Some particles have a hexagonal-rod shape. Fig. 3.10D shows an EDX spectrum and the composition of the small particles on the top surface of the ingot, in which Si and N elements are detected together with C elements. The small particles attached on the top surface of the ingot are Si$_x$N$_y$, and these particles do not seriously affect the ingot quality even though the grains in the area are smaller than those in other cross-sectional areas because these perticles prevent the contamination from the gas phase [133].

3.4.2 Identification of Si$_3$N$_4$ particles

α-Si$_3$N$_4$ and β-Si$_3$N$_4$ particles are formed in the reaction between N gas and Si [154,155]. Both α- and β-Si$_3$N$_4$ have a hexagonal structure [156,157], but the lattice constant of α-Si$_3$N$_4$ (a = 7.766 and c = 5.615 Å)

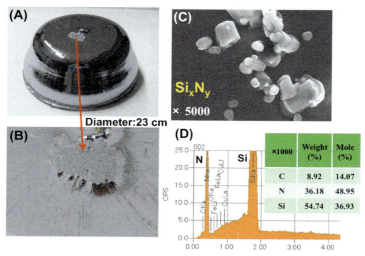

Fig. 3.10 (A) p-Type ingot with a diameter of 23 cm grown using a crucible coated with a mixture of α- and β-Si₃N₄ particles. (B) Small particles on the top surface of the ingot. (C) Si$_x$N$_y$ particles on the top surface of the ingot observed through an SEM. (D) EDX spectrum of the small particles on the top surface of the ingot, in which Si and N elements are detected along with C elements. *(Referred from Fig. 3 in K. Nakajima et al. J. Cryst. Growth 389 (2014) 112.)*

is longer than that of β-Si₃N₄ (c = 2.911 Å) [156–158]. The α-Si₃N₄ structure (c = 5.6193 Å) has trigonal symmetry and the β-Si₃N₄ structure (c = 2.9061 Å) has hexagonal symmetry [159,160]. Regarding the morphology, α-Si₃N₄ particles have a trigonal shape with several morphologies including whiskers, elongated particles, and equiaxed particles, and β-Si₃N₄ particles have a hexagonal shape [55,161]. α-Si₃N₄ in powder form has been observed to transform into β-Si₃N₄ at temperatures of approximately 1300 °C [162,163].

N dissolves from the surface of the Si₃N₄ coating material into a Si melt while the Si feedstock is melted and the Si melt is maintained at a high temperature. Dissolved N precipitates from the Si melt as other Si₃N₄ particles during the cooling of the Si melt [163]. Such precipitated Si₃N₄ particles float on the surface of the Si melt and act as nucleation sites during growth. When a Si ingot is grown using silica crucibles coated with a mixture of α- and β-Si₃N₄ particles, inclusions in the grown ingot are β-Si₃N₄ with a hexagonal shape [138]. In the growth, β-Si₃N₄ is formed through a reaction between O or oxides and a N-saturated Si melt because β-Si₃N₄ particles are more easily formed through the first reaction between

O and Si and the second reaction between N and silicon oxide [154]. Amorphous Si_3N_4 floating on the surface of a Si melt is transformed into $\beta\text{-}Si_3N_4$ crystalline rods with a hexagonal shape [150]. The wetting behavior depended on the Si_3N_4 coating was reported by several researchers [136,164]. The precipitated Si_3N_4 particles seriously affect on the quality of Si ingots grown by the cast method because such Si_3N_4 particles generate small angle grain boundaries and dislocations in the ingots grown by the cast method [165,166].

Impurities such as Fe diffuse from the Si_3N_4 coating materials to a grown ingot, and the quality at the periphery of the ingot is reduced [25,167,168]. Buonassisi et al. [169] measured impurities in commercially available $\alpha\text{-}Si_3N_4$ particles. The particles contain many kinds of impurities such as O, C, Al, Fe, Cu, Ca, Co, Mn, sodium (Na), Ni, Mo, magnesium (Mg), Cr, potassium (K), Zn, Ag, Sn, Ga, hafnium (Hf), As, Sb and Au, which are arranged in descending order. Among them, similar species such as Fe, Cu, Co, Mo, Cr, Zn, Ag, Sn, Ga, Hf, As and Sb can be detected in an Si ingot grown using the $\alpha\text{-}Si_3N_4$ coated crucible. The metal silicide nanoprecipitates detected in the ingot typically contain Fe, Cu, Co and Ni, which have the highest solubilities and diffusivities in Si. Abundant Fe elements distribute homogeneously as iron di-silicide ($FeSi_2$) and concentrate in discrete particles as di-iron trioxide (Fe_2O_3). Cu-rich particles are copper silicide (Cu_3Si) precipitates, which are identified at grain boundaries and dislocations [170]. A complex intermetallic alloy phase with fluorite-type structure containing Ni, Fe, Cu and Si is observed as precipitates in multicrystalline Si which accompanies mixed occupancies of Fe on the Ni site and Cu on the Si site [171].

The O, N, and C concentrations of ingots can be measured from longitudinal and horizontal cross sections of ingots through the FTIR analysis. The FTIR measurement in ingots is performed to confirm whether the N dissolves from the Si_3N_4 coating material into the Si melt. The ingots are grown by the NOC method using crucibles without and with a Si_3N_4 coating with a mixture of α- and $\beta\text{-}Si_3N_4$ particles. The O concentration in the ingots grown using crucibles without a Si_3N_4 coating is significantly higher than that in the ingots grown using crucibles with a Si_3N_4 coating [133]. The Si_3N_4 coating is effective for preventing a reaction between the silica crucible and the Si melt. The N concentration of the ingots grown using crucibles with a Si_3N_4 coating is significantly higher than that of ingots grown using crucibles without a Si_3N_4 coating. N clearly dissolves from the Si_3N_4 coating material into the Si melt, where it is saturated and incorporates in the ingots as inclusions during crystal growth [133]. The dissolved N from the Si_3N_4 coating material also acts as n–type dopants [172].

Reimann et al. and Trempa et al. [135,142] studied the N concentration in ingots in detail. The N concentration at the center of ingots is lower than that at the periphery in the ingots grown by the cast method. The N concentration depends on the growth rate and it has a low and uniform distribution by increasing a convective flow. On the other case, the N concentration at the center of the ingots is higher than that at the periphery [133]. N dissolved in the Si melt continuously evaporates from the surface of the melt. Therefore, the N concentration is lowest near the melt surface and increases toward the bottom of the melt when the convection is suppressed. As the segregation coefficient of N is very small, the N concentration in the melt increases near and at the center of the ingot during growth (see Section 3.3.4). Moreover, as the ingot becomes larger, it becomes more difficult for N to diffuse from the coating and evaporate from the melt near the center of the ingot than near the periphery. Therefore, the N concentration in the melt rapidly increases near the center of the ingot. This is the reason why the N concentration at the center of the ingot is higher than that at the periphery in this case. The same distribution is observed for the O concentration for the similar reason.

Fig. 3.11A shows a p-type ingot with a diameter of 10.3 cm grown using a crucible coated with a mixture of α- and β-Si_3N_4 particles. A large number of Si_xN_y particles appears at the top surface of the ingot. The orientation and structure of the small particles can be analyzed on basis of their EBSD pattern. To carry out the SEM and EBSD measurements, an area on the top surface of the ingot is used as a sample, as indicated by the red arrow in Fig. 3.11A. Fig. 3.11B and C show the Si_xN_y particles with a columnar structure observed by SEM at 400x and 2500x magnifications, respectively. Fig. 3.11D, E, and F show orientation image maps analyzed on basis of EBSD patterns for Si, α-Si_3N_4, and β-Si_3N_4 taken from a cross section of the sample, respectively. The Si sample contains many Si_xN_y particles as shown in Fig. 3.11B, which correspond to the black dots in Fig. 3.11D. Although most of the Si_xN_y particles do not act as nucleation sites, some of them acts as nucleation sites from which grains and grain boundaries appear, but it cannot be distinguished which Si_xN_y particles act as nucleation sites. Very few signals from α-Si_3N_4 are detected as shown in Fig. 3.11E, while a strong signal from β-Si_3N_4 is detected as indicated in Fig. 3.11F. Therefore, there are very few α-Si_3N_4 particles in the sample

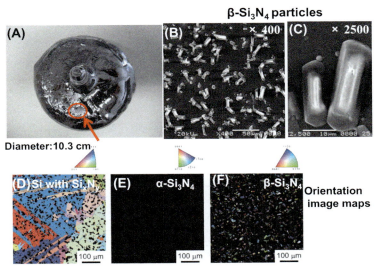

Fig. 3.11 (A) *p*-Type ingot with a diameter of 10.3 cm grown using a crucible coated with a mixture of α- and β-Si$_3$N$_4$ particles as in case I. (B) Si$_x$N$_y$ particles with a columnar structure observed by SEM at 400x magnification. (C) Si$_x$N$_y$ particles with a columnar structure observed by SEM at 2500x magnification. (D) Orientation image map analyzed on basis of EBSD pattern for Si. The black dots correspond to Si$_x$N$_y$ particles in the Si ingot. (E) Orientation image map analyzed on the basis of EBSD pattern for α-Si$_3$N$_4$. Very few α-Si$_3$N$_4$ particles were detected. (F) Orientation image map analyzed on the basis of EBSD pattern for β-Si$_3$N$_4$. β-Si$_3$N$_4$ particles are clearly detected. *(Referred from Fig. 5 in K. Nakajima et al. J. Cryst. Growth 389 (2014) 112.)*

and most of the Si$_x$N$_y$ particles on the surface of the ingot shown in Fig. 3.11B and C are β-Si$_3$N$_4$ [133]. This result is consistent with Søiland's result [138]. The phase of the small particles can be identified by X-ray diffraction (XRD) using Cu-Kα radiation. It is also confirmed by XRD that a large number of the Si$_x$N$_y$ particles on the top surface of the same ingot as shown in Fig. 3.7A are β-Si$_3$N$_4$ [133].

3.4.3 Process to form Si$_3$N$_4$ transformers or precipitates from Si$_3$N$_4$ coating materials

The transformation and precipitation processes can be considered to occur from the analysis of Si$_3$N$_4$ particles on the surface of ingots. Fig. 3.12 shows a schematic illustration summarizing the behavior of Si$_3$N$_4$ particles coated on a crucible wall during cooling and crystal growth [133]. Silica crucibles coated with a mixture of α- and β-Si$_3$N$_4$ particles are used. In case I, a large number of Si$_3$N$_4$ particles are removed from the Si$_3$N$_4$ coating material and

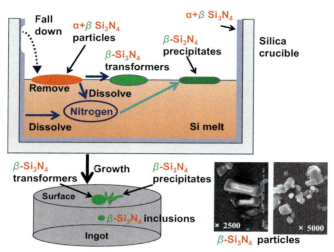

Fig. 3.12 Schematic illustration summarizing the behavior of Si$_3$N$_4$ particles coated on a crucible wall. *(Referred from Fig. 7 in K. Nakajima et al. J. Cryst. Growth 389 (2014) 112.)*

initially float on the surface of the Si melt as a mixture of α- and β-Si$_3$N$_4$ particles. Then, they transform into β-Si$_3$N$_4$ particles through a reaction with the Si melt. Some of the floating Si$_3$N$_4$ particles dissolve in the Si melt as N, and β-Si$_3$N$_4$ particles then precipitate from the Si melt oversaturated by N during the cooling. A few of the α-Si$_3$N$_4$ particles removed from the Si$_3$N$_4$ coating material continue to float and remain as α-Si$_3$N$_4$ particles on the surface of the Si melt without undergoing a transformation or dissolution. In case II, N dissolves in the Si melt from the surface of the Si$_3$N$_4$ coating material and becomes saturated in the Si melt at a higher temperature than the melting point of Si. The solubility of N in the Si is expressed by the composition ratio expressed as (the composition of N/the composition of Si) at the melting point of Si (1414 °C). The composition ratio of ~10^{-4} corresponds to the N solubility of 10^{19} cm^{-3} [173]. The solubility of N is strongly reduced by cooling the melt temperature resulting in precipitation of Si$_3$N$_4$. The β-Si$_3$N$_4$ particles from the N-oversaturated Si melt precipitate on the Si melt while it is cooled from above 1450 °C to the melting point of the Si. The number of precipitated Si$_3$N$_4$ particles is limited by the saturated concentration of N in the Si melt and depends on the temperature reduction. N dissolved from the Si$_3$N$_4$ coating material also acts as n-type dopants [172].

For the NOC growth shown in Chapter 7, such β-Si$_3$N$_4$ particles float near the center of the surface of the Si melt before crystal growth because of the low-temperature region in the melt. Finally, in both cases I and II, floating β-Si_3N_4 particles appear on the surface of the ingot and have several morphologies depending on the growth conditions. Some of the β-Si_3N_4 particles attached on the ingot act as nucleation sites, forming grains and grain boundaries. N in the Si melt is also incorporated in the ingots as β-Si_3N_4 inclusions during crystal growth [138]. Buonassisi et al. reported that Si_3N_4 particles can be a significant source of contaminants, such as metal point defects, in Si ingots [169]. The Si_3N_4 formation process is the same in Si crystal growth using crucibles with a Si_3N_4 coating such as directional solidification cast growth. No difference is found in the Si_3N_4 formation process for the p- and n-type ingots.

3.4.4 Wetting properties of Si_3N_4 coating

Non-wetting behavior of crucible materials is favorable to obtain spontaneous detachment of a solidified Si ingot from crucible wall as well as negligible reactivity. A larger interface energy between a Si melt and a substrate leads to stronger non-wetting behavior. The contact angle or wetting angle between a Si melt and the crucible materials should be higher than $90°$ for the favorable non-wetting behavior. The wetting angle of Si on SiC, SiO_2 and Si_3N_4 is lower than $90°$ and that on SiC is lowest [174,175]. The wetting angle of Si on SiC is $8°$ and molten Si spreads over the SiC plate [174]. The wetting angle of Si on SiO_2 is $83 \pm 4°$ which is close to $90°$ and results in sticking at the Si/SiO_2 interface [136]. The wetting angle of Si on Si_3N_4 is $82-89°$ and molten Si wets well the deoxidized Si_3N_4 coating materials [136,164,174]. The spreading rate of Si on Si_3N_4 is controlled by deoxidation of the substrate by reaction between Si and the oxide film with formation of the volatile SiO [176]. The wetting angles of Si on Al_2O_3 and magnesium oxide (MgO) are $86°$ and $88°$, respectively [174]. The wetting angle of Si on boron nitride (BN) is $117°$ or $145°$ [174,176]. The reactivity in the Si/BN system is so weak that chemical equilibrium or saturation of B in Si melt is established well before the equilibrium contact angle on the reaction product (Si_3N_4) is attained [176]. Si does not well wet BN, but B contamination leads to over-doping in Si [136]. The discontinuous Si_3N_4 layer is formed at the interface between Si and BN and it retards the dissolution of BN into the Si melt to a certain extent [174]. The spreading of Si on BN is limited by the rate of formation of such Si_3N_4 at the interface due reaction between Si and BN [176].

The wetting properties of a coating layer of Si_3N_4 particles depend on oxidation of the layer. The presence of a poorly wettable SiO_2 skin layer on

Si_3N_4 particles is effective as a barrier to wetting. The wetting angle of Si on Si_3N_4 coated SiO_2 decreases as oxidized Si_3N_4 coated particles transform to deoxidized ones [136]. The wetting angle of Si on Si_3N_4 changes from $92 \pm 3°$ to $49 \pm 3°$ by the deoxidation of Si_3N_4 [164]. The deoxidation can take place ahead of the infiltration front by the reaction between the Si vapor and the silica skin [146]. The deoxidation occurs both by lateral propagation on the coating surface and by in–depth penetration onside the coating, and the complete deoxidation can occur for 7 h at 1480 °C above which the deoxidation dramatically starts under Ar flow [177]. The possible reactions of the deoxidation as follows [146,177],

$$SiO_2 \ (s) + Si \ (g) = 2SiO \ (g), \quad or \tag{3.4}$$

$$SiO_2 \ (s) = SiO \ (g) + 1/2O_2 \ (g), \quad or \tag{3.5}$$

$$2SiO_2 \ (s) + Si_3N_4 \ (s) = 6SiO \ (g) + 2N_2 \ (g), \tag{3.6}$$

The reaction shown by Eq. (3.1) also has some possibility for the deoxidation of Si_3N_4 coating materials. The propagation of the deoxidation front inside the coating under Ar flow is due to the reaction of Eq. (3.6) [177]. Significant losses of O can be produced by this reaction [146]. Thus, the deoxidation can be explained by the reduction of the SiO_2 skin layer by Si_3N_4. The ingot/crucible/vapor triple lines are week points on sticking where deoxidation easily occurs by rapid SiO evaporation, and they lead to wetting and infiltration of Si in the coating layer which are responsible for cracking of an ingot [136]. The infiltration of Si likely occurs when the wetting angle is below 90°.

Brynjulfsen et at [164]. studied the relation between wetting and oxidation in detail. The wetting angle of Si on the Si_3N_4 coating layer increases with the O concentration in the coating layer. A high O content in the Si_3N_4 coating layer on SiO_2 enhances non–wettable properties and the oxidation forms a sinoite (Si_2N_2O) layer as follows,

$$Si_3N_4 \ (s) + 3O_2(g) = 3SiO_2 \ (s) + 2N_2 \ (g), \tag{3.7}$$

$$Si_3N_4 \ (s) + 3/4O_2 \ (g) = 3/2 \ Si_2N_2O \ (s) + 1/2N_2 \ (g), \tag{3.8}$$

Before formation of a SiO_2 layer on the Si_3N_4 particles, the Si_2N_2O layer is formed on the surface of the Si_3N_4 particles because Si_3N_4 is un-stable. Atomspheric moisture promotes formation of SiO_2 over Si_2N_2O. Thus, the oxidation of Si_3N_4 toward SiO_2 goes through an intermediate Si_2N_2O phase. The composition gradually varies from an outer SiO_2 layer toward an inner Si_3N_4 layer. The wetting angle of Si on Si_2N_2O with O concentration higher than 14 wt% is around $100°$. Deoxidation of this layer occurs at high temperature. The superheating above 150 K increases the wettability and leads to penetration of Si melt into the coating.

3.5 Electrical properties and solar cells of Si multicrystalline ingots

3.5.1 Electrical properties

The high diffusion length can be obtained for an intra-grain of the p-type wafer grown by the cast method after P gettering and hydrogenation [178]. The main factors to limit the solar-cell performance are recombination at crystal defects. Decoration of impurities makes dislocations, sub-grain boundaries and grain boundaries electrically active. The area of recombination active dislocations is large in the lower part of an ingot, but it normally becomes smaller in the middle part of the ingot. The recombination activity of grain boundaries varies depending on the grain boundary geometry or type, which has an impact on the ability to aggregate metal impurities [130]. Most of dislocations in p- and n-type crystals are already recombination active in as-grown state before any processing [179]. Grain boundaries in p-type crystals are more recombination active than those in n-type crystals [179]. Recombination active grain boundaries and dislocations lead to a local reduction of lifetime and excess carrier concentration at the defects, resulting in a local reduction of both the open circuit voltage and the short circuit current of solar cells [179].

The P gettering increases the recombination strength for grain boundaries in the p- and n-type wafers due to a redistribution of impurities near the grain boundaries caused by high-temperature processes. The high diffusion length can be obtained for an intra-grain of p- and n-type wafers after P gettering and hydrogenation [130]. On the other hand, the main advantage of n-type wafers is their superior electrical properties in the intra-grain regions and grain boundaries, especially after both P gettering and hydrogenation. Grain boundaries in n-type wafers are less recombination active than those in p-type wafers. Hydrogenation is most effective on gettered n-type wafers.

The minority carrier lifetime is largely reduced near the top part of an ingot because many aggregated dislocations appear in this part as shown in Fig. 5.3 [180,181]. To assure high V_{oc} values, it is mandatory to use wafers with a higher minority carrier lifetime than 1 ms, which is not limited by bulk recombination [182]. The external P gettering is efficient in reducing the metallic impurities but its effect strongly depends on the initial metallic concentration. For Fe contaminated multicrystalline Si, the minority carrier lifetime is strongly improved by higher temperature of P diffusion gettering, especially in areas of low dislocation density [183]. Moreover, lower-temperature P diffusion following a high-temperature P diffusion produces further lifetime benefits, leaving a moderate Fe_i concentration [183]. On the other hand, dislocations act as an internal gettering at high temperature to trap metals and decrease the bulk quality. However, the external P gettering is much stronger than the internal gettering [109].

3.5.2 Solar cells

The front surface texturing is performed for preparation of multicrystalline Si solar cells to reduce the light reflectance from the solar-cell surface. The wet texturing such as acidic etching and alkaline etching is commonly used owing to its easiness and low-cost. Usually, multicrystalline Si wafers are textured by the isotropic acidic texturing method because of their randomly oriented grains, while <100> oriented CZ single wafers are textured by the alkaline texturing method to obtain a pyramidal structure with low reflectance [184]. The average reflectance of the acidic texturing for a bare multicrystalline Si wafer is 30% in the wavelength range of 600–700 nm [185]. Submicron surface textures can be effectively reduced using submicron Si surface grating with appropriate aspect ratios [186]. The reactive ion etching (RIE) texturing can be used for the submicron surface texturing. The RIE texturing obtains the good reliability and reproducibility, and it has high rates of isotropic etching. The average reflectance of RIE is 20% at 700 nm for a bare multicrystalline Si wafer [187]. The novel technology to use the maskless RIE in combination with acidic etching can reduce the average reflectance to 8.5% in the whole wavelengths for a bare multicrystalline Si wafer, and the conversion efficiency of solar cells increases by 0.5% [185]. Subwavelength structure with much smaller dimensions (~ 100 nm period) than the wavelength of light is a method to obtain a wide-band antireflection effect and to offer a low reflectivity over a wide range of incident angles [188]. Ion implantation doping is tried for multicrystalline Si solar cells together with flash lamp annealing within a millisecond annealing time as a low thermal technology and its good effects on electrically activation of implanted P and recrystallization of Si are demonstrated [189].

For Si multicrystalline wafers, the typical average conversion efficiency of solar cells is 17.2−17.6% and the scattering width of the conversion efficiency is 1.6% [190]. The scattering width is defined as the difference between the best and worst conversion efficiencies (see Section 7.5.2). Fe_i plays a major role concerning efficiency losses of Si multicrystalline solar cells [100]. The losses are significantly reduced by using a high-purity crucible for the ingot growth [100]. Especially, for lower injection levels, Fe_i becomes the most dominant recombination channel. For the higher injection levels, Fe precipitates play a more important role compared with Fe_i and the total impact of Fe is even stronger [100]. For Fe contaminated solar cells, annealing at appropriate times and temperature can allow the transformation from supersaturated more detrimental Fe point defects to distributed less detrimental iron silicide precipitates [191]. Independent of the as-grown Fe distribution, relative short annealing times at low temperatures after P diffusion gettering lead to an appreciable increase in solar cell efficiency [192]. On the other hand, the conversion efficiency of solar cells is not affected by the initially large Cu concentration due to the complementary actions of the external gettering effect developed by the P diffusion and hydrogenation [120].

3.6 Growth of large-scale ingots in industry

The industrial scale of Si ingots grown by the cast method increases from G5 to G6 by upgrading design of the hot zone area of a furnace. The G5 and G6 ingots are cut to 25 and 36 bricks, respectively. For the growth of such large-scale ingots, a normally used cast furnace has two zone carbon heaters such as the side and top ones which is similar to the furnace shown in Fig. 3.1. Inside a stainless chamber, the furnace has several insulating materials such as graphite insulators, a graphite case, a graphite partition block, a heat-exchanger, a gas shield, a graphite block for a crucible and a protecting graphite cover-plate against contamination. For the growth of large ingots, the crucible is placed on the graphite block cooled by thermal radiation as the insulation basket moves upwards [193]. In the furnace, Ar cooling is used for heat extraction in a closed loop and the heat is exchanged by cold water. The reduced pressure (~ 0.6 atm) of Ar is flushed to protect graphite materials in the hot zone from oxidation during growth [181]. The Ar flow is effective to carry away SiO evaporated from a melt and prevent the back diffusion of CO from graphite materials. The furnace is designed to realize a crystal-melt interface with slightly convex in the

146 Crystal Growth of Si Ingots for Solar Cells Using Cast Furnaces

growth direction to move impurities outward, reduce nucleation on the crucible wall and reduce stress in an ingot by increasing the melt temperature near the side wall of the crucible. At the early stage of the growth, there are two counter-rotated vortices in a Si melt owing to the temperature difference between the side and center of the Si melt, which can be known by three-dimensional (3-D) simulation of casting process [194].

The Si feedstocks used for the growth of G5 and G6 ingots are usually 450 and 800 kg, respectively. The heights of the G5 and G6 ingots are 27 and 35 cm, respectively. The solidification rate increases with the ingot size increases even through the total growth time is longer for the larger ingot. The total growth time is usually between 30 and 40 h for the G5 and G6 ingots and the whole growth cycle for the G6 ingot takes 70 h including the meltdown, growth, annealing and cooling procedures [181]. The yield of high-quality parts obtained from these ingots is usually around 70% for each ingot. From a p-type G6 ingot, about 27,000 wafers with 200 µm thick and a resistivity range between 1.2 and 1.8 Ω-cm can be obtained [193]. The resistivity of the p-type ingot increases from the top (1 Ω-cm) to the bottom (2 Ω-cm). The larger ingot can be more efficiently manufactured and contribute the low–impurity and low cost. The in-diffusion depth of Fe increases with the crucible size, whereas the Fe concentration in the center of the ingot decreases at the same time [4]. Recently, much larger ingots such as G7 and G8 are manufactured in the industries. The G7 and G8 ingots are cut to 49 and 64 bricks, respectively. The Si feedstocks used for the growth of G7 and G8 ingots are usually 1,200 and 1,600–2,000 kg, respectively. The heights of the G7 and G8 ingots are 40 and 42 cm, respectively. To increase the height of the ingot, additional Si are sometimes recharged during meltdown stage.

The ingot growth starts by moving the insulation basket or the temperature gradient upward in the furnace to cool the melt temperature down (see section 3.1.1). The initial cooling rate for solidification is about 5–20 °C h^{-1} and the solidified ingot is cooled to RT at a cooling rate of 10 °C min^{-1} [68]. Generally, the low growth rate (<0.07 mm min^{-1}) is used during the early stage of the ingot growth to suppress initial formation of dislocations caused by thermal stress, and the high growth rate (>0.16 mm min^{-1}) is used during the later stage of the growth to increase the production efficiency [63,68,181]. After the ingot is solidified completely, the insulation basket is lowered down again for annealing. The annealing period is a common part of the industrial process and it intends to relieve stress by movement of dislocations and sub-grain boundaries to minimize the stored stress energy.

Regions with defects such as dislocations, dislocation clusters and sub-grain-boundaries normally increase from the bottom to the top of an ingot. As much as 10% of the surface area of commercially available multicrystalline wafers is covered with dislocation clusters with a dislocation density between 10^6 and 10^8 cm^{-2} [86]. The size of dislocation clusters varies from 0.1 mm to 1 mm. An ingot grown by the cast method has low lifetime regions at the bottom and sides of the ingot where they directly contact with the crucible wall as shown in Fig. 5.3. The ingot also has a low lifetime region at the top of the ingot where impurities segregated from the growing interface are finally frozen in place. The ingot generally contains inclusions of SiC and Si_3N_4 particles which mainly come from carbon heaters, Si_3N_4 coating material and crucible. The SiC particles has a chance to wind up to the upper part of the Si melt due to the two counter-rotated vortices and they act as nucleation sites of grains [169].

3.7 Key points for improvement

For a Si ingot grown by the cast method, the grain structure approaches to a stable state as the ingot height increases, in which the average grain size is constant and grains with preferential orientations remain. It should be studied why the stable state appears or what are the main factors to determine the average grain size. The suitable grain size to reduce stress, grain boundaries and dislocations and to largely increase the quality of an ingot should be much more studied for the advanced cast method. By making clear the mechanism for the appearance of the final stable state, the main factors to determine the average grain size and the suitable grain size to obtain the higher quality of the ingot, novel technologies will be invented for improvement of the quality and structure of Si multicrystalline ingots.

If the remaining stress in an ingot will be largely reduced, the density of crystal defects such as dislocations can be also largely reduced and the quality of the ingot can be largely improved. The adhesion between the ingot and crucible should be largely reduced using a nonwetting crucible or nonwetting coating materials. The wetting research between the ingot and crucible will be more important to improve the quality of the cast ingot [195,196].

The impurities in ingots are basic problems for the cast method. Their main origins are the graphite materials, quartz, Si feedstock and Si_3N_4 coating which mainly contains B. P, Al, Fe, Co, Ni and Cr of several ppm or less. A nonwetting crucible such as a Liquinert quartz crucible has a potential to reduce impurities [195—197]. Such a new technology will give a chance to largely improve the quality of Si multicrystalline ingots (see Section 8.2.2).

References

[1] C.W. Lan, W.C. Lan, T.F. Lee, A. Yu, Y.M. Yang, W.C. Hsu, B. Hsu, A. Yang, J. Cryst. Growth 360 (2012) 68.

[2] Y.M. Yang, A. Yu, B. Hsu, W.C. Hsu, A. Yang, C.W. Lan, Prog. Photovoltaics Res. Appl. 23 (2015) 340.

[3] Q. Yu, L. Liu, G. Zhong, X. Huang, in: Proceeding of 7th Intern. Symp. On multiphase flow, Heat Mass Transfer and Energy Conversion 1547, 2013, p. 652.

[4] J. Wei, H. Zhang, L. Zheng, C. Wang, B. Zhao, Sol. Energy Mater. Sol. Cells 93 (2009) 1531.

[5] M.C. Schubert, J. Schön, F. Schindler, W. Kwapil, A. Abdollahinia, B. Michl, S. Riepe, C. Schmid, M. Schumann, S. Meyer, W. Warta, IEEE J. Photovolt. 3 (2013) 1250.

[6] X. Ma, L. Zheng, H. Zhang, B. Zhao, C. Wang, F. Xu, J. Cryst. Growth 318 (2011) 288.

[7] Q. Yu, L. Liu, W. Ma, G. Zhong, X. Huang, J. Cryst. Growth 358 (2012) 5.

[8] L. Chen, B. Dai, J. Cryst. Growth 354 (2012) 86.

[9] Z. Li, L. Liu, X. Liu, Y. Zhang, J. Xiong, J. Cryst. Growth 385 (2014) 9.

[10] W. Ma, G. Zhong, L. Sun, Q. Yu, X. Huang, L. Liu, Sol. Energy Mater. Sol. Cells 100 (2012) 231.

[11] X. Qi, W. Zhao, L. Liu, Y. Yang, G. Zhong, X. Huang, J. Cryst. Growth 398 (2014) 5.

[12] H. Miyazawa, L. Liu, K. Kakimoto, Cryst. Growth Des. 9 (2009) 267.

[13] C. Ding, M. Huang, G. Zhong, L. Liu, X. Huang, Cryst. Res. Technol. 49 (2014) 405.

[14] Z. Wu, G. Zhong, Z. Zhang, X. Zhou, Z. Wang, X. Huang, J. Cryst. Growth 426 (2015) 110.

[15] F. Liu, C.-S. Jiang, H. Guthrey, S. Johnston, M.J. Romero, B.P. Gorman, M.M. Al-Jassim, Sol. Energy Mater. Sol. Cells 95 (2011) 2497.

[16] M.A. Martorano, J.B. Ferreira Neto, T.S. Oliveira, T.O. Tsubaki, Metall. Mater. Trans. A 42 (2011) 1870.

[17] I. Brynjulfsen, K. Fujiwara, N. Usami, L. Arnberg, J. Cryst. Growth 356 (2012) 17.

[18] H.J. Möller, C. Funke, M. Rinio, S. Scholz, Thin Solid Fims 487 (2005) 179.

[19] R. Buchwald, S. Würzner, H.J. Möller, A. Ciftja, G. Stokkan, E. Øvrelid, A. Ulyashin, Phys. Solidi A 212 (2015) 25.

[20] S. Würzner, R. Buchwald, H.J. Möller, Phys. Status Solidi C 12 (2015) 1119.

[21] A.L. Endrös, Sol. Energy Mater. Sol. Cells 72 (2002) 109.

[22] M. Kivambe, D.M. Powell, S. Castellanos, M.A. Jensen, A.E. Morishige, K. Nakajima, K. Morishita, R. Murai, T. Buonassisi, J. Cryst. Growth 407 (2014) 31.

[23] P.S. Ravishankar, J.P. Dismukes, W.R. Wilcox, J. Growth 71 (1985) 579.

[24] S. Dumitrica, D. Vizman, J.-P. Garandet, A. Popescu, J. Cryst. Growth 360 (2012) 76.
[25] F.-M. Kiessling, F. Büllesfeld, N. Dropka, C. Frank-Rotsch, M. Müller, P. Rudolph, J. Cryst. Growth 360 (2012) 81.
[26] I. Buchovska, N. Dropka, S. Kayser, F.M. Kiessling, J. Cryst. Growth 507 (2019) 299.
[27] E. Scheil, Z. Metallkd. 34 (1942) 70.
[28] M.P. Bellmann, G. Stokkan, A. Ciftja, J. Denafas, T. Kaden, J. Cryst. Growth 504 (2018) 51.
[29] G. Stokkan, M.D. Sabatino, R. Søndenå, M. Juel, A. Autruffe, K. Adamczyk, H.V. Skarstad, K.E. Ekstrøm, M.S. Wiig, C.C. You, H. Haug, M. M'hamdi, Phys. Status Solidi 214 (2017) 1700319.
[30] X. Huang, T. Hoshikawa, S. Uda, J. Cryst. Growth 306 (2007) 422.
[31] T.F. Ciszek, G.H. Schwuttke, K.H. Yang, J. Cryst. Growth 46 (1979) 527.
[32] B. Gao, S. Nakano, K. Kakimoto, J. Electrochem. Soc. 157 (2010) H153.
[33] Y.-Y. Teng, J.-C. Chen, C.-W. Lu, C.-Y. Chen, J. Cryst. Growth 360 (2012) 12.
[34] C. Schmid, M. Schumann, F. Haas, S. Riepe, The 28th european photovoltaic solar energy conference and exhibition, Paris, France, September 30—October 4, 2013.
[35] B. Gao, X.J. Chen, S. Nakano, K. Kamimoto, J. Cryst. Growth 312 (2010) 1572.
[36] V.V. Kalaev, I.Y. Evstratov, Y.N. Makarov, J. Cryst. Growth 249 (2003) 87.
[37] S. Yuan, D. Hu, X. Yu, F. Zhang, H. Luo, L. He, D. Yang, Silicon, vol. 16, Springer, January 2018 (On line).
[38] G. Wang, D. Yang, D. Li, Q. Shui, J. Yang, D. Que, Physica B 308—310 (2001) 450.
[39] D. Yang, R. Fan, L. Li, D. Que, K. Sumino, Appl. Phys. Lett. 68 (1996) 487.
[40] H. Zhang, M. Stavola, M. Seacrist, J. Appl. Phys. 114 (2013) 093707.
[41] H. Kusunoki, T. Ishizuka, A. Ogura, H. Ono, Appl. Phys. Exp. 4 (2011) 115601.
[42] X. Ma, L. Fu, D. Tian, D. Yang, J. Appl. Phys. 98 (2005) 084502.
[43] K. Fujiwara, K. Nakajima, T. Ujihara, N. Usami, G. Sazaki, H. Hasegawa, S. Mizoguchi, K. Nakajima, J. Cryst. Growth 243 (2002) 275.
[44] K. Fujiwara, Y. Obinata, T. Ujihara, N. Usami, G. Sazaki, K. Nakajima, J. Cryst. Growth 262 (2004) 124.
[45] K. Fujiwara, Y. Obinata, T. Ujihara, N. Usami, G. Sazaki, K. Nakajima, J. Cryst. Growth 266 (2004) 441.
[46] K. Fujiwara, Growth of multicrystalline silicon for solar cells: dendritic cast method, in: Yang (Ed.), Chapter 9, Section Two Crystalline Silicon Growth, Hand Book of "Photovoltaic Silicon Material", Springer, Berlin Heidelberg, 2017, pp. 1—22 (On line).
[47] K. Fujiwara, R. Maeda, K. Maeda, H. Morito, Scripta Mater. 133 (2017) 65.
[48] K. Fujiwara, K. Kutsukake, H. Kodama, N. Usami, K. Nakajima, J. Cryst. Growth 312 (2010) 3670.
[49] M. Tokairin, K. Fijiwara, K. Kutsukake, N. Usami, K. Nakajima, Phys. Rev. B 80 (2009) 174108.
[50] K. Fujiwara, R. Gotoh, X.B. Yang, H. Koizumi, J. Nozawa, S. Uda, Acta Meter 59 (2011) 4700.
[51] K. Fujiwara, W. Pan, N. Usami, K. Sawada, M. Tokairin, Y. Nose, A. Nomura, T. Shishido, K. Nakajima, Acta Mater. 54 (2006) 3191.
[52] K. Fujiwara, K. Maeda, N. Usami, G. Sazaki, Y. Nose, A. Nomura, T. Shishido, K. Nakajima, Acta Mater. 56 (2008) 2663.
[53] S. Rajendran, W.R. Wilcox, P.S. Ravishankar, J. Cryst. Growth 75 (1986) 353.
[54] M. Tokairin, K. Fujiwara, K. Kutsukake, H. Kodama, N. Usami, K. Nakajima, J. Cryst. Growth 312 (2010) 3670.
[55] M. Beaudhuin, K. Zaidat, T. Duffer, M. Lemiti, J. Cryst. Growth 336 (2011) 77.

[56] M. Beaudhuin, G. Chichignoud, P. Bertho, T. Duffer, M. Lemiti, K. Zaidat, Mater. Chem. Phys. 133 (2012) 284.

[57] K. Fujiwara, K. Maeda, N. Usami, K. Nakajima, Phys. Rev. Lett. 101 (2008) 055503.

[58] K. Nakajima, N. Usami, K. Fujiwara, K. Kutsukake, S. Okamoto, in: Proceedings of the 24th European Photovoltaic Solar Energy Conference, 2009, p. 1219.

[59] M. Trempa, I. Kupka, C. Kranert, T. Lehmann, C. Reimann, J. Friedrich, J. Cryst. Growth 459 (2017) 67.

[60] T. Strauch, M. Demant, P. Krenckel, R. Riepe, S. Rein, J. Cryst. Growth 454 (2016) 147.

[61] K. Kutsukake, T. Abe, N. Usami, K. Fujiwara, K. Morishita, K. Nakajima, Scripta Mater. 65 (2011) 556.

[62] H.T. Nguyen, M.A. Jensen, L. Li, C. Samundsett, H.C. Sio, B. Lai, T. Buonassisi, D. MacDonald, IEEE J. Photovol. 7 (2017) 772.

[63] Z. Wu, G. Zhong, X. Zhou, Z. Zhang, Z. Wang, W. Chen, X. Huang, J. Cryst. Growth 441 (2016) 58.

[64] R.E. Thomson, D.J. Chadi, Phys. Rev. B 29 (1984) 889.

[65] A. Voigt, E. Wolf, H.P. Strunk, Mater. Sci. Eng., B 54 (1998) 202.

[66] N. Usami, K. Kutsukake, T. Sugawara, K. Fujiwara, W. Pan, Y. Nose, T. Shishido, K. Nakajima, Jpn. J. Appl. Phys. 45 (2006) 1734.

[67] M. Trempa, C. Reimann, J. Friedrich, G. Müller, A. Krause, L. Sylla, T. Richter, Cryst. Res. Technol. 50 (2015) 124.

[68] H.W. Tsai, M. Yang, C. Chuck, C.W. Lan, J. Cryst. Growth 363 (2013) 242.

[69] T. Lehmann, M. Trempa, E. Meissner, M. Zschorsch, C. Reimann, J. Friedrich, Acta Mater. 69 (2014) 1.

[70] D. Oriwol, E.R. Carl, A.N. Danilewsky, L. Sylla, W. Seifert, M. Kittler, H.S. Leipner, Acta Mater. 61 (2013) 6903.

[71] M. Kitamura, N. Usami, T. Sugawara, K. Kutsukake, K. Fujiwara, Y. Nose, T. Shishido, K. Nakajima, J. Cryst. Growth 280 (2005) 419.

[72] H.K. Lin, M.C. Wu, C.C. Chen, C.W. Lan, J. Cryst. Growth 439 (2016) 40.

[73] J. Chen, T. Sekiguchi, D. Yang, F. Yin, K. Kido, S. Tsurekawa, J. Appl. Phys. 96 (2004) 5490.

[74] J. Chen, T. Sekiguchi, Jpn. J. Appl. Phys. 46 (2007) 6489.

[75] K. Kutsukake, N. Usami, T. Ohtaniuchi, K. Fujiwara, K. Nakajima, J. Appl. Phys. 105 (2009) 044909.

[76] G. Stokkan, Acta Mater. 58 (2010) 3223.

[77] S. Woo, M. Bertoni, K. Choi, S. Nam, S. Castellanos, D.M. Powell, T. Buonassisi, H. Choi, Sol. Energy Mater. Sol. Cells 155 (2016) 88.

[78] M.M. Kivambe, G. Stokkan, T. Ervik, B. Ryningen, O. Lohne 110 (2011) 063524.

[79] D. You, J. Du, T. Zhang, Y. Wan, W. Shans, L. Wang, D. Yang, Proceedings of 35th IEEE Photovol. Specialists Conf. (2010) 2258−2261.

[80] S. Würzner, R. Helbig, C. Funke, H.J. Möller, J. Appl. Phys. 108 (2010) 083516.

[81] F. Secco d'Aragona, J. Electrochem. Soc. 119 (1972) 948.

[82] B.L. Sopori, J. Electrochem. Soc. 131 (1984) 667.

[83] D. Oriwol, M. Trempa, L. Sylla, H.S. Leipner, J. Cryst. Growth 463 (2017) 1.

[84] T. Ervik, G. Stokkan, T. Buonassisi, Ø. Mjøs, O. Lohne, Acta Mater. 67 (2014) 199.

[85] E.B. Yakimov, Metal impurities and gettering in crystalline silicon, in: D. Yang (Ed.), Crystalline Silicon Growth, Hand Book of "Photovoltaic Silicon Material", Springer, Berlin Heidelberg, 2018, pp. 1−46 (On line).

[86] B. Ryningen, G. Stokkan, M. Kivambe, T. Ervik, O. Lohne, Acta Mater. 59 (2011) 7703.

[87] M. Kivambe, G. Stokkan, T. Ervik, S. Castellanos, J. Hofstetter, T. Bounassisi, Solid State Phenom. 205−206 (2014) 71.

[88] Y. Hayama, T. Matsumoto, T. Muramatsu, K. Kutsukake, H. Kudo, N. Usami, Sol. Energy Mater. Sol. Cells 189 (2019) 239.

[89] K. Arafune, T. Sasaki, F. Wakabayashi, Y. Terada, Y. Ohshita, M. Yamaguchi, Physica B 376—377 (2006) 236.

[90] M.I. Bertoni, D.M. Powell, M.L. Vogl, S. Castellanos, A. Fecych, T. Bounassisi, Phys. Status Solidi RRL 5 (2011) 28.

[91] G. Stokkan, Y. Hu, Ø. Mjøs, M. Juel, Sol. Energy Mater. Sol. Cells 130 (2014) 679.

[92] K. Kutsukake, T. Abe, N. Usami, K. Fujiwara, I. Yonenaga, K.M.K. Nakajima, J. Appl. Phys. 110 (2011) 083530.

[93] G. Stokkan, J. Cryst. Growth 384 (2013) 107.

[94] C. Reimann, J. Friedrich, E. Meissner, D. Oriwol, L. Sylla, Acta Mater. 93 (2015) 129.

[95] K. Hartman, M.I. Bertoni, J. Serdy, T. Buonassisi, Appl. Phys. Lett. 93 (2008) 122108.

[96] K. Choi, S. Castellanos, D.M. Powell, T. Buonassisi, H.J. Choi, Phys. Status Solidi 212 (2015) 2315.

[97] H.J. Choi, M.I. Bertoni, J. Hofstetter, D.P. Fenning, D.M. Powell, S. Castellanos, T. Buonassisi, IEEE J. Photovol. 3 (2013) 189.

[98] S. Pizzini, A. Sandrinelli, M. Beghi, D. Narducci, F. Allegretti, S. torchio, J. Electrochem. Soc. 135 (1988) 155.

[99] A.A. Istratov, H. Hieslmair, E.R. Weber, Appl. Phys. A 70 (2000) 489.

[100] F. Schindler, B. Michl, J. Schön, W. Kwapil, W. Warta, M.C. Schubert, IEEE J. Photovol. 4 (2014) 122.

[101] K. Arafune, E. Ohishi, H. Sai, Y. Ohshita, M. Yamaguchi, J. Cryst. Growth 308 (2007) 5.

[102] L. Liu, S. Nakano, K. Kakimoto, J. Cryst. Growth 292 (2006) 515.

[103] K. Kakimoto, L. Liu, S. Nakano, Mater. Sci. Eng. B 134 (2006) 269.

[104] Y. Patrick, B. Mouafi, B. Herzog, G. Hahn, in 29th EU PVSEC, Amsterdam, September 22—26, 2014.

[105] L. Arnberg, M. Di Sabatino, E. Øvrelid, Electron. Mater. Solid. 63 (2011) 38.

[106] Y. Boulfrad, G. Stokkan, M. M'hamdi, E. Øvrelid, L. Arnberg, Solid State Phenom. 178—179 (2011) 507.

[107] J. Schön, H. Habenicht, M.C. Schubert, W. Warta, Solid State Phenom. 156—158 (2010) 223.

[108] G. Coletti, R. Kvande, V.D. Mihailetchi, L.J. Geerligs, L. Arnberg, E.J. Øvrelid, J. Appl. Phys. 104 (2008) 104913.

[109] M. Rinio, A. Yodyunyong, S. Keipert-Colberg, D. Borchert, A. Montesdeoca-Santana, Phys. Status Solidi 208 (2011) 760.

[110] S. Wu, L. Wang, X. Li, P. Wang, D. Yang, D. You, J. Du, T. Zhang, Y. Wan, Cryst. Res. Technol. 47 (2012) 7.

[111] T. Jiang, X. Yu, L. Wang, X. Gu, D. Yang, J. Appl. Phys. 115 (2014) 012007.

[112] E.R. Weber, Appl. Phys. A 30 (1983) 1.

[113] A.A. Istratov, T. Buonassisi, M.D. Pickett, M. Heuer, E.R. Weber, Mater. Sci. Eng. B 134 (2006) 282.

[114] T. Buonassisi, A.A. Istratov, M.D. Pickett, M. Heuer, J.P. Kalejs, G. Hahn, M.A. Marcus, B. Lai, Z. Cai, S.M. Heald, T.F. Ciszek, R.F. Clark, D.W. Cunningham, A.M. Gabor, R. Jonczyk, S. Narayanan, E. Sauar, E.R. Weber, Prog. Photovoltaics Res. Appl. 14 (2006) 513.

[115] A.A. Istratov, T. Buonassisi, R.J. McDonald, A.R. Smith, R. Schindler, J.A. Rand, J.P. Kalejs, E.R. Weber, J. Appl. Phys. 94 (2003) 6552.

[116] H.J. Möller, L. Long, M. Werner, D. Yang, Phys. Status Solidi. (a) 171 (1999) 175.

[117] O.V. Feklisova, X. Yu, D. Yang, E.V. Yakimov, Semiconductors 47 (2013) 232.

[118] X. Li, D. Yang, X. Yu, D. Que, Trans. Nonferrous Metals Soc. China 21 (2011) 691.

[119] Z. Xi, D. Yang, H.J. Möller, Infrared Phys. Technol. 47 (2006) 240.

[120] T. Turmgambetov, S. Dubois, J.P. Garandet, B. Martel, N. Enjalbert, J. Veirman, E. Pihan, Phys. Status Solidi C 11 (2014) 1697.

[121] J. Lindroos, D.P. Fenning, D.J. Backlund, E. Verlage, A. Gorgulla, S.K. Estreicher, H. Savin, T. Buonassisi, J. Appl. Phys. 113 (2013) 20906.

[122] T. Buonassisi, M. Heuer, O.F. Vyvenko, A.A. Istratov, E.R. Weber, Z. Cai, B. Lai, T.F. Ciszek, R. Schindler, Physica B 340−342 (2003) 1137.

[123] O.F. Vyvenko, T. Buonassisi, A.A. Istratov, H. Hieslmair, A.C. Thompson, R. Schindler, E.R. Weber, J. Appl. Phys. 91 (2002) 3614.

[124] T. Buonassisi, A.A. Istratov, M. Heuer, M.A. Marcus, R. Jonczyk, J. Isenberg, B. Lai, Z. Cai, S. Heald, W. Warta, R. Schindler, G. Willeke, E.R. Weber, J. Appl. Phys. 97 (2005) 074901.

[125] J. Schön, A. Haarahiltunen, H. Savin, D.P. Fenning, T. Buonassisi, W. Warta, M.C. Schubert, IEEE J. Photovol. 3 (2013) 131.

[126] J. Hofstetter, D.P. Fenning, J.F. Lelièvre, C. del Cañizo, T. Buonassisi, Phys. Status Solidi 209 (2012) 1861.

[127] T. Buonassisi, A.A. Istratov, S. Peters, C. Ballif, J. Isenberg, S. Riepe, W. Warta, R. Schindler, G. Willeke, Z. Cai, B. Lai, E.R. Weber, Appl. Phys. Lett. 87 (2005) 121918.

[128] T. Buonassisi, A.A. Istratov, M.D. Pickett, M.A. Marcus, G. Hahn, S. Riepe, J. Isenberg, W. Warta, G. Willeke, T.F. Ciszek, E.R. Weber, Appl. Phys. Lett. 87 (2005) 044101.

[129] T. Buonassisi, M. Heuer, A.A. Istratov, M.D. Pickett, M.A. Marcus, B. Lai, Z. Cai, S.M. Heald, E.R. Weber, Acta Mater. 55 (2007) 6119.

[130] T. Buonassisi, A.A. Istratov, M.D. Pickett, M.A. Marcus, T.F. Ciszek, E.R. Weber, Appl. Phys. Lett. 89 (2006) 042102.

[131] M.M. Mandurah, K.C. Saraswat, C.R. Heims, T.I. Kamins, J. Appl. Phys. 51 (1980) 5755.

[132] T. Orellana, E.M. Tejado, C. Funke, S. Riepe, J.Y. Pastor, H.J. Möller, J. Mater. Sci. 49 (2014) 4905.

[133] K. Nakajima, K. Morishita, R. Murai, N. Usami, J. Cryst. Growth 389 (2014) 112.

[134] Z. Xi, J. Tang, H. Deng, D. Yang, D. Que, Sol. Energy Mater. Sol. Cells 91 (2007) 1688.

[135] C. Reimann, M. Trempa, T. Jung, J. Friedrich, G. Müller, J. Cryst. Growth 312 (2010) 878.

[136] B. Drevet, O. Pajani, N. Eustathopoulos, Sol. Energy Mater. Sol. Cells 94 (2010) 425.

[137] I. Takahashi, S. Joonwichien, S. Matsushima, N. Usami, J. Appl. Phys. 117 (2015) 095701.

[138] A.K. Søiland, E.J. Øvrelid, T.A. Engh, O. Lohne, J.K. Tuset, Ø. Gjerstad, Mater. Sci. Semicond. Process. 7 (2004) 39.

[139] J. Li, R.R. Prakash, K. Jiptner, J. Chen, Y. Miyamura, H. Harada, K. Kakimoto, A. Ogura, T. Sekiguchi, J. Cryst. Growth 377 (2013) 37.

[140] G. Du, L. Zhou, P. Rossetto, Y. Wan, Sol. Energy Mater. Sol. Cells 91 (2007) 1743.

[141] V. Ganapati, S. Schoenfelder, S. Castellanos, S. Oener, R. Koepge, A. Sampson, M.A. Marcus, B. Lai, H. Morhenn, G. Hahn, J. Bagdahn, T. Buonassisi, J. Appl. Phys. 108 (2010) 063528.

[142] M. Trempa, C. Reimann, J. Friedrich, G. Müller, J. Cryst. Growth 312 (2010) 1517.

[143] H. Aufgebauer, J. Kundin, H. Emmerich, M. Azizi, C. Reimann, J. Friedrich, T. Jauß, T. Sorgenfrei, A. Cröll, J. Cryst. Growth 446 (2016) 12.

[144] J. Friedrich, C. Reimann, T. Jauß, A. Cröll, T. Sorgenfrei, J. Cryst. Growth 447 (2016) 18.

[145] Y. Tao, T. Sorgenfrei, T. Jauß, A. Cröll, C. Reimann, J. Friedrich, J.J. Derby, J. Cryst. Growth 448 (2017) 24.

[146] D. Camel, B. Drevet, V. Brizé, F. Disdier, E. Cierniak, N. Eustathopoulos, Acta Mater. 129 (2017) 415.

[147] A. Selzer, V. Brizé, R. Voytovych, B. Drevet, D. Camel, N. Eustathopoulos, J. Eur. Ceram. Soc. 37 (2017) 69.

[148] K.E. Ekstrøm, E. Undheim, G. Stokkan, L. Arnberg, M.D. Sabatino, Acta Mater. 109 (2016) 267.

[149] K. Nakajima, R. Murai, K. Morishita, K. Kutsukake, N. Usami, J. Cryst. Growth 344 (2012) 6.

[150] K. Nakajima, K. Morishita, R. Murai, K. Kutsukake, J. Cryst. Growth 355 (2012) 38. Appl. Phys. Lett. 89 (2006) 042102.

[151] K. Nakajima, R. Murai, K. Morishita, K. Kutsukake, J. Cryst. Growth 372 (2013) 121.

[152] K. Nakajima, R. Murai, K. Morishita, Jpn. J. Appl. Phys. 53 (2014) 025501.

[153] Y. Yatsurugi, N. Akiyama, Y. Endo, J. Electrochem. Soc. 120 (1973) 975.

[154] W. Kaiser, C.D. Thurmond, J. Appl. Phys. 30 (1959) 427.

[155] H.M. Jennings, J. Mater. Sci. 18 (1983) 951.

[156] E.T. Turkdogan, P.M. Bills, V.A. Tippett, J. Appl. Chem. 8 (1958) 296.

[157] F.L. Riley, J. Am. Ceram. Soc. 83 (2000) 245.

[158] I. Kohatsu, J.W. McCauley, Mater. Res. Bull. 9 (1974) 917.

[159] C.-M. Wang, X. Pan, M. Rühle, F.L. Riley, M. Mitomo, J. Mater. Sci. 31 (1996) 5281.

[160] H. Fujimori, N. Sato, K. Ioku, S. Goto, T. Yamada, J. Am. Ceram. Soc. 83 (2000) 2251.

[161] D.R. Messier, F.L. Riley, R.J. Brook, J. Mater. Sci. 13 (1978) 1199.

[162] C. Creskovich, S. Prochazka, J. Am. Ceram. Soc. 60 (1977) 471.

[163] H. Suematsu, M. Mitomo, T.E. Mitchell, J.J. Petrovic, O. Fukunaga, N. Ohashi, J. Am. Ceram. Soc. 80 (1997) 615.

[164] I. Brynjulfsen, A. Bakken, M. Tangstad, L. Arnberg, J. Cryst. Growth 312 (2010) 2404.

[165] T. Tachibana, T. Sameshima, T. Kojima, K. Arafune, K. Kakimoto, Y. Miyamura, H. Harada, T. Sekiguchi, Y. Ohshita, A. Ogura, J. Appl. Phys. 111 (2012) 074505.

[166] T. Tachibana, T. Sameshima, T. Kojima, K. Arafune, K. Kakimoto, Y. Miyamura, H. Harada, T. Sekiguchi, Y. Ohshita, A. Ogura, Jpn. J. Appl. Phys. 51 (2012) 02BP08.

[167] H. Sugimoto, M. Tajima, T. Eguchi, I. Yamaga, T. Saitoh, Mater. Sci. Semicond. Process. 9 (2006) 102.

[168] T.F. Li, H.C. Huang, H.W. Tsai, A. Lan, C. Chuck, C.W. Lan, J. Cryst. Growth 340 (2012) 202.

[169] T. Buonassisi, A.A. Istratov, M.D. Pickett, J.−P. Rakotoniaina, O. Breitenstein, M.A. Marcus, S.M. Heald, E.R. Weber, Cryst. Growth 287 (2006) 402.

[170] S. Castellanos, M. Kivambe, J. Hofstetter, M. Rinio, B. Lai, T. Buonassisi, J. Appl. Phys. 115 (2014) 183511.

[171] M. Heuer, T. Buonassisi, M.A. Marcus, A.A. Istratov, M.D. Pickett, T. Shibata, E.R. Weber, Phys. Rev. B 73 (2006) 235204.

[172] K. Arafune, M. Nohara, Y. Ohshita, M. Yamaguchi, Sol. Energy Mater. Sol. Cells 93 (2009) 1047.

[173] L. Alphei, A. Braun, V. Becker, A. Feldhoff, J.A. Becker, E. Wulf, C. Krause, F.-W. Bach, J. Cryst. Growth 311 (2009) 1250.

[174] Z. Yuan, W.L. Huang, K. Mukai, Appl. Phys. A 78 (2004) 617.

[175] I. Kupka, T. Lehmann, M. Trempa, C. Kranert, C. reimann, J. Friedrich, J. Cryst. Growth 465 (2017) 18.

[176] B. Drevet, R. Voytovych, R. Israel, N. Eustathopoulos, J. European Soc. 29 (2009) 2363.

[177] B. Drevet, A. Selzer, V. Brizé, R. Voytovych, D. Camel, N. Eustathopoulos, J. Eur. Ceram. Soc. 37 (2017) 75.

[178] H.C. Sio, S.P. Phang, P. Zheng, Q. Wang, W. Chen, H. Jin, D. Macdonald, Jpn. J. Appl. Phys. 56 (2017) 08MB16.

[179] H.C. Sio, D. Macdonald, Sol. Energy Mater. Sol. Cells 144 (2016) 339.

[180] D. Zhu, L. Ming, M. Huang, Z. Zhang, X. Huang, J. Cryst. Growth 386 (2014) 52.

[181] C.W. Lan, A. Lan, C.F. Yang, H.P. Hsu, M. Yang, A. Yu, B. Hsu, W.C. Hsu, A. Yang, J. Cryst. Growth 468 (2017) 17.

[182] F. Jay, D. Muñoz, T. Desrues, E. Pihan, V.A. de Oliveira, N. Enjalbert, A. Jouini, Sol. Energy Mater. Sol. Cells 130 (2014) 690.

[183] D.P. Fenning, A.S. Zuschlag, J. Hofstetter, A. Frey, M.I. Bertoni, G. Hahn, T. Buonassisi, IEEE J. Photovol. 4 (2014) 866.

[184] G. Zhong, Q. Qinghua, Yu, X. Huang, L. Liu, Sol. Energy 111 (2015) 218.

[185] S. Liu, X. Niu, W. Shan, W. Lu, J. Zheng, Y. Li, H. Duan, W. Quan, W. Han, C.R. Wronski, D. Yang, Sol. Energy Mater. Sol. Cells 127 (2014) 21.

[186] H. Sai, Y. Kanamori, K. Arafune, Y. Ohshita, M. Yamaguchi, Prog. Photovoltaics Res. Appl. 15 (2007) 415.

[187] B.T. Chan, E. Kunnen, M. Uhlig, J.-F. de Marneffe, K. Xu, W. Boullart, B. Rau, J. Poortmans, Jpn. J. Appl. Phys. 51 (2012) 10NA01.

[188] H. Sai, H. Fujii, K. Arafune, Y. Ohshita, Y. Kanamori, H. Yugami, M. Yamaguchi, Jpn. J. Appl. Phys. 46 (2007) 3333.

[189] S. Prucnal, B. Abendroth, K. Krockert, K.K.]S. Prucnal, B. Abendroth, K. Krockert, K.K.]S. Prucnal, B. Abendroth, K. Krockert, K. König, D. Henke, A. Kolitsch, H.J. Möller, W. Skorupa, J. Appl. Phys. 111 (2012) 123104.

[190] X. Zhang, L. Gong, B. Wu, M. Zhou, B. Dai, Sol. Energy Mater. Sol. Cells 139 (2015) 27.

[191] M.D. Pickett, T. Buonassisi, Appl. Phys. Lett. 92 (2008) 122103.

[192] J. Hofstetter, D.P. Fenning, M.I. Bertoni, J.F. Lelièvre, C. del Cañizo, T. Buonassisi, Prog. Photovoltaics Res. Appl. 19 (2011) 487.

[193] C.W. Lan, C.F. Yang, A. Lan, M. Yang, A. Yu, H.P. Hsu, B. Hsu, C. Hsu, Cryst. Eng. Comm. 18 (2016) 1474.

[194] B. Wu, R. Clark, J. Cryst. Growth 318 (2011) 200.

[195] Y. Horioka, S. Sakuragi, Coating Method of Quartz Crucibles for Si Crystal Growth, November 11, 2011. Japanese Patent Number: JP-4,854,814.

[196] T. Fukuda, Y. Horioka, N. Suzuki, M. Moriya, K. Tanahashi, S. Simayi, K. Shirasawa, H. Takato, J. Cryst. Growth 438 (2016) 76.

[197] K. Fujiwara, Y. Horioka, S. Sakuragi, Energy Sci. Eng. 3 (2015) 419.

CHAPTER 4

Dendritic cast method

4.1 Motivation to develop the dendritic cast method

A conventional Si multicrystalline ingot has many grains with random orientations and random sizes as shown in the upper left figure of Fig. P.3. To reduce the scattering effect of grain boundaries on diffusion of minority carriers, the unidirectional growth is performed to obtain a Si ingot with a columnar structure with parallel grain boundaries in the growth direction as shown in the lower left figure of Fig. P.3. The ingot with the columnar structure is vertically cut in the growth direction to obtain wafers with minimum grain boundaries. Generally, random grain boundaries and small angle grain boundaries decrease the electrical properties together with impurities of an ingot. To reduce grain boundaries and prepare efficient texture structure, grains with almost same orientations and proper sizes is required in a multicrystalline ingot for solar cells as shown in the right figure of Fig. P.3. The grain boundaries preferably have electrically inactive characteristics to improve the quality of the multicrystalline ingot as that of a CZ single ingot. Such a concept has not been tried yet for the growth of Si multicrystalline ingots for solar cells. To largely improve the quality of multicrystalline ingots, Nakajima et al. proposed the new concept namely the structure control, and they realized the structure control by inventing the dendritic cast method [1—6]. In the dendritic cast method, dendrite crystals are used as as–grown seed crystals on the bottom of the crucible as shown in Fig. P.4. The structure control is based on controlling nucleation sites on the crucible bottom using several types of seed at the initial stage of the growth. This method is effective to control the density of random grain boundaries and the coherency of grain boundaries by controlling grain orientation, grain size and arrangement of dendrite crystals.

The target of the dendritic cast method is the development of a Si multicrystalline ingot to increase the yield of high-efficiency solar cells using a low-cost method [1—6]. To grow the high-quality Si multicrystalline ingot using a cast furnace, the Si ingot should be grown by controlling the crystal

Crystal Growth of Si Ingots for Solar Cells Using Cast Furnaces
ISBN 978-0-12-819748-6
https://doi.org/10.1016/B978-0-12-819748-6.00004-9

Copyrigh: © 2020 Elsevier Inc.
All rights reserved.

structure, because the ingot should ideally have grains with same orientations, proper sizes, electrically inactive grain boundaries and few crystal defects such as dislocations. The proper grain size is required to reduce both stress and random grain boundaries because there is a trade-off between reduction of stress and reduction of random grain boundaries as shown in Fig. 4.24 (see Section 4.8.1). The stress in the ingot acts as a driving force to generate dislocations. In this method, dendrite crystals are grown along the bottom of a crucible at the initial stage of growth by locally controlling the supercooling of a Si melt. The dendritic crystals are used as seed crystals naturally grown on the crucible bottom to control the grain orientation and grain size of the Si multicrystalline ingot. The quality and uniformity of the Si multicrystalline ingot are largely improved by the structure control.

4.2 Growth and behavior of dendrite crystals using the in-situ observation system

4.2.1 Growth of dendrite crystals

To develop the dendritic cast method, the growth mechanism of multicrystalline Si should be studied in detail. The in-situ observation system is usefully for such study as shown in Fig. 3.3 [7,8]. By controlling the supercooling along the bottom of a crucible, a thin layer grows along the crucible bottom at the initial stage of the growth, which can be directly observed by the in-situ observation system [2,3]. Then, Si multicrystals grow on the upper surface of the thin layer. By direct observation on the surface of a very thin Si melt, the thin layer is known to be a dendrite crystal which appears along the bottom of the crucible under a cooling rate of 1 K min^{-1} as shown in Fig. 4.1 [2,3]. For this observation, the thickness of the sample melt is only 0.1 cm. The dendrite crystal starts to grow from the side-wall of a small crucible in one direction. Many branched grain boundaries dendritically appear from the center of the dendrite crystal as shown in Fig. 4.3. Grains also dendritically and symmetrically grow to widen the dendrite crystal during lengthening the crystal. Large size grains in dendrite regions can be confirmed by EBIC [9]. This type of growth behavior is called as facet dendrite growth related to twin boundaries [10]. Extensive researches have been reported on Si or Ge facet dendrites [10−17]. In metal castings, many equiaxed dendrites nucleate and grow simultaneously in supercooled melt. The Si facet dendrite varies morphology in the growth direction with increasing supercooling because the dendrite selects atomically rough interface in order to promote the

Dendritic cast method 157

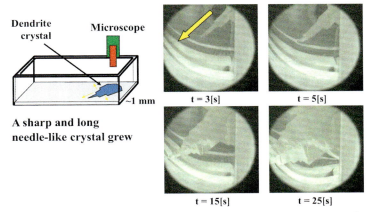

Fig. 4.1 Direct observation on the surface of a very thin Si melt using the in-situ observation system. A dendrite crystal grows along the bottom of the crucible a thin layer. *(Referred from Fig. 4 in K. Fujiwara et al. Acta Mater. 54 (2006) 3191.)*

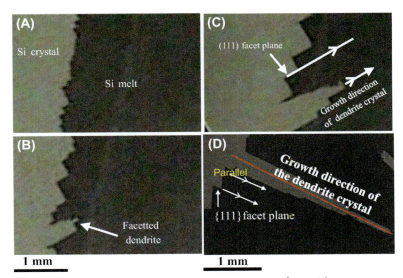

Fig. 4.2 Dendrite crystal appeared from the growing interface with many steps with the (111) facet when the supercooling of the Si melt is larger than 10 K. The growth direction of the dendrite crystal is exactly parallel to the {111} facet. *(Referred from Fig. 3 in K. Fujiwara et al. Scripta Mater. 57 (2007) 81.)*

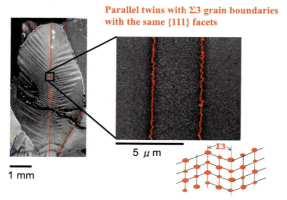

Fig. 4.3 Typical dendrite crystal of Si, which has two parallel twins with Σ3 grain boundaries. *(Referred from Fig. 1 in K. Fujiwara et al. Scripta Mater. 57 (2007) 81.)*

incorporation of atoms at high supercooling [17]. The simulation model for growth of equiaxed dendrites was proposed by Steinbach et al. [18]. The model is valid for the steady-state growth of dendrites.

A twin-related dendrite (facet dendrite) used for the dendritic cast method appears at a supercooling smaller than 100 K [17]. This type of dendrite crystals usually appears in Si multicrystalline ingots grown by the cast method. A twin-free dendrite crystal which preferentially grows to the <100> direction appears at the supercooling larger than 100 K [17]. The degree of supercooling required to generate a dendrite crystal can be determined using the in-situ observation system with a thermocouple set close to the wall of the small crucible. The growing interface has many steps with {111} facets which have an angle of 70.53° at the steady state which corresponds to the angle between {111} faces as shown in Fig. 1.12 [19]. When the supercooling of a melt is larger than 10 K by changing the temperature gradient in the melt, the growing interface varies from a flat interface to a zigzag faceted interface. Then, parallel twin grain boundaries or parallel twins are formed and a dendrite crystal originated from the parallel twins appears with a sharp and long needle-like crystal at its top as shown in Figs. 4.2 and 4.4 [20–22]. Then, the {111} facets grow slowly on the both sides of the dendrite crystal and widen the dendrite crystal. The growth direction of the dendrite crystal is exactly parallel to the {111} facet plane. Fig. 4.3 shows is a typical dendrite crystal of Si, which has parallel

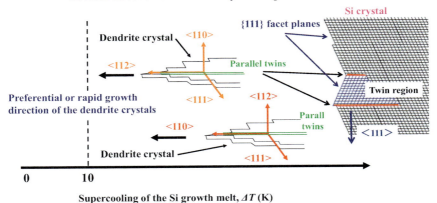

Fig. 4.4 Preferential or rapid growth direction of the dendrite crystals with two parallel twins with S3 grain boundaries. The growth direction is <112> or <110>. The two parallel twins form on a growing interface with many steps with (111) facet. *(K. Fujiwara, K. Maeda, N. Usami, G. Sazaki, Y. Nose, K. Nakajima, Formation mechanism of parallel twins related to Si-facetted dendrite growth, Scripta Materialia, 57 (2) (2007) 81–84; Kazuo Nakajima et al., Development of Textured High-Quality Si Multicrystal Ingots with Same Grain Orientation and Large Grain Sizes by the New Dendritic Casting Method, 2006 IEEE 4th World Conference on Photovoltaic Energy Conference, © 2006 IEEE.)*

twins with {111} Σ3 grain boundaries at the center of the dendrite crystal [20]. Such a twin-related dendrite crystal is a facet dendrite crystal which is bounded by {111} facet planes. The growth direction of the parallel twins is exactly same as that of the dendrite crystal. The {111} Σ3 grain boundaries have the same {111} facets as the grain boundary planes. The atomic arrangement near the parallel twins with the Σ3 grain boundaries is shown in the lower right figure of Fig. 4.3. Crystals on both sides of the twin region have the same crystallographic orientation. The formation mechanism of the parallel twins can be known by the right figure of Fig. 4.4. In this figure, the growing interface has many steps with {111} facets and different shapes depending on the growth orientation [20,21]. At first, a twin stochastically nucleates and forms on an {111} facet because the formation energy of the twin boundary is very low [23]. Then, another twin forms exactly parallel to the first one because crystals on both sides of the twin region have the same crystallographic orientation. The distance between these two twins depends on the wavelength of the zigzag faceted interface, and it decreases as the growth rate or supercooling on the facet increases because the probability of twin formation increases as the supercooling increases [22,24]. The parallel twins are formed between grain boundaries and do not generate at grain boundaries because crystals on both sides of the twin region have same crystallographic orientation [21]. Precisely speaking, the parallel twins nucleate from long ridges between <111> facets on the growing interface [25]. A dendrite crystal appears from the growing interface using the parallel twins as an origin. Thus, the dendrite crystal keeps parallel twins inside it and the growth direction of the dendrite crystal is exactly parallel to an {111} facet plane because the growth direction of the twins is also exactly parallel to the {111} facet plane [20]. These parallel twins can be observed by EBIC as dark contrast at RT because of the presence of minority carrier recombination centers [9].

Several types of dendrite crystal appear in the growth direction of <112>, <110> and <100> under supercooling from 5.7 to 157 K [15]. The transition supercooling value for <110> to <112> is 22 K and that for <112> to <100> is 68 K [15]. However, it is well known that the growth direction of a dendrite crystal varies from <112> to <110> to <100> with increasing degree of supercooling, and the <112> or <110> dendrites are observed under the supercooling lower than 100 K [17]. A dendrite crystal with the <100> growth direction appears under the supercooling of a melt above 100 K [16,17]. Thus, the transition supercooling for <110> to <100> is about 100 K [17]. A dendrite crystal with the <110> or <112> direction has twins, but an <100> dendrite crystal is twin-free which cannot have parallel twins because the <100> direction is not parallel to the {111} plane [16,17]. The twin-free <100> dendrite crystals have two types such as an $<100>_{<100>}$ dendrite and an $<100>_{<110>}$ dendrite which have the <100> and <110> secondary arms, respectively [17]. The surface of the <112> and <110> dendritic crystals is bounded by {111} facet planes, so all planes of the <112> and <110> dendrite crystals are covered by atomically smooth {111} facets [17,26]. The $<100>_{<100>}$ dendrite crystal is composed of atomically rough {110} and {100}, and the $<100>_{<110>}$ dendrite crystal is composed of {111}, {110} and {100} [17]. The <100> dendrite crystal breaks up into small pieces under the supercooling of a melt above 200 K [27].

Under the supercooling of a melt above 10 K, Fujiwara et al. [2,3,28] determined the supercooling dependence on the growth direction of dendrite crystals. The preferential or rapid growth direction of the facet dendritic crystal is <112> or <110> which is parallel to the {111} plane, and the growth direction of both sides of the facet dendritic crystal is <111> as shown in Fig. 4.4. Thus, the dendrite crystal rapidly grows in the <112> or <110> direction and propagates in the <111> direction perpendicular to the <112> or <110> growth direction. The growth rate in the rapid direction of the facet dendrite crystal is much larger than that of a crystal grown on the {100}, {110} and {111} faces. The <112> and <110> dendritic crystals have parallel twins with Σ3 grain boundaries, and the Σ3 grain boundaries have the same {111} facets, but there are no twins inside the secondary arms as shown in Fig. 4.3. The upper surface of the dendritic crystals is fixed to be {110} or {112} for the <112> or <110> dendrite crystals, respectively. The {110} and {112} planes are parallel to the <111> direction of the twin boundary planes. The dendrite crystal with the <112> growth direction appears under slightly smaller supercooling than that with the <110> growth direction does. In actual ingot growth, however, dendrite crystals with both <112> and <110> growth directions often appear together with each other, and the dominant orientations in the upper surface of an ingot grown by the dendritic cast method are usually both {110} and {112} [3]. Dendrite crystals can be formed by a higher cooling rate than 10 K min^{-1} [29], but it is very difficult to control the growth of the dendrite crystals along the bottom of a crucible by the cooling rate. The orientation of dendrite crystals has a possibility to strongly depend on the relative orientation of the growing interface. Moreover, the initial structure such as the dendrite crystals not only depends on the supercooling, but also has a possibility to depend on the size of the coated Si_3N_4 particles in contact with the Si melt [30] (see Section 1.2.1.2).

4.2.2 Growth mechanisms of dendrite crystals

The in-situ observation system can be effectively used to observe the growth mechanisms of dendrite crystals. A sheet crystal is set inside the furnace and melted to remain some edge part as seed. Crystallization starts from the seed part during rapid cooling. By such observation, the shapes and growth mechanisms of the <112> and <110> facet dendrite crystals can be well known to be markedly different. The tips of the <112> and <110> facet dendrite crystals are wide and narrow, respectively [26,28].

By the precise observation of dendrite crystals using the in-situ observation system, the growth mechanism of dendrite crystal with <112> direction can be made clear [28,31]. The <112> dendritic crystal is bounded by {111} faces and contains parallel twins. Fig. 4.5 shows a schematic image of a model Si crystal with parallel twins surrounded by {111} faces, which was proposed by Hamilton and Seidensticker [11]. The dendrite crystal grows only in one <112> direction, for simplicity. In this case, the crystal growth starts on one face as growing interface on the

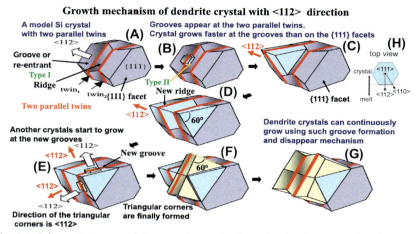

Fig. 4.5 Schematic image of the growth mechanism of a dendrite crystal with <112> direction from the initial stage to the second stage. *(Referred from Fig. 4 in K. Fujiwara et al. Phys. Rev. Lett. 101 (2008) 055503.)*

model Si crystal as shown in Fig. 4.5H. At the initial stage, grooves or re-entrants and ridges appear at the parallel two twins as shown in Fig. 4.5A [28]. The two twins are distinguished by labeling them twin₁ and twin ₂. A groove with an external angle of 141° (type I) appears at the growth surface only at twin₁. A ridge with an external angle of 219° also appears at the growth surface only at twin₂. Then crystals grow faster at the grooves than on the {111} faces as shown in Fig. 4.5B. Triangular corners with an <112> direction and an angle of 60° are formed due to the rapid growth from the type I grooves and finally form new ridges from the ruins of the grooves as shown in Fig. 4.5C [26,28]. The rapid growth is inhibited after the growth of the triangular corners is finished. Crystals continue to grow on the {111} faces of the newly grown triangular corners to make the triangular corners thicker. The thicker parts finally become new triangular crystals as shown in Fig. 4.5D. Then, the triangular crystals form new type I grooves made on the previous ridges with the triangular crystals, on the growth surface at twin₂. On the second stage, other crystals start to grow at the new grooves as shown in Fig. 4.5E. The triangular corners with an <112> direction and an angle of 60° are newly formed from the new grooves in the same manner as before as shown in Fig. 4.5F. Then, the triangular corners stop growing and crystals continue to grow only on the {111} faces of the newly grown triangular corners. Finally, the new triangular crystals are formed and new other grooves are also formed as shown in Fig. 4.5G. Then crystals start growing at these new grooves again to form other triangular corners in <112> direction. The direction of the triangular corners changes during growth. The facet dendrite crystal can continuously grow by repeating the same processes and forming the 60° corner at the growth tip [28]. The tip of the dendrite crystal becomes wider during growth as shown in Fig. 4.5G.

By the in-situ observation system, the growth mechanism of dendrite crystal with <110> direction can be made clear [26]. The <110> dendritic crystal is bounded by {111} faces and contains parallel twins. Fig. 4.6A shows a schematic image of a model Si crystal with parallel twins surrounded by {111} faces. The initial shape of the crystal is the same as that shown in Fig. 4.5A. The dendrite crystal grows only along the <110> direction, for simplicity. In this case, the crystal growth starts on two faces as growing interface on the model Si crystal as shown in Fig. 4.6F. At the initial stage, grooves with an external angle of 141° (type I) and ridges with an external angle of 219° appear at the parallel two

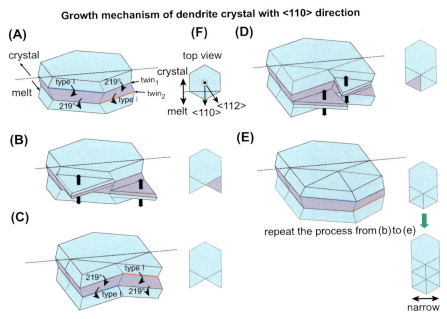

Fig. 4.6 Schematic image of the growth mechanism of a dendrite crystal with <110> direction from the initial stage to the second stage. *(Referred from Fig. 6 in K. Fujiwara et al. Phys. Rev. B 81 (2010) 224106.)*

twins, and the type I grooves exist at both twin$_1$ and twin $_2$ as shown in Fig. 4.6A [26]. Then crystals grow faster at the two type 1 grooves simultaneously as shown in Fig. 4.6B. Triangular corners with an <112> direction and an angle of 60° are formed due to the rapid growth from the type I grooves and finally form new ridges from the ruins of the grooves [26]. The rapid growth is inhibited after the growth of the triangular corners is finished. Crystals continue to grow on the {111} faces of the newly grown triangular corners to make the triangular corners thicker as shown in Fig. 4.6B and C. The thicker parts finally become new triangular crystals as shown in Fig. 4.6C. Then, the triangular crystals form new type I grooves made on the previous ridges with the triangular crystals, on the growth surface at twin$_1$ and twin$_2$. On the second stage, other crystals start to grow at the new grooves as shown in Fig. 4.6D. The triangular corners with an <112> direction and an angle of 60° are newly formed from the new grooves in the same manner as before. When the triangular crystals propagate across the other twin, two type I groves appear as shown in Fig. 4.6E. Finally, the new triangular crystals are closely packed and the type I groves appear at both twins simultaneously. Then crystals start growing at these new grooves again and the facet dendrite crystal can continuously grow by repeating the same processes. The tip of the dendrite crystal remains narrow during growth and the growth direction of the dendrite crystal is <110> as shown in Fig. 4.6E.

From the above growth mechanisms for each dendrite crystal, the opportunity whether a dendrite crystal starts to grow as the <112> or <110> facet dendrite crystal strongly depends on the initial shape of growing interface of the model Si crystal or the opportunity strongly depends on whether the initial dendrite crystal starts to grow as Fig. 4.5H with one face or as Fig. 4.6F with two faces. The opportunity whether the initial interface has one face or two faces of the model crystal depends on the shape of growing interface which depends on the cooling rate or the supercooling. The growth mechanism of a dendrite crystal of Si is simulated using a phase-field model by considering the growth rates at a re-entrant or grove and a ridge [24,32]. In this simulation, a new grove is considered, which has an external angle of 109.5° (type II: Fig. 4.5B) formed by a crystal grown at a grove and an initial side face of a ridge. When the growth rate at a re-entrant or grove, V_{re} is faster than that at the type II grove formed at the ridge, V_{ri}, the facet dendrite prefers the <112> dendritic growth, and when V_{re} is equal to V_{ri}, the facet dendrite prefers the <110> dendritic growth [24].

4.3 Ingot growth controlled by dendrite crystals grown along the bottom of a crucible

Fig. 4.7 shows the EBSD orientation maps of two crystals grown by the conventional and dendritic cast methods [2,5]. The cooling rates for the conventional and dendritic growth are 1 K min^{-1} and 50 K min^{-1}, respectively. For the conventional cast method, inhomogeneous nucleation occurs at random points on the bottom of the crucible, and many grains with random orientations grow from them. The grain structure develops from few independent nuclei [33]. The grain structure approaches to a stable state as the ingot height increases. In a short distance, large grains with preferential orientations such as the {112} and {111} stable faces with low surface energy remain during grain competition. The structure of Si multicrystalline ingots grown by the conventional cast method cannot be controlled at all. Using dendrite crystals as seeds grown along the bottom of the crucible, the grain orientation and grain size can be controlled by the surface orientation and length of the dendrite crystals, respectively, as shown in Fig. 4.7 [3]. For this observation, the thickness of the sample melt is 0.7 cm. The growth direction and upper surface of the dendrite crystal is <110> and {112}, respectively. A grain grown on the dendrite crystal is larger than other grains.

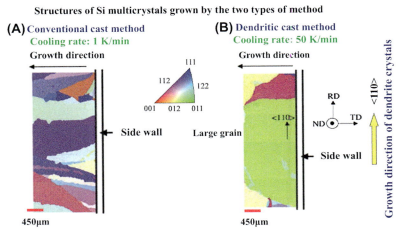

Fig. 4.7 EBSD orientation maps of two crystals grown by (A) the conventional cast method and (B) the dendritic cast method to compare between two structures of the crystals. The cooling rates for the conventional and dendritic growth are 1 K min^{-1} and 50 K min^{-1}, respectively. *(Referred from Fig. 3 K. Fujiwara et al. Acta Mater. 54 (2006) 3191.)*

Thus, the dendrite crystals have been used for the structure control to control nucleation sites on the bottom of the crucible for the dendritic cast method. In this growth, the dendrite crystals are used as-grown seed crystals. Fig. 4.8A shows the concept for controlling the grain orientation and the grain size using the dendritic cast method [2,3,28]. At first, dendritic crystals are rapidly grown in the <110> or <112> direction only along the bottom of the crucible using a local supercooling larger than 10 K to prevent the vertical growth of the dendritic crystals. Such growth conditions are obtained by rapidly cooling only the crucible bottom at 50 K min^{-1}. A large temperature gradient to the melting isotherm confines the vertical growth of the dendritic crystals and only the most favorable dendritic crystals grow forth along the crucible bottom [3,30]. Such dendritic crystals have the upper surface in the <110> or <112> direction. Then, Si multicrystals are grown on the upper surface of the dendritic crystals in the form of same unidirectional growth as the conventional cast growth. The upper surface acts as an oriented seed crystal with a large grain. Fig. 4.8B shows the distributions of the dendritic

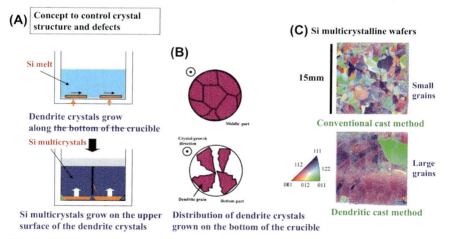

Fig. 4.8 (A) Concept for controlling the grain orientation and the grain size using the dendritic cast method. At first, dendritic crystals are grown in the <110> or <112> direction along the bottom of a crucible, then Si multicrystals are grown on the upper surface of the dendritic crystals. (B) Distribution of the dendrite crystals on the bottom of the crucible and grains grown on the dendrite crystals. (C) EBSD orientation maps of two ingots grown by the conventional and dendritic cast methods. They are smaller grains with random orientations and larger grains with the same <112> orientation, respectively. *(Referred from Figs. 5 and 6 in K. Fujiwara et al. Acta Mater. 54 (2006) 3191.)*

crystals grown on the bottom of the crucible and that of grains grown on the dendritic crystals [2,5]. The resulting Si multicrystalline ingot has larger grains with the <112> and <110> orientations, comparing with the conventional Si ingot that have smaller grains with random orientations as shown in the EBSD maps of Fig. 4.8C. In this case, the dislocation density in the ingot grown by the present method is very low [25]. The fraction of large grain areas to the total area is largest comparing with those of the other methods [34]. The dominating grain face is {112} which is obtained by a dendrite crystal with the <110> vector or a shallow angle to this vector [25]. The {112} face is favorable for solar cells applications. The surface area of grains with the {110} or {112} face increases by controlling the degree of supercooling to be almost same at any parts of the melt along the crucible bottom because dendrite crystals with the same surface orientation increase. Thus, an ingot with almost same orientation will be obtained by controlling uniformity of the degree of supercooling all over the melt along the crucible bottom. Random grain boundaries are a few but twins inevitably appear for a Si multicrystalline ingot with larger grains grown using the dendritic cast method [5]. When the supercooling of the Si melt is not well controlled along the bottom of a crucible, the dendrites crystals grow inside the melt and grains grown on them have not the {110} or {112} planes to the growth direction. In this case, the dislocation density is very high [25].

The dendritic cast method is proved to be useful for obtaining structure-controlled multicrystalline Si ingots [25,35]. Such controlled nucleation creates large grains with a favorable crystal orientation and parallel grains in the growth direction [25]. Lan et al. [35−37] used an active cooling spot using a graphite rod below the bottom of a crucible to make a high thermal gradient for growth of dendrite crystals. The other end of the rod is connected to a water-cooled pad. They also used a zirconia insulation for the crucible to make a convex growing interface for growth of large grains due to outward growth. Even though large dendrite crystals are not grown by this method, the dislocation density is very low (about $10^3-10^4 \text{ cm}^{-2}$) in a region with high twin boundaries. The similar results are obtained using notched crucibles with the cooling spot [38]. However, the dislocation density is high (about 10^5 cm^{-2}) in a region with random grain boundaries. Random grain boundaries do not increase during growth once a large number of twins are introduced. The minority carrier lifetime and quantum efficiency of solar cells are largely improved comparing with those of the conventional cast method. Controlling the

Fig. 4.9 Distribution of dendrite crystals appeared on the bottom of a Si multi-crystalline ingot.

arrangement of dendritic crystals is effective in decreasing the density of random grain boundaries which are one of origins of dislocations as shown in Fig. 4.11 (see Section 4.4.1).

Distribution of dendrite crystals which appear on the bottom of a Si multicrystalline ingot can be observed from an overview of the ingot as shown in Fig. 4.9. The diameter and height of the ingot are 30 cm and 7 cm, respectively. A large dendrite crystal appears along the bottom of the ingot. These dendrite crystals usually appear with a random distribution and a random direction at the bottom of an ingot when a graphite plate with a uniform thermal conductivity is placed under crucibles as shown in Fig. 4.10A[6]. The red lines show the dendrite crystals that appear first, and the light blue lines show the dendrite crystals that appear second by branching off from the dendrite crystals that first appear. Many dendrite crystals appear from the edge of the ingot because the crucible wall strongly acts as a nucleation site of dendrite crystals. Long and short dendrite crystals are mixed together. Most of the dendrite crystals that appear first and second distribute almost randomly within an angle of $\pm 180°$, which can be known by measuring distribution of dendrite crystals as a function of the angle of dendrite crystals inclined from the y-axis shown in Fig. 4.10. In such distribution of dendrite crystals, dendrite crystals often collide with each other. The growth behavior just after collision can be observed using the in-situ observation system [4]. Before collision with another dendrite crystal, the tip of a dendrite crystal

Distribution of dendrite crystals grown along the bottom of an ingot (30 cm$^\varphi$)

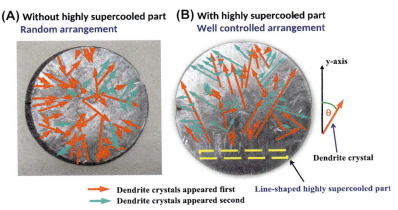

Fig. 4.10 Distribution of dendrite crystals grown on the bottom of a Si ingot with a diameter of 30 cm. The red arrows show the dendrite crystals that appear first and the light blue arrows show the dendrite crystals that appear second. (A) Random arrangement of dendrite crystals without a highly supercooled part. (B) Well controlled arrangement of dendrite crystals with the line-shaped highly supercooled part. *(Referred from Figs. 2 and 5 in K. Nakajima et al. J. Cryst. Growth 319 (2011) 13.)*

is needlelike. As the dendrite crystal continues to grow toward another dendrite crystal and the distance between the two crystals becomes shorter, the tip of the dendrite crystal becomes flat because the latent heat of crystallization is kept between them and the melt temperature locally becomes higher [4]. In this way, flat grain boundaries are naturally formed between dendrite crystals where the dendrite crystals collide with each other. A large grain appears from the upper surface of the dendrite crystal during growth. There are small grains between large grains.

4.4 Arrangement of dendrite crystals
4.4.1 Effect of angle between dendrite crystals with several orientations

To obtain a high-quality Si multicrystalline ingot, the ingot should be manufactured on the viewpoint of the structure control because crystal defects such as dislocations and random grain boundaries can be reduced by controlling the crystallographic structure. Random grain boundaries are origins of dislocations in a Si multicrystalline ingot, which are mainly formed at the bottom of the ingot [3]. Moreover, a Si multicrystalline ingot grown by the dendritic cast method have high-dense dislocations in small grain regions, which are local areas surrounded by large grains grown on dendrite crystals as seeds on the crucible bottom as shown in Fig. 4.17. In these areas, many small grains directly grow on Si_3N_4 coating on the crucible bottom (see Section 4.6). Dislocations should be reduced as much as possible because highly-dense dislocations ($>10^5$ cm^{-2}) in a Si multi-crystalline ingot strongly affect the minority carrier lifetime [39] and the efficiency of solar cells [40,41]. In the dendritic cast method, controlling arrangement of dendrite crystals grown along the bottom of a crucible is effective to prevent the generation of such areas and reduce dislocations in the ingot.

To measure the effect of the contact angle between dendrite crystals with several orientations, many dendrite crystals with several orientations and several grain boundary structures should be obtained. To determine the effect of the contact angle on the dislocation density and the co-herency of grain boundaries, the dendritic cast method is effectively used to prepare such an ingot with dendrite crystals with several orientations [42]. In this method, the dendrite crystals can be arranged by controlling the thermal conductivity under crucibles and the heat flow from the crucible bottom [42]. The floating cast method can be more effectively used to grow such a Si ingot from the melt surface, which is fully covered by dendrite crystals with several orientations on the ingot surface [43−45]. In this method, the distribution of dendrite crystals on the surface matches the microstructure of the mid-ingot wafer because crystal growth starts from the surface to the crucible bottom, and the measured contact angles on the surface are valid to describe the coherency of grain boundaries [43]. The adjacent dendrite crystals can be arranged in parallel to realize the contact angle of $0°$ using insulators to realize a unidirectional temperature gradient on the melt surface [46]. Fig. 4.11 shows the effect of the contact angle between dendrite crystals with several orientations on the shear stress and dislocation density [4,42,43]. The contact angle is defined as the angle between growth directions of each $\Sigma3$ twin boundary in two adjacent dendrite crystals as known by Figs. 4.4 and 4.11. It can be determined by measuring each growth direction or each twin boundary orientation of the adjacent dendrite crystals using the EBSD analysis. The angle can be employed as a structural parameter to describe the quality or coherency of a grain boundary formed by the adjacent dendrite crystals. This means that when the angle is zero, the dendrite crystals arrange completely parallel to each other and the grain boundary is coherent.

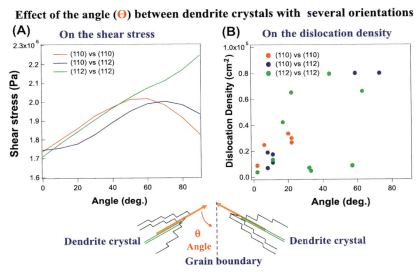

Fig. 4.11 Effect of the angle between dendrite crystals with several orientations on the shear stress and dislocation density. *(Referred from Fig. 4 in N. Usami et al. J. Appl. Phys. 107 (2010) 013511.)*

The relation between the shear stress around a grain boundary and the angle can be calculated by finite element analysis at 0 K as shown in Fig. 1.13 [47,48]. The shear stress decreases as the angle becomes smaller and the minimum shear stress can suppress generation of dislocations around the grain boundary as shown in Fig. 4.11. The dislocation density can be experimentally determined by counting EPD on cross sections of the ingots. In Fig. 4.11, three combinations of the upper surfaces of the adjacent dendrite crystals are plotted such as <110> versus <110>, <110> versus <112> and <112> versus <112>. To obtain these data, 6 ingots and 13 grain boundaries are investigated. The dislocation density decreases as the angle becomes smaller regardless of the orientation of dendrite crystals, and the combination of dendrite types has a minor impact [43]. This means that the dislocation density can be drastically reduced by controlling the arrangement of dendrite crystals to be fairly parallel to each other. Moreover, the dislocation density does not largely increase during crystal growth when the contact angle is smaller because the shear stress is lower around the grain boundary, and the lower shear stress corresponds to a lower dislocation density [43]. The dislocation density has a maximum point around 90° [43].

Electrical properties of wafers with grain boundaries can be measured by the PL imaging. Crystal defects acted as non-radiative recombination centers can be detected as dark contrast in the PL image. Dark regions can be observed in random grain boundaries and strongly in dislocation clusters. $\Sigma3$ grain boundaries have no dark regions and cannot be detected at all. The contrast of the PL intensity varies with the contact angle between dendrite crystals with several orientations. The PL contrast is defined by the ration which is determined by the difference between the bright and dark intensities divided by the bright intensity. The PL contrast becomes weaker as the angle approaches to zero [47]. This means that the parallel arrangement of dendrite crystals weakens the PL contrast from a grain boundary formed by the dendrite crystals and suppresses generation of dislocations. Grain boundaries with lower interface energy do not act as sink for impurities. They have a potential to efficiently remove impurities or suppress recombination activity by gettering.

4.4.2 Control of arrangement of dendrite crystals

Nucleation sites of dendrite crystals must be on the bottom surface of a crucible as heterogeneous nucleation sites. To control the arrangement of dendrite crystals grown along the bottom of an ingot, it is necessary to intentionally make nucleation sites at the bottom of a crucible. Controlling heat flow from the crucible bottom is very effective for the arrangement of dendrite crystals, which can be determined by controlling the thermal conductivity under crucibles [49]. Nakajima et al. [6] proposed a method of controlling thermal conductivity under a crucible to locally control supercooling and nucleation sites. Fig. 4.12A shows the temperature distribution of a Si melt all over the bottom surface of a crucible with a line-shaped highly supercooled part. The dark blue and light blue show a highly supercooled part and slightly supercooled parts in the Si melt. As shown in Fig. 4.12B, such temperature distribution can be realized by combining graphite plates with high and low thermal conductivities of 40 W m^{-1} K^{-1} and 0.35 W m^{-1}K^{-1} at the melting point of Si, respectively. The line-shaped highly cooled part is created from a graphite plate with a high thermal conductivity with a thickness of 0.5 cm, and a width of 3 cm. It is located 3 cm from the edge of the circular graphite plate with a diameter of 32 cm. To realize a line-shaped highly supercooled part for an actual crystal growth, these graphite plates with high and low thermal conductivities are combined into one circular plate. Then, it is set between a crucible and a holder as shown in Fig. 4.12C.

Dendritic cast method 173

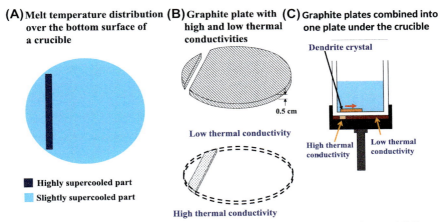

Fig. 4.12 Circular graphite plate to make a line-shaped highly cooled part. (A) Temperature distribution all over the bottom surface of a crucible with a line-shaped highly supercooled part. (B) Circular graphite plate with a line-shaped highly cooled part made by the combining of graphite plates with high and low thermal conductivities (40 W m^{-1}K^{-1} and 0.35 W m^{-1}K^{-1}, respectively, at the melting point of Si). (C) Layout drawing of the combined graphite plate in a cast furnace. It is set between a crucible and a holder. Dendrite crystals start to grow from the highly supercooled part and grow along the bottom of the crucible. *(Referred from Figs. 1 and 4 in K. Nakajima et al. J. Cryst. Growth 319 (2011) 13.)*

The combined plate can completely cover the bottom surface of the crucible. Dendrite crystals start to grow from the highly supercooled part and arrange fairly parallel to each other. Then, Si multicrystals grow on the upper surface of the arranged dendrite crystals to obtain an ingot.

Fig. 4.10B shows the distribution of dendrite crystals grown along the bottom of an ingot with a diameter of 30 cm and a height of 4.5 cm. The ingot was grown using the circular graphite plate with the line-shaped highly cooled part as shown in Fig. 4.12B. In Fig. 4.10B, the red lines show the dendrite crystals that appear first (total = 27), and the light blue lines show the dendrite crystals that appear second (total = 41) by branching off from the dendrite crystals that appear first. Most of the dendrite crystals that appear first are occurred from the line-shaped highly supercooled part. Their distribution is well arranged fairly parallel to each other. There are few dendrite crystals that appear from the edge of the ingot because the highly supercooled part in the Si melt strongly acts as nucleation sites of dendrite crystals. The supercooling in the Si melt must be larger than 10 K on the high conductivity part when dendrite crystals appear along the bottom of the ingot, because the dendrite crystals can be formed when the supercooling in the Si melt is larger than 10 K [19]. From Fig. 4.10B, nucleation easily occurs near the highly supercooled part, so many small grains appear inside and near this region. Dendrite crystals appear from growing interfaces of the small grains at the edge of this region. The y-axis is defined as the direction perpendicular to the line-shaped highly supercooled part as shown in Fig. 4.10B. Most of the dendrite crystals that appear first is controlled within an angle of $\pm50-60°$, which can be known by measuring distribution of dendrite crystals as a function of the angle of dendrite crystals inclined from the y-axis [15]. The dendrite crystals are not perfectly arranged parallel to each other because the size of the crucible is not enough large (30 cm$^{\Phi}$) to reduce the effect of the crucible wall on heat distribution in the center of the melt. The distribution of most of the dendrite crystals that appear second is within an angle of $\pm120°$ [6]. The dendrite crystals that appeared second are more widely spread than the dendrite crystals that appear first. However, as most of the dendrite crystals that appear first are well arranged within the small angle range, the distribution of dendrite crystals will be more significantly controlled by locally arranging the highly supercooled part in a Si melt using a much larger crucible.

Thus, graphite plates with different thermal conductivities placed all over the bottom surface of a crucible are effective to control the arrangement of dendrite crystals. Any type and structure (e.g. width, thickness, combination and thermal conductivity) of graphite plates can be designed, and any arrangement of dendrite crystals can be realized by selecting the size of a crucible and by controlling the balance of supercooling between two types of graphite plate, namely a highly supercooled one with a high thermal conductivity and a uniformly supercooled one with a low thermal conductivity.

In an ingot grown by the conventional cast method, grain size shows no large variation during crystal growth. In an ingot grown by the dendritic cast method, grains grow on dendrite crystals grown along the ingot bottom and the distribution of grains basically corresponds to the distribution of dendrite crystals. The grain distribution with the ingot height can be known using wafers cut from the ingot as a function of the distance from the bottom of the ingot. At 0.1 cm from the ingot bottom, dendrite crystals grown along the bottom of the ingot can be observed. At 0.3 cm, the dendrite crystals leave no originally shaped traces and grains grow correspondingly to the distribution of dendrite crystals. This means that the initial thickness of dendrite crystals just after their lateral growth is only

0.2 cm. The grain size becomes larger during ingot growth. At 0.6 cm, some large grains with surface areas of 8—10 cm × 4 cm appear fairly parallel to each other. For a total number of 27 of dendrite crystals, their average length is 6.2 cm and the maximum length is 18.2 cm. Most of grains grown on dendrite crystals grow vertically in the growth direction and continue from the bottom to the top of the ingot as shown in Fig. 4.21 [48]. Thus, the dendritic cast method can be effectively used to prepare Si multicrystalline ingots with large grains with almost same orientation and few random grain boundaries. The low density of random grain boundaries that act as recombination centers of photo-carriers is preferable for performance of solar cells [50—52]. The grain boundaries density also strongly affects the precipitation of metal impurities [53].

4.5 Generation of dislocations

A sub-grain boundary is a small angle grain boundary which comprises a dislocation array with many dislocations. Sub-grain boundaries deteriorate solar cell performance especially when the grain size is much larger than the diffusion length of minority carriers [54,55]. The diffusion length of minority carriers (μm) strongly depends on the sub-grain boundary density (mm^{-1}) in an ingot as shown in Fig. 2.1 [56]. Buonassisi et al. reported that the diffusion length of minority carriers was drastically improved by controlling metal nano-defects [57]. The sub-grain boundaries are electrically active comparably with random grain boundaries [56]. The reduction of sub-grain boundaries or dislocations is very important to obtain high-efficiency solar cells as explained in Section 2.5.

Dislocations can be generated from random grain boundaries during crystal growth [58]. Low density of random grain boundaries is important on the view point of suppressing sub-grain boundaries or dislocations [4,58]. Fig. 4.13 shows a schematic illustration of the relation between random grain boundaries and dislocations during crystal growth using the conventional cast method and the dendritic cast method [4,5,15]. For Si multicrystalline ingots, dislocations are usually generated from random grain boundaries as origins of dislocations during crystal growth. Si multicrystals grown by the dendritic cast method have few random grain boundaries especially at the initial stage of the growth because the grain size is large and the grains have the <112> and <110> same orientations. The suitable density of random grain boundaries can be determined by trade-off between reduction of dislocations and reduction of stress in an ingot.

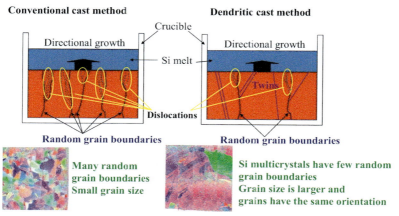

Fig. 4.13 Schematic illustration of the relation between random grain boundaries and dislocations during crystal growth for the conventional casting method and the dendritic cast method. *(Referred from Fig. 5 in K. Nakajima et al. Proceedings of the 24th European Photovoltaic Solar Energy Conference, 2009, p. 1219.)*

Thus, the dendritic cast method is effective to reduce origins of dislocations because the grain size and grain boundary character can be determined by controlling the length and relative growth direction of dendrite crystals on the bottom of the ingot.

Many dislocations near a random grain boundary can be confirmed by the spatially resolved X-ray rocking analysis which has the angle resolution of better than 0.01° [59]. Fig. 4.14 shows a photo image of a Si wafer cut from a Si ingot parallel to the growth direction. Straight lines are twin boundaries and a single curve is a random grain boundary. Dislocations with small angle difference in a Si multicrystalline ingot can be detected by the rocking curve measurement of X-ray diffraction. The local broadening of the rocking curve is caused by a small rotation of crystal lattice by dislocations or strain relaxation near dislocations. Dislocations with relative angle between 0.1 and 10° can be observed by this method. The X-ray rocking curve profiles around the twin boundaries show only one peak because there are no dislocations in the twin regions. However, the X-ray rocking curve profiles around the random grain boundary show split peaks because there are many dislocations with small angular difference near the random grain boundary. The Bragg angle changes abruptly from one grain to another one near random grain boundaries because the lattice planes are locally tilted against each other. The reason why dislocations appear on only one side of the grain boundaries can be known by calculation of the

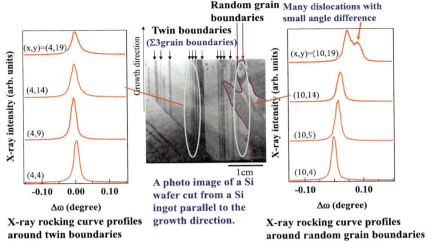

Fig. 4.14 Photo image and x-ray rocking curve profiles of a Si wafer cut from a Si ingot parallel to the growth direction. Straight lines are twin boundaries and a single curve is a random grain boundary. There are many dislocations with small angular difference near the random grain boundary. *(Referred from Figs. 4, 6 and 8 in N. Usami et al. J. Appl. Phys. 102 (2007)103504.)*

distribution of shear stress near artificially manipulated grain boundaries as shown in Fig. 1.13 [5,42,58,60].

Dislocations more easily generate near a rounded grain boundary than near a straight grain boundary [42]. The high dislocation density results from a multiplication of dislocations caused by the stress concentration at sharp angles of the grain boundary [61]. Thus, generation of dislocations is strongly affected by the shape of grain boundaries. Generation of dislocations is also strongly affected by the coherency of grain boundaries. The coherency can be expressed by the contact angle between two adjacent grains grown on two dendrite crystals and it increase as the contact angle becomes smaller, which corresponds to the results shown in Fig. 4.11. To clarify the relationship between the coherency and the dislocation density, multicrystalline ingots grown by the dendritic cast method is useful because they have many grain boundaries with different contact angles. Fig. 4.15 shows optical micrographs around straight grain boundaries with different contact angles of (A) 2°, (B) 63° and (C) 73°, which are obtained from different ingots with the almost same height between 0.5 and 0.67 cm [42]. In this figure, dark dots around the straight grain boundaries show EPD of dislocations. The straight grain boundaries are chosen to

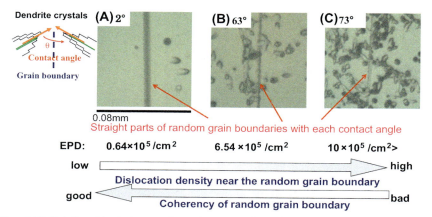

Fig. 4.15 Relationship between the coherency of random grain boundary and the dislocation density near some straight parts of random grain boundaries. Optical micrographs show straight grain boundaries with different contact angles of (A) 2°, (B) 63° and (C) 73°. *(Referred from Fig. 3 in N. Usami et al. J. Appl. Phys. 107 (2010) 013511.)*

exclude the effect of the shape of grain boundaries. Fig. 4.15 also shows the relationship between the coherency of random grain boundary and the dislocation density near each straight part of the random grain boundaries [42]. As the coherency of the random grain boundary increases, the dislocation density near the random grain boundary becomes lower. The dislocation density strongly depends on the coherency of random grain boundary. The density of origins of dislocations and the stress in an ingot depend on the density of coherent grain boundaries, and they strongly depend on the density of random grain boundaries.

Difference in local shear stress due to anisotropic elastic constants and/or variation of grain boundary energy depends on the coherency [5,42,58]. The coherent grain boundaries contain little stress and cannot largely reduce remained stress in an ingot. As the coherency of grain boundary increases, the driving force to generate dislocations such as local stress decreases, the origins of dislocations decrease and the stress in an ingot increases. The coherent grain boundaries cannot prevent movement of dislocations, but do not badly affect the solar cell performance. Therefore, the existence of the coherent grain boundaries is not bad when the remained stress is small or the small grain regions in an ingot are suppressed. Usami et al. [62] intentionally prepared a Si multicrystalline ingot which contains many Σ3 grain boundaries with electrically inactive characteristics, using an artificial multicrystalline seed with random grain boundaries and a high growth rate of 1.0 mm min^{-1}. Under these conditions, most of the random grain boundaries are spontaneously modified to $\Sigma 3$ grain boundaries. This result supports that the initial grain boundary configuration is important to realize Si multicrystalline ingots with electrically inactive $\Sigma 3$ grain boundaries. To realize such Si multicrystalline ingots with electrically inactive $\Sigma 3$ grain boundaries, graphite plates with different thermal conductivities as shown in Fig. 4 12 are also effective on the control of the initial grain boundary configuration by controlling the arrangement of dendrite crystals. The coherency of random grain boundaries is also one of the important factors to reduce dislocations. The arrangement of dendrite crystals grown along the bottom of an ingot is effective to realize a Si ingot with many coherent grain boundaries by controlling the growth of dendrite crystals to be fairly parallel to each other [6]. The fraction of twins is slightly larger than that for the other cast methods.

On the other hand, the reduction of remained stress all over an ingot becomes more difficult as coherent grain boundaries largely increases because of their small effect to reduce stress. The proper density of coherent grain boundaries strongly depends on the amount of remained stress in the ingot. When the large stress remains in the ingot, the stress should be largely reduced by introduction of random grain boundaries to reduce the dislocation density. When the little stress remains in the ingot, the origins of dislocations should be largely reduced by increasing coherent grain boundaries and decreasing random grain boundaries. Therefore, to improve the quality of a Si multicrystalline ingot, the suitable density of random grain boundaries must be carefully understood, which is determined by the trade-off between reduction of dislocations and reduction of remained stress in the ingot. The nature of random grain boundaries such as the shape and coherency should be also considered. Moreover, dislocations nucleated from precipitates or particles usually spread in cascades with the ingot height toward the top of ingot. The dislocation density always increases toward the top of ingot [25]. The dislocation density in ingot should by largely reduced by developing future novel technologies to obtain a higher-quality Si ingot with lower remained stress and lower crystal defects.

4.6 Quality and solar-cell performance of Si ingots using the dendritic cast method

The quality of Si ingots grown by the dendritic cast method can be known by the minority carrier lifetime and the conversion efficiency of solar cells.

Diffusion length and resistivity distributions over Si multicrystalline ingots

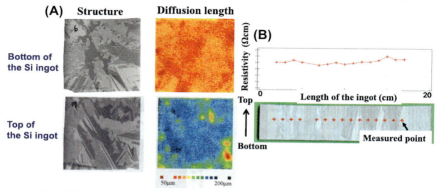

The diffusion length distribution is not affected by dendrite crystals except for the bottom

Fig. 4.16 (A) Crystalline structure and the distribution of diffusion length of minority carriers over a Si multicrystalline ingot. (B) Distribution of resistivity over a Si multicrystalline ingot. The ingots are grown by the dendritic cast method.

Fig. 4.16A shows the multicrystalline structures and the distributions of diffusion length of minority carriers over a Si multicrystalline ingot. The structure and electrical property of the bottom of the ingot are affected by rapidly grown dendrite crystals. The diffusion length at the bottom is lower than 100 μm. The structure of the top of the Si ingot is not affected by dendrite crystals, and the diffusion length is improved to be higher than 150 μm all over the Si wafer. Even through the initial structure is somewhat affected be the dendrite crystals, the diffusion length is largely improved in directionally grown parts of the ingot with its height, and the distribution of the diffusion length is not affected by dendrite crystals except for the bottom part. Some local regions with low diffusion length in the maps contain many dislocations. Especially, the Si wafer cut from the bottom part of the ingot contains many such regions which are called as small grain regions as shown in Fig. 4 17A. Fig. 4.16B shows the distribution of resistivity over a Si multicrystalline ingot. The resistivity is almost constant over a side of the ingot doped by B.

The small grain-regions usually appear in an ingot because dendrite crystals cannot cover all over the bottom of the crucible and many small grains directly grow on a Si_3N_4 coating. To compare the quality of a large grain grown on a dendrite crystal with many small grains directly grown on the Si_3N_4 coating, a Si crystal mixed with the large and small grain regions

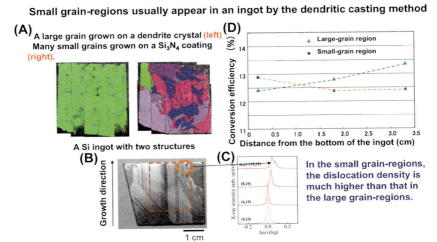

Fig. 4.17 Quality of a large grain region grown on a dendrite crystal comparing with that of a small grain region directly grown on a Si₃N₄ coating. (A) Si crystals with the large and small grain regions prepared by the dendritic cast method. (B) Photo image of the crystal with the large and small grain regions. (C) X-ray measurements which show the quality of the small grain-region is lower than that of the large grain-region because of the multiple and broad peaks of the rocking curve observed in the small grain-region. (D) Conversion efficiency of p-type solar cells prepared using the Si crystal with the small and large grain-regions. *(Referred from Figs. 6 and 8 in N. Usami et al. J. Appl. Phys. 102 (2007) 103504.)*

was prepared by the dendritic cast method as shown in Fig. 4.17A. Fig. 4.17B shows the photo image of the crystal with the large and small grain regions. Multiple twins are observed as parallel straight lines. The X-ray measurement shows the quality of the small grain-region is lower than that of the large grain-region because the multiple and broad peaks of the rocking curve are observed in the small grain-region as shown in the top curve in Fig. 4.17C [59]. The line width of the rocking curve in the multiple twins is narrow and uniform. In the small grain-regions, the dislocation density is much higher than that in the large grain-regions. Fig. 4.17D shows the conversion efficiency of p-type solar cells prepared using the Si crystal with the small or large grain-regions. Obviously, the conversion efficiency of the solar cell with the large grain-region is higher than that of the solar cell with the small-grain region except for the initially grown part. Especially, even though the quality of the small grain-region is higher at the initial stage of the growth, it largely decreases as the ingot grows. This behavior is very similar to that of ingots grown by the HP cast method (see Section 5.4.1). On the other hand, the quality of the large grain-region is remarkably improved in the finally grown part of the ingot. Therefore, the most important point to improve the dendritic cast method is reduction of the small grain-regions.

Even at the initial stage of the development of the dendritic cast method, the conversion efficiency of solar cells prepared by a Si multi-crystalline ingot grown by the dendritic cast method is higher than that of solar cells prepared by the conventional cast method and is kept high even at its finally grown part of the ingot [1,4]. The conversion efficiency of solar cells prepared using the dendritic cast method depends on the crucible size used [4]. The conversion efficiency increases with increasing the crucible size because the volume of the high-quality part in the center of an ingot becomes larger with the crucible size. Fig. 4.18 shows the conversion efficiency of p-type solar cells prepared by a Si multicrystalline ingot as a function of the distance from the bottom of the ingot [1,5]. The ingot was prepared by the dendritic cast method using a 15 cm diameter crucible. The diameter and height of the ingot are 15.0 cm and 4.7 cm, respectively. The average grain size is more than 3 cm. The solar cells were prepared by Sharp using Si wafers with 10×10 cm^2. For comparison, the average level of the conversion efficiency of solar cells prepared using a CZ single crystal is shown as 18.0% in Fig. 4.18. The level of the conversion efficiency of

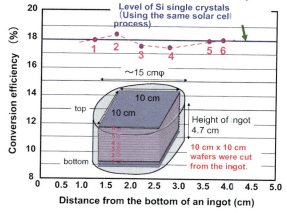

Fig. 4.18 Conversion efficiency of solar cells prepared by a Si multicrystalline ingot grown using the dendritic cast method and a 15 cm-diameter crucible as a function of the distance from the bottom of the ingot. High-efficiency solar cells are obtained from more than 70% of this ingot. (Referred from Fig. 4 in K. Nakajima et al. Proceedings of the 24th European Photovoltaic Solar Energy Conference, 2009, p. 1219.)

solar cells prepared by the dendritic cast method is very closed to that of the solar cells prepared by CZ single wafers using same solar cell processes. The conversion efficiency of 18.2% was in the highest world record group for the large-scale solar cells at that time. Especially, even though the conversion efficiency of the solar cells prepared by the conventional cast method usually dropped down from 60% of the ingot height at that time, the conversion efficiency of the solar cells prepared by the present method keeps high even at the top of the ingot. The dendritic cast method has a possibility to obtain such high-efficiency solar cells from more than 70%–80% of the ingot height except for the initially and finally grown parts. These data showed for the first time that high-quality and uniform Si multicrystalline ingots could be obtained even using the cast method as long as the structure control was executed. These data gave a strong impact to promote the structure control in this field and its effectiveness was proved in many institutes and companies. Fig. 4.19 shows the conversion efficiency of p-type solar cells prepared by larger Si multicrystalline ingots as a function of the distance from the bottom of the ingots [4]. The ingots were grown using the dendritic cast method and the solar cells were prepared by Sharp. The data shown by pink dots are the conversion efficiency of the solar cells with 10×10 cm^2 shown in Fig. 4.19. The data shown by dark

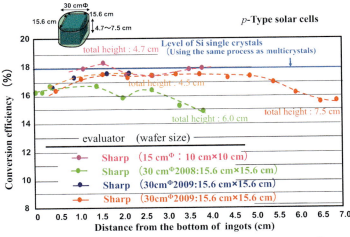

Fig. 4.19 Conversion efficiency of solar cells prepared by Si multicrystalline ingots as a function of the distance from the bottom of the ingots, together with the results in Fig. 4.18. The conversion efficiency of the solar cells keeps high in a large part of the ingot with a height of 7.5 cm. *(From K Nakajima et al., High efficiency solar cells obtained from small size ingots with 30 cmΦ by controlling the distribution and orientation of dendrite crystals grown along the bottom of the ingots, 2010 35th IEEE Photovoltaic Specialists Conference, © 2010 IEEE.)*

blue, red and green dots are the conversion efficiency of the present solar cells with 15.6 × 15.6 cm². The diameter of the crucible used for the crystal growth is 30 cm. The conversion efficiency of solar cells prepared at 2009 becomes higher around 17.7% due to roughly control of the arrangement of dendrite crystals, which is almost comparable with the conversion efficiency (18%) of solar cells prepared by CZ single wafers using the same solar cell processes. The conversion efficiency of the solar cells keeps high in a large part of the present ingot with a height of 7.5 cm even though the ingot size is small.

4.7 Pilot furnace for manufacturing industrial scale ingots

High-quality Si ingots can be obtained even using a small crucible by the dendritic cast method. A pilot furnace was developed by Dai-ichi Kiden Co., Ltd. (KDN) for industrial scale ingots by controlling the structure of the ingots as shown in Fig. 4.20. An 80 cm-diameter crucible can be set in the furnace. Large scale ingots with a mass of 400–450 kg can be manufactured using the furnace. Using the pilot furnace, the structure of Si ingots can be freely designed for the growth of dendrite crystals, the ingot growth using seeded crystals, the control of grain structure and so on. A very long dendrite crystal with 70 cm appears from the edge of an ingot with a mass of 250 kg prepared by the furnace. A high-quality Si ingot with a mass of 420 kg was grown by KDN using the pilot furnace without controlling the

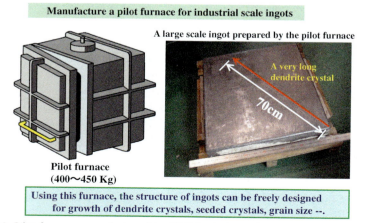

Fig. 4.20 Pilot furnace manufactured by KDN for industrial scale ingots by controlling the structure of the ingots. A large-scale ingot with a mass of 400–450 kg can be grown using a 80 cm-diameter crucible set in the furnace. A very long dendrite crystal with 70 cm appears from the edge of an ingot with a mass of 250 kg.

arrangement of dendrite crystals. Fig. 4.21 shows the distribution of grains from the bottom to the top of an ingot grown using the pilot furnace. Fig. 4.21A shows a cross section of an ingot parallel to the growth direction. Fig. 4.21B shows a wafer cut at 0.48 cm from the bottom of the ingot. Four large grains grow on dendrite crystals. Two grains marked by red and green in the bottom of the ingot shown in Fig. 4.21A correspond to the grains marked by red and green at 0.48 cm from the bottom of the ingot shown in Fig. 4.21B. Thus, the grains grown on dendrite crystals clearly continue to the top of the ingot from the bottom. The average grain size normally increases with the ingot height. The distribution of the minority carrier lifetime in a cross section of the ingot is uniform in the lower half part of the ingot, but its upper half part should be still improved by controlling the arrangement of dendrite crystals as shown in Fig. 4.22. Fig. 4.23 shows the distribution of conversion efficiency of p-type solar cells prepared by 3,000 wafers selected from every parts of this ingot [63]. The size and thickness of the wafers are 15.6 × 15.6 cm and 180 μm, respectively. The peak value of the distribution is 17.2%. The peak value will be higher by controlling the arrangement of dendrite crystals. The solar cells prepared from 92% to 59% of the total wafers show the conversion efficiencies above 16.5% and 17%, respectively. The scattering width of the conversion efficiencies is about

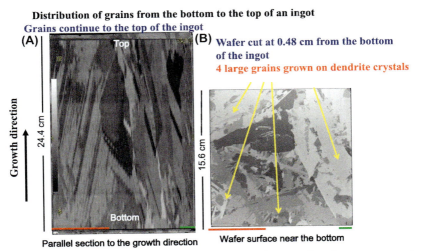

Fig. 4.21 Distribution of grains from the bottom to the top of an ingot grown using the pilot furnace. (A) Cross section of an ingot parallel to the growth direction. (B) Wafer cut at 4.8 mm from the bottom of the ingot. The grains grown on dendrite crystals clearly continue to the top of the ingot from its bottom.

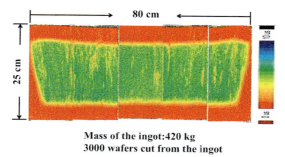

Mass of the ingot: 420 kg
3000 wafers cut from the ingot

Fig. 4.22 Distribution of the minority carrier lifetime in a cross section of the ingot. It is uniform in the lower half part of the ingot except for its upper half part.

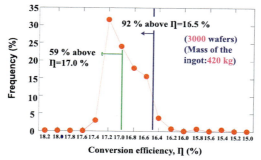

Fig. 4.23 Distribution of conversion efficiency of p-type solar cells prepared by 3,000 wafers selected from every parts of an ingot with a mass of 420 kg grown by KDN using the pilot furnace without controlling the arrangement of dendrite crystals. The peak value of the distribution is 17.2%. The scattering width of the conversion efficiencies is about 1.6%. (Referred from Fig. 10.14 in C.W. Lan, C. Hsu, K. Nakajima, in: T. Nishinaga, P. Rudolph (Eds.), Handbook of Crystal Growth, Bulk Crystal Growth: Basic Techniques Vol. II, Part A, Elsevier, Amsterdam, 2015, pp. 373–411.)

1.6% between the highest and lowest ones, which is quite narrow. This value is important for comparing with the scattering width using other methods such as the HP cast and NOC methods. The yield of the solar cells with high conversion efficiency is high for the ingot grown by the dendritic cast method. Thus, the dendritic cast method has a possibility to increase the yield of high-quality Si multicrystalline ingots. Using small ingots, Lan et al. [64] reported that a very high percentage of $\Sigma 3$ grain boundaries and very high minority carrier lifetime could be obtained by using dendrite crystals as seeds. Even in industry-scale wafers, the twining area has rather low defects and high lifetime.

4.8 Key points for improvement and impact

4.8.1 Key points for improvement

The dendritic cast method has a potential to prepare an ingot with fewer dislocations by properly growing dendrite crystals only along the crucible bottom because nucleation on dendrite crystals can provide an environment to generate fewer crystal defects and improve electrical properties. Stokkan [25] pointed out that low-dislocation density ingots were most likely caused by nucleation on dendrite crystals grown along the crucible bottom and high dislocation density ingots were grown on dendrite crystals grown away from the crucible bottom at a wider angle. In this case, the grain orientation of the high dislocation density ingots is not close to {211} or {110}, which is expected for grains grown on the upper surface of dendrite crystals. The most important points to use this method as an advanced industrial cast method are to grow dendrite crystals only along the bottom of a crucible as seed crystals and largely reduce the small grain-regions as shown in Fig. 4.17. The dendrite crystals can be surely controlled to grow only along the crucible bottom by creating the local and sharp supercooling only along the crucible bottom.

The small grain-regions usually appear in an ingot because dendrite crystals cannot cover all over the crucible bottom and many small grains directly grow on the Si_3N_4 coating on the crucible bottom. These small grains appear due to random nucleation on the coating materials. The quality of the small grain-regions is lower than that of the large grain-regions. The dislocation density is much higher in the small grain-regions than that in the large grain-regions. Dislocation clusters also easily appear in the small grain-regions. High-quality ingots should contain only large grain-regions with lower defects, without containing such small grains.

To largely reduce such small-grain regions in an ingot, dendrite crystals should be arranged completely parallel to each other. The line-shaped highly supercooled part shown in Fig. 4.12A and B is effective to arrange dendrite crystals completely parallel to each other because the highly supercooled part in the Si melt strongly acts as a nucleation site of dendrite crystals as shown in Fig. 4.10B. In this case, a much larger crucible than 30 cm$^{\varphi}$ should be used for the ingot growth to prevent dendrite crystals appeared from the edge of the ingot. When dendrite crystals are arranged completely parallel to each other, dislocation density can be largely reduced as shown in Fig. 4.11. Growth of dendrite crystals with some angles to the crucible bottom should be avoided by precisely establishing locally-limited supercooling only along the crucible bottom for the method. To use the dendritic cast method for the practical use, the larger supercooling than 10 K should be uniformly controlled only along the Si_3N_4 coated bottom of a large crucible in an industrial-scale furnace. The local supercooling should be effectively created along the bottom of the crucible through the thick quartz bottom-wall.

The dendritic cast method can control the crystallographic structure of a Si multicrystalline ingot. The structure control is effective to reduce stress and dislocations and increase uniformity of the ingot. For the structure control, the grain size and the density of random grain boundaries are the most important factors to reduce dislocations in the multicrystalline ingot and increase the yield of high-performance solar cells using the ingot. As shown in Fig. 4.24, an ingot with many small grains has low stress and many random grain boundaries. In the ingot, many small grains can effectively reduce stress by activating many types of slip systems, rearrangement of grains during crystal growth and introduction of the high-density of random grain boundaries. This concept is used for the HP cast method. An ingot with few large grains has high stress and few random grain boundaries, and the stress is mainly relaxed by introduction of dislocations in the ingot and formation of many small grains appeared from the crucible wall. This concept is used for the mono-like cast method. The average dislocation density in a multicrystalline ingot is higher than in a mono-like cast

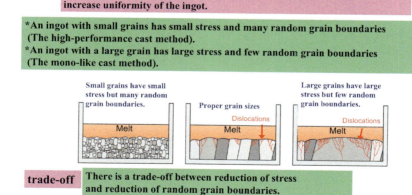

Fig. 4.24 Trade-off between reduction of stress and introduction of random grain boundaries or dislocations. There is also a trade-off between the densities of random grain boundaries and dislocations, and the grain size strongly concerns in this trade-off.

ingot [65]. In the multicrystalline ingot, the stress is much reduced and the distribution of residual strain is more homogeneous comparing with those in the mono-like cast ingot [65].

On the view point of crystallinity, it is best to grow large grains without the electrically active small angle grain boundaries and the mechanically weak multi-twin boundaries [66]. Moreover, Si wafers with larger grains and fewer grain boundaries are stronger and better from the mechanical point of view [67]. To reduce impurities and stress and to obtain an efficient textural surface, an ingot with large-size oriented grains is proposed as a future structure of an ingot grown by the dendritic cast method using a non-wetting crucible [22,68]. The random grain boundaries can reduce stress in the ingot but act as origins of dislocations in a Si multicrystalline ingot together with local high stress. There is a trade-off between reduction of stress and reduction of random grain boundaries or dislocations. There is also a trade-off between reduction of stress and reduction of grain density (increase of large grains), therefore the grain size strongly concerns in this trade-off. Proper grain sizes will be required to reduce both dislocations and random grain boundaries. To determine such proper grain sizes, it is effective to use the dendritic cast method because it can control the grain size and the density of random grain boundaries. The proper fraction between twins and random grain boundaries in an ingot grown by the dendritic cast method is not clear. As a dendrite crystal grows using parallel twins by a local supercooling larger than 10 K, the fraction of twins may be larger than that for the other methods. This point should be much more studied to improve the quality of the ingot.

4.8.2 Key points for impact on development of seed-assisted cast methods

As a new concept, the dendritic cast method was proposed for the structure control to obtain the best crystallographic structure to realize high-quality Si multicrystalline ingots [1,2]. At the early stage of the structure-controlled ingot growth, the conversion efficiency of solar cells prepared by this method was found to be kept high and uniform almost all over the ingot [5]. It was proved for the first time that the quality and uniformity of Si multicrystalline ingots could be largely improved even using the cast method as long as the structure control was executed. This result gave a large motivation for the worldwide researchers to start the ingot growth on the basis of the structure control. The dendritic cast method was also widely tried by several methods and its effectiveness was widely recognized [29,37,64].

The largely improved results obtained by the structure control gave a significant impact on the development of seed-assisted ingot-growth technologies using several types of artificially manipulated seeds such as the mono-like and seeded cast methods. Among them, a new practical cast method was developed in Taiwan and China in the processes of trial and error for the dendritic cast method or the structure control method. This method was named as the HP cast method by Prof. Lan [69]. The HP cast method using Si or non-Si seed particles are used for manufacturing most of Si multicrystalline ingots for solar cells in the present market. Stokkan et al. [70] reported that such pioneering work was performed by the dendritic cast method, focusing on the nucleation and initial growth which showed how grain size and grain orientation could be controlled using the initial supercooling.

In future, the dendritic cast method will be much more effective to obtain a novel Si multicrystalline ingot with lager grains and few dislocations by controlling the parallel arrangement of dendrite crystals only along the crucible bottom and largely reducing the small grain regions. The impurity concentration in the ingot and the stress between the ingot and the crucible will be largely reduced using the non-wettable crucible. Although the density of random grain boundary is much smaller comparing to the HP ingot, the density of dislocation will be also much smaller controlling the parallel arrangement of dendrite crystals to obtain much coherent grain boundaries and to reduce the small grain regions. Even though the coherent grain boundary cannot largely reduce the stress in the ingot, the coherent grain boundaries and the large grains do not badly affect the solar cell performance by themselves comparing with the random grain boundaries and many small grains under the reduced stress in the ingot. The development of such novel structure of Si multicrystalline ingot will be desired.

References

[1] K. Nakajima, K. Fujiwara, W. Pan, M. Tokairin, Y. Nose, N. Usami, in: Conference Record of the 2006 IEEE 4th World Conference on Photovoltaic Energy Conversion, Hilton Waikoloa Village, Waikoloa, Hawaii, USA, May 7–12, 2006, vol. 1–2, 2006, pp. 964–967.

[2] K. Fujiwara, W. Pan, N. Usami, K. Sawada, M. Tokairin, Y. Nose, A. Nomura, T. Shishido, K. Nakajima, Acta Mater. 54 (2006) 3191.

[3] K. Fujiwara, W. Pan, K. Sawada, M. Tokairin, N. Usami, Y. Nose, A. Nomura, T. Shishido, K. Nakajima, J. Cryst. Growth 292 (2006) 282.

[4] K. Nakajima, K. Kutsukake, K. Fujiwara, N. Usami, S. Ono, I. Yamasaki, Proceedings of the 35th IEEE Photovoltaic Specialists Conference, Hawaii Convention Center, Honolulu, Hawaii, USA, June 20-25, 2010, pp. 817–819.

[5] K. Nakajima, N. Usami, K. Fujiwara, K. Kutsukake, S. Okamoto, Proceedings of the 24th European Photovoltaic Solar Energy Conference and Exhibition, CCH-Congress Center and International Fair Hamburg, Germany, September 21-24, 2009, pp. 1219–1221.

[6] K. Nakajima, K. Kutsukake, K. Fujiwara, K. Morishita, S. Ono, J. Cryst. Growth 319 (2011) 13.

[7] K. Fujiwara, Y. Obinata, T. Ujihara, N. Usami, G. Sazaki, K. Nakajima, J. Cryst. Growth 262 (2004) 124.

[8] K. Fujiwara, Y. Obinata, T. Ujihara, N. Usami, G. Sazaki, K. Nakajima, J. Cryst. Growth 266 (2004) 441.

[9] W. Yi, J. Chen, S. Ito, K. Nakazato, T. Kimura, T. Sekiguchi, K. Fujiwara, Crystals 8 (2018) 317.

[10] E. Billig, Proc Roy Soc A229 (1955) 346.

[11] D.R. Hamilton, R.G. Seidensticker, J. Appl. Phys. 31 (1960) 1165.

[12] R.S. Wagner, Acta Metall. 8 (1960) 57.

[13] G. Devaud, D. Turnbull, Acta Metall. 35 (1987) 765.

[14] C.F. Lau, H.W. Kui, Acta Mater. 41 (1993) 1999.

[15] K.K. Leung, H.W. Kui, J. Appl. Phys. 75 (1994) 1216.

[16] K. Nagashio, H. Murata, K. Kuribayashi, Acta Mater. 52 (2004) 5295.

[17] K. Nagashio, K. Kuribayashi, Acta Mater. 53 (2005) 3021.

[18] I. Steinbach, H.–J. Diepers, C. Beckermann, J. Cryst. Growth 275 (2005) 624.

[19] K. Fujiwara, K. Nakajima, T. Ujihara, N. Usami, G. Sazaki, H. Hasegawa, S. Mizoguchi, K. Nakajima, J. Cryst. Growth 243 (2002) 275.

[20] K. Fujiwara, K. Maeda, N. Usami, G. Sazaki, Y. Nose, K. Nakajima, Scripta Mater. 57 (2007) 81.

[21] K. Fujiwara, K. Maeda, N. Usami, G. Sazaki, Y. Nose, A. Nomura, T. Shishido, K. Nakajima, Acta Mater. 56 (2008) 2663.

[22] K. Fujiwara, Growth of multicrystalline silicon for solar cells: dendritic cast method, in: D. Yang (Ed.), Chapter 9, Section Two, Crystalline Silicon Growth, Hand Book of "Photovoltaic Silicon Material", Springer, Berlin Heidelberg, 2017, pp. 1–22 (On line).

[23] M. Kohyama, R. Yamamoto, M. Doyama, Phys. Status Solidi B138 (1986) 387.

[24] G.Y. Chen, C.W. Lan, Scripta Mater. 125 (2016) 54.

[25] G. Stokkan, Acta Mater. 58 (2010) 3223.

[26] K. Fujiwara, H. Fukuda, N. Usami, K. Nakajima, S. Uda, Phys. Rev. B 81 (2010) 224106.

[27] K. Nagashio, H. Okamoto, K. Kuribayashi, I. Jimbo, Metal. Mater. Trans. A 36 (2005) 3407.

[28] K. Fujiwara, K. Maeda, N. Usami, K. Nakajima, Phys. Rev. Lett. 101 (2008) 055503.

[29] H.W. Tsai, M. Yang, C. Chuck, C.W. Lan, J. Cryst. Growth 363 (2013) 242.

[30] K.E. Ekstrøm, E. Undheim, G. Stokkan, L. Arnberg, M.D. Sabatino, Acta Mater. 109 (2016) 267.

[31] K. Nakajima, K. Fujiwara, N. Usami, High-quality Si multicrystals with same grain orientation and large grain size by the newly developed dendritic casting method for high-efficiency solar cell applications, in: Y. Fujikawa, K. Nakajima, T. Sakurai (Eds.), Advances in Materials Research, Frontiers in Materials Research Part II, vol. 10, Springer, Berlin Heidelberg, 2008, pp. 123–140.

[32] G.Y. Chen, H.K. Lin, C.W. Lan, Acta Mater. 115 (2016) 324.

[33] A. Voigt, E. Wolf, H.P. Strunk, Mater. Sci. Eng., A B54 (1998) 202.

[34] T. Strauch, M. Demant, P. Krenckel, R. Riepe, S. Rein, J. Cryst. Growth 454 (2016) 147.

[35] T.Y. Wang, S.L. Hsu, C.C. Fei, K.M. Yei, W.C. Hsu, C.W. Lan, J. Cryst. Growth 311 (2009) 263.

[36] C.W. Lan, W.C. Lan, T.F. Lee, A. Yu, Y.M. Yang, W.C. Hsu, B. Hsu, A. Yang, J. Cryst. Growth 360 (2012) 68.

[37] K.M. Yeh, C.K. Hseih, W.C. Hsu, C.W. Lan, Prog. Photovoltaics Res. Appl. 18 (2010) 265.

[38] T.F. Li, K.M. Yeh, W.C. Hsu, C.W. Lan, J. Cryst. Growth 318 (2011) 219.

[39] K. Arafune, T. Sasaki, F. Wakabayashi, Y. Terada, Y. Ohshita, M. Yamaguchi, Physica B 376—377 (2006) 236.

[40] C.A. Dimitriadis, J. Phys. D Appl. Phys. 18 (1985) 2489.

[41] M. Rinio, A. Yodyungyong, S. Keipert-Colberg, D. Borchert, A. Montesdeoca-Santana, Phys. Status Solidi 208 (2011) 760.

[42] N. Usami, R. Yokoyama, I. Takahashi, K. Kutsukake, K. Fujiwara, K. Nakajima, J. Appl. Phys. 107 (2010) 013511.

[43] I. Takahashi, S. Joonwichien, S. Matsushima, N. Usami, J. Appl. Phys. 117 (2015) 095701.

[44] Y. Nose, I. Takahashi, W. Pan, N. Usami, K. Fujiwara, K. Nakajima, J. Cryst. Growth 311 (2009) 228.

[45] N. Usami, I. Takahashi, K. Kutsukake, K. Fujiwara, K. Nakajima, J. Appl. Phys. 109 (2011) 083527.

[46] T. Hiramatsu, I. Takahashi, S. Matsushima, N. Usami, Jpn. J. Appl. Phys. 55 (2016) 091302.

[47] I. Takahashi, N. Usami, H. Mizuseki, Y. Kawazoe, G. Stokkan, K. Nakajima, J. Appl. Phys. 109 (2011) 033504.

[48] K. Nakajima, R. Murai, K. Morishita, K. Kutsukake, K. Fujiwara, N. Usami, S. Ono, 21th International Photovoltaic Science and Engineering Conference (PVSEC-21), Fukuoka Sea Hawk, Fukuoka, Japan, November 28—December 2, 2011.

[49] G. Stokkan, J. Cryst. Growth 384 (2013) 107.

[50] J. Palm, J. Appl. Phys. 74 (1993) 1169.

[51] Z.J. Wang, S. Tsurekawa, K. Ikeda, T. Sekiguchi, T. Watanabe, Interface Sci. 7 (1999) 197.

[52] J. Chen, T. Sekiguchi, D. Yang, F. Yin, K. Kido, S. Tsurekawa, J. Appl. Phys. 96 (2004) 5490.

[53] S. Bernardini, S. Johnston, B. West, T. U. Nærlanc, M. Stuckelberger, and M. I. Bertoni, The 43th IEEE Photovoltaic Specialists Conference (43th PVSC), Portland, Oregon, US, June 5—10, 2016.

[54] J. Chen, T. Sekiguchi, R. Xie, P. Ahmet, T. Chikyo, D. Yang, S. Ito, F. Yin, Scripta Mater. 52 (2005) 1211.

[55] H. Sugimoto, M. Inoue, M. Tajima, A. Ogura, Y. Ohshita, Jpn. J. Appl. Phys. 45 (2006) L641.

[56] K. Kutsukake, N. Usami, T. Ohtaniuchi, K. Fujiwara, K. Nakajima, J. Appl. Phys. 105 (2009) 044909.

[57] T. Buonassisi, A.A. Istratov, M.A. Marcus, B. Lai, Z. Cai, S.M. Heald, E.R. Weber, Nat. Mater. 4 (2005) 676.

[58] I. Takahashi, N. Usami, K. Kutsukake, G. Stokkan, K. Morishita, K. Nakajima, J. Cryst. Growth 312 (2010) 897.

[59] N. Usami, K. Kutsukake, K. Fujiwara, K. Nakajima, J. Appl. Phys. 102 (2007) 103504.

[60] I. Takahashi, N. Usami, K. Kutsukake, K. Morishita, K. Nakajima, Jpn. J. Appl. Phys. 49 (2010) 04DP01.

[61] V.A. Oliveira, B. Marie, C. Cayron, M. Marinova, M.G. Tsoutsouva, H.C. Sio, T.A. Lafford, J. Baruchel, G. Audoit, A. Grenier, T.N. Tran Thi, D. Camel, Acta Mater. 121 (2016) 24.

[62] N. Usami, K. Kutsukake, T. Sugawara, K. Fujiwara, W. Pan, Y. Nose, T. Shishido, K. Nakajima, Jpn. J. Appl. Phys. 45 (2006) 1734.

[63] C.W. Lan, C. Hsu, K. Nakajima, Multicrystalline silicon crystal growth for photovoltaic applications, in: T. Nishinaga, P. Rudolph (Eds.), Handbook of Crystal Growth, Bulk Crystal Growth: Basic Techniques, vol. II, Elsevier, Amsterdam, 2015, pp. 373−411. Part A.

[64] C.W. Lan, A. Lan, C.F. Yang, H.P. Hsu, M. Yang, A. Yu, B. Hsu, W.C. Hsu, A. Yang, J. Cryst. Growth 468 (2017) 17.

[65] S. Nakano, B. Gao, K. Jiptner, H. Harada, Y. Miyamura, T. Sekiguchi, M. Fukuzawa, K. Kakimoto, J. Cryst. Growth 474 (2017) 130.

[66] J. Chen, B. Chen, T. Sekiguchi, M. Fukuzawa, M. Yamada, Appl. Phys. Lett. 93 (2008) 112105.

[67] V.A. Popovich, A. Yunus, M. Janssen, I.M. Richardson, I.J. Bennett, Sol. Energy Mater. Sol. Cells 95 (2011) 97.

[68] K. Fujiwara, Y. Horioka, S. Sakuragi, Energy Sci. Eng. 3 (2015) 419.

[69] C.W. Lan, C.F. Yang, A. Lan, M. Yang, A. Yu, H.P. Hsu, B. Hsu, C. Hsu, Cryst. Eng. Comm. 18 (2016) 1474.

[70] G. Stokkan, Y. Hu, Ø. Mjøs, M. Juel, Sol. Energy Mater. Sol. Cells 130 (2014) 679.

CHAPTER 5

High performance (HP) cast method

5.1 Concept of the HP cast method

The structure control using dendrite crystals as-grown seeds for growth of a Si multicrystalline ingot is effective to reduce stress and dislocations in it and increase the uniformity of the ingot (see Section 4). The grain size and the density of random grain boundaries are the most important factors to reduce stress and dislocations in the ingot. Small grains have an effect to reduce stress by rearrangement of grains during growth and many random grain boundaries formed between them. The random grain boundaries are essentially recombination active in as-grown state and very harmful for solar cells [1]. As shown in Fig. 4.24, an ingot with small grains has small stress and quite high-density of random grain boundaries. The random grain boundaries can also reduce stress in the ingot, but if they have local high stress, they also act as one of origins of dislocations in a Si multicrystalline ingot. In this sense, the introduction of random grain boundaries increases the dislocation density near the random grain boundaries as shown in Fig. 4.15. Generally, there is a trade-off between reduction of stress and reduction of random grain boundaries (see Section 4.5). On the other hand, many random grain boundaries can prevent the movement or multiplication of dislocations and dislocation clusters because dislocations cannot across the random grain boundaries [2]. The HP cast method was proposed to obtain uniform ingot structure introducing small grains by Lan at 2011 [3].

For the HP cast method, to reduce the dislocation density in the initial part of an ingot, the interference of the movement or multiplication of dislocations and dislocation clusters was tried by intentionally introducing random grain boundaries at the initial stage of the growth. Large amount of random grain boundaries is effective to relax thermal stress and prevent movement of dislocations, resulting in reduction of the dislocation density

Crystal Growth of Si Ingots for Solar Cells Using Cast Furnaces
ISBN 978-0-12-819748-6
https://doi.org/10.1016/B978-0-12-819748-6.00005-0

Copyright © 2020 Elsevier Inc.
All rights reserved.

in the bottom and middle parts of the ingot. To pull out such effects, the HP cast method has a feature to grow a Si multicrystalline ingot with many small grains and many random grain boundaries in its bottom part by the directional solidification using a particle seed layer set on the bottom of the crucible [4—6]. The fraction of $\Sigma 3$ grain boundaries which can reduce only little stress is very low in the ingot. Twin planes with such CSL grain boundaries also have effects to prevent movement of dislocations when they have a large angle in the growth direction [2]. To increase the uniformity of the ingot and reduce the dislocation density, this approach is based on the concept of four factors, such as growth of uniform small grains by uniform nucleation, rearrangement of small grains to reduce stress, introduction of many random grain boundaries to reduce stress and suppression of the formation, movement and multiplication of dislocations using the random grain boundaries. Even though the ingot has huge amount of random grain boundaries which are very detrimental for the performance of solar cells, the hydrogenation is very much effective to make the random grain boundaries quite harmless [1,7]. The hydrogenation is a typical process to prepare solar cells after P gettering.

The concept of the HP cast method was found during an ingot growth by controlling the supercooling to induce large grains using the dendritic cast method [6,8]. Lan et al. [6] said that during the development of the dendritic cast method, they obtained very small grains at the beginning incidentally. It turned out that with the small grains, the multiplication of dislocation clusters was significantly reduced, and the ingot lifetime was uniform. Thus, small grains occasionally grown during this process were found to be a much better uniformity. As a result, a high fraction of random grain boundaries is effective to realize the uniform distribution of the conversion efficiency of solar cells. The dendritic cast method is the fully-melted cast method, but the HP cast method is the half-melted cast method because some seed layers should be remained without melting.

5.2 Control of grain size, grain orientation and grain boundaries using assisted seeds

5.2.1 Si particles or chips placed on the bottom of crucibles

If a Si ingot which has good uniformity for distribution of small grains can be prepared only by controlling the initial supercooling of a Si melt, the ingot can be reused as multicrystalline seed plates for another new growth [4]. However, the reuse of multicrystalline seed plates will not become a suitable production technology because it will increase the cost of the ingot even though the yield of ingot growth will be high. Azuma et al. [9] used Ge seed particles with random orientation for directional solidification of the SiGe and Ge ingots and found that the grain size increased with the ingot height, the area fraction of the preferred orientation of <110> increased with the ingot height, and the {110} face was finally remained during growth. They further used a {110} Ge seed plate with the preferred orientation to effectively retain polycrystallization of the ingot. These pioneering results have given an important implication for the development of the HP cast method using Si seed particles as shown in Fig. 1.5 (see Section 1.3.1). Wong et al. [10] used spherical Si beads with twins to determine the preferred orientation for the Si ingot growth, and they found that the {112} and {111} faces were dominant and a high fraction of random grain boundaries initially appeared thanks to the initial nucleation from the beads and to the new grains with different orientations generated by twining.

In the HP cast method as a seed-assisted growth method, granular Si particles or small Si chips can be used as assisted seeds or nucleation agents placed on the bottom of a quartz crucible to control the size and density of grains of a Si ingot. The granular shaped or feedstock particles with the size of 0.015−0.5 cm are paved on the entire bottom of a crucible [4,6,8,11]. Smaller particles can be easily contaminated due to larger surface area. The thickness of the layer of the paved seed particles is usually 2 cm. The temperature gradient near the seed layer is selected to be as large as 26 K cm^{-1} to melt the surface part of the seed layer and remain its un−melted height of about 1 cm. The remaining thickness of the seed layer is proportional to the so-called red zone with high impurities and low minority carrier lifetime [8]. The use of granular seed particles (the size: 0.15−4 mm) results in a longer bottom red zone comparing with the red zone caused by block seeds like the mono-like cast seeds because of contaminated gaps between the seed particles [12]. A slow growth rate of 0.07 mm s^{-1} is used at the initial stage of the growth even through the average growth rate is 0.2 mm s^{-1}.

There are large gaps between the irregularly shaped poly-Si chips used as seed particles, which allow for melt penetration into the seeding structure. The growth initiates on the top of the chips, but the average grain size of 0.0706 cm is relatively smaller due to the gaps between the chips even though the chip size is 0.6−1.5 cm [13]. The voids or gaps between particles are effective for the growth of uniform small grains because the grain growth can avoid further interaction with seed particles thanks to the voids. When fluidized bed reactor granules are used as seeding materials with quite small size, the average grain size has a similar value of about 0.0727 cm because of a considerable coarsening of the grain structure grown on quite small granules [13]. The coarsening of genuine non–melted seed particles can be observed. Thus, the size of particles does not essentially affect the average grain size or the grain density (the density of random grain boundary) because of the coarsening. The grain size in the region of infiltrated melt inside the seed particles becomes larger with the height of the genuine non–melted seed layer and these grains grow longer due to their wider grain average width resulting in representing the final upper structure of the HP ingot [14]. A wafer cut from the top part of a conventional multicrystalline ingot has almost 33% of the wafer area covered in dislocation clusters. A wafer cut from the top part of the HP ingot grown on the poly-Si chips has 7%−13% of the wafer area covered in dislocation clusters. This value is approximately 35% more than that grown on the small granules [13]. The poly-Si chips appear to generate more dislocations than small granules.

As growth takes place on all grains intersecting the initial growing interface, a strong correlation between the seed layer and the first growth area occurs resulting in affecting on the grain size and grain orientation [14]. To know the grain structure in detail, a typical HP multicrystalline Si wafer is compared with a typical conventional multicrystalline Si wafer as shown in Fig. 5.1 [3]. The conventional wafer has large and non–uniform grains (a), but the HP wafer has small and uniform grains (b). When the microstructure or grain size in seed particles or feedstock particles is smaller than the particle size of Si feedstock materials, small grains with a homogeneous orientation distribution are obtained near the bottom of an ingot because each grain in seed particles acts as a seed to form a grain [15]. In this case, the mean grain size is slightly smaller than that for the smallest feedstock single crystal as a seed particle [16]. Wider grains in the seed particles at the growing interface have an advantage during grain competition and the largest grains in seed particles are more prone to generate the final grain structure after further growth [14]. When feedstock particles with a single grain are used for a seeding layer, the initial grain size depends on the size of the feedstock particles and decreases with decreasing the size of the feedstock particles [16].

Reimann et al. [15] studied the effect of the random grain boundary density on the defect reduction. When the initial grain size becomes

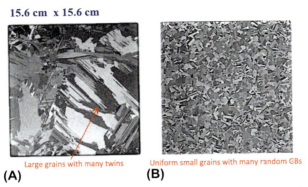

Fig. 5.1 (A) Typical conventional multicrystalline Si wafer with large and non-uniform grains. (B) Typical HP multicrystalline Si wafer with small and uniform grains. *(Referred from Fig. 2 in C.W. Lan, in: D. Yang (Eds.), Chapter 10, Section Two Crystalline silicon growth, Hand Book of "Photovoltaic Silicon Material", Springer, Berlin Heidelberg, 2018, On line.)*

smaller, the initial length fraction or density of random grain boundary slightly increases and the dislocation density becomes lower. If Si single crystalline particles are used, the small particle size is preferable to obtain initial small grains with a homogeneous orientation distribution. In the HP ingot, the length fraction or density of random grain boundaries is generally high and that of the $\Sigma 3$ grain boundaries is low. At 2.5 cm above the seed layer, the defected area fraction of high EPD ($>10^5$ cm^{-2}) decreases as the length fraction or density of random grain boundaries increases which is an important factor to determine the defect density. When the length fraction of random grain boundaries is near 80%, the defected area fraction is as low as 10%.

Trempa et al. [17] reported that the mean grain size at the bottom of an ingot grown by the HP method using Si small chips is smaller than that of an ingot grown by the conventional cast method because the HP ingot contains very small and nearly isometric gains at its bottom part as shown in Fig. 5.2A. At the lower part of the HP ingot, the mean grain size rapidly increases with increasing the ingot height [17–19]. At the middle and top parts of both ingots, the mean grain size is almost constant regardless of the ingot height, but the mean grain size still continuously increases with the ingot height for the industrially grown HP ingot. The grain size distribution can be described by the coefficient of variation CV_{GS} which is defined by Trempa et al. as the ratio between the standard deviation of the mean grain size and the mean grain size itself [17]. This value means that the grain size

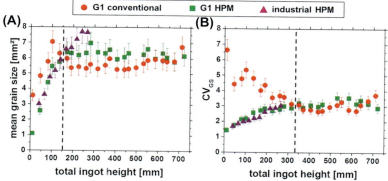

Fig. 5.2 Mean grain size and grain size distribution of ingots grown by the conventional and HP cast methods as a function of the ingot height. The mean grain size at the bottom of the HP ingots is smaller than that of the conventional cast ingot. As the CV becomes smaller, the grain size distribution becomes more homogeneous. In the lower part of the HP ingots, the grain size distribution is very homogeneous. In the middle and upper parts of the ingots, the mean grain size and the grain size distribution become very similar for both methods. (Referred from Fig. 2 in M. Trempa et al. J. Cryst. Growth 459 (2017) 67.)

distribution is more homogeneous as the CV_{GS} is smaller. Fig. 5.2B shows CV_{GS} as a function of the total ingot height for the ingots grown by the conventional and HP cast methods. The CV_{GS} values are distinctly different at the bottom part of both ingots. The initial grain size distribution in the conventional ingot is quite inhomogeneous due to dendrites and twins, but that in the HP ingot is more homogeneous due to very small and nearly isometric grains. At the middle and top parts of the ingots, the distribution and uniformity of the grain size become very similar for both ingot types [16,19]. The shape of grains changed from an initial spherical shape to a columnar shape with the ingot height [18]. At the bottom of the HP ingot, the grain orientation also homogeneously distributes and many orientations evenly appear [17,18]. However, at the upper part of the ingot, the distribution of grain orientation becomes quite similar to that of grain orientation in the ingot grown by the conventional cast method because the distribution naturally reaches a stable state with low energy at the final stage as shown in Fig. 1.6. Thus, the difference in grain structure properties and in the recombination active area between both types of ingot is not significant in the top region. Therefore, the advantage of the HP ingot is most evident in the lower parts of the ingot, where the difference in grain structure between both types of ingot is largest [16].

5.2.2 Other assisted Si seeds with different shapes

For the growth of the seed assisted HP cast method, several shapes or types of assisted seed are tried to be used and the quality of grown ingots are investigated. Among them, for manufacturing industrial low-cost HP multicrystalline Si ingots, polycrystalline (poly-Si) planar slabs are used as assisted seeds for nucleation sites [20]. The 3—5 cm thick slab seeds are cut from Siemens poly-Si rods with diameters of 16—20 cm and are paved on the bottom of crucible. The gaps between the slabs are filled with poly-Si nuggets. Grains in a slab-assisted ingot are much uniform compared to those in a nugget-assisted ingot, and the density of dislocation cluster in the slab-assisted ingot is much lower than that in the nugget-assisted ingot, especially in the middle and top parts of the ingot [20]. The bottom edge of the slab-assisted ingot can be fully recycled for reducing the manufacturing cost because the ingot has dense structure and the clear cutting can be preformed [20]. The poly-Si slabs are used for manufacturing Si ingots as the half-melted cast method instead of the fully melted cast method in which some mixture of Si particles and other particles such as Si_3N_4 or SiO_2 is used.

Si waste wafer flakes with diameters of 0.5—2 cm and a thickness of 180 μm are tried to be used in industry as low-cost assisted seeds for nucleation sites [21]. These flake plates are horizontally and orderly stacked along the bottom of crucible to make multilayer seeds. Crystals are epitaxially grown on them and the size of grains is determined by the flake size. A lot of small dislocation clusters appear in this case [21]. When the wafer flakes are piled up intentionally to form some interspaces, the size of grains (~ 0.05 cm diameter) becomes relatively smaller and more uniform, and dislocation cluster are almost invisible [21]. The distribution of minority carrier lifetime on the vertical cross-section of the ingot grown using the pileup flakes is more uniform comparing with that of the ingot grown using the stacked flakes [21]. This phenomenon is very similar to the case of the voids between Si particles.

Single-layer Si beads coated with Si_3N_4 are tried as assisted seeds for nucleation sites [22—24]. The purpose of this method is to reduce the precise temperature control of the Si seed particles which is required to prevent the particles from completely melting. The Si beads are spread on the bottom of crucible and coated with Si_3N_4. Smaller grain size, more uniform grain distribution and lower density of dislocation cluster are obtained comparing with those of conventional cast wafers because of their small bead size (<0.1 cm) and uniform seed shapes [23,24]. However, the dislocation density is higher and the grains are larger than those in an ingot grown using Si seed particles [25]. Nucleation occurs on deep concave portions of the bead nucleation-layer and the nuclei expands over large areas of the nucleation-layer resulting in larger grains and few random grain boundaries. This mechanism can be known by the application of the Voronoi diagram [22] (see Section 1.3.2).

5.2.3 Non-Si particles used as coating material on the bottom of crucibles

Generally, the nucleation energy becomes lower as the wetting angle becomes lower. To avoid increasing the processing time and the yield loss of the unmolten seed layer for Si particles, a non-Si nucleation layer is used for the HP cast method. Even for such non-Si nucleation layer, a nucleation layer should be stable at high temperatures and be wettable by a Si melt to reduce nucleation energy. Thus, for the HP cast method, the wetting angle of Si on such a nucleation layer should be lower than on a Si_3N_4 coating used for the conventional cast method because the nucleation rate increases as the wetting angle decreases for the same supercooling. Kupka et al. [26] reported effects of several non-Si particles on the nucleation of initial grains grown on them. For both SiO_2 and SiC layers, the initial grain size decreases as the surface roughness of particle layer increases because the rough layer has more steps, edges and corners with low nucleation energy. The thermal conductivity is another factor to determine the grain size because the high thermal conductivity increases the cooling rate and supercooling locally at the crucible bottom and grow large grains with same orientation thanks to dendrite crystals. The thermal conductivity of SiO_2 is lower than that of SiC, and the grain size on SiO_2 layers is smaller than that on SiC layers. On the view point of the homogeneity of grain orientation distribution and the introduction of random grain boundaries, SiO_2 layers have a larger effect than SiC layers because of preventing dendrite growth owing to low thermal conductivity of SiO_2. However, the density of random grain boundaries is not as high as that in the case of Si seed particles [3].

Several kinds of materials such as fused quarts, SiO_2 particles and a mixed coating of the Si, SiO_2 and Si_3N_4 powders are used as seed or nucleation agent on the bottom of a quartz crucible to control the size and density of grains [27,28]. Among them, fused quarts particles attain high yield and high quality for the ingot growth because they are effective on the nucleation and grain control of Si multicrystalline ingots [27,29]. Therefore, the fused quarts particles are used as seed materials on the bottom of a quartz crucible. They are uniformly fixed on the crucible bottom by fused quartz slurry, then a mixture of silica sol, Si_3N_4 powder and water is brushed on it as a coating material. After adherence of them, another mixture of silica sol, Si_3N_4 powder and water is sprayed above the brushed coating to retard erosion of molten Si to the coating [30]. Usually, Si_3N_4, SiC, and C materials are added to SiO_2 particles.

The main factors to control the nucleation and grain are the density and size of fused quartz seed particles acted as seed. The nucleation occurs from both the quarts particles and the gap area between them. The high density of particles increases the ratio between particle and gap area. The grain size generally becomes smaller and uniform as the ratio becomes larger or the nucleation from quartz particles increases. The size of fused quartz seed particles affects the nucleation area on the particles, and the nucleation ratio between the nucleation area and the bottom area of a crucible decreases as the particle size becomes smaller. Thus, smaller particles form smaller gap areas and increase the nucleation area which causes smaller grains and uniform distribution of them. However, the too large density of particles or too smaller particles causes sticking of particles at high temperature. Ding et al. reported that an ingot with the most uniform small grains and lowest density of dislocation clusters was obtained at the intermediate density of 220 particles cm^{-2} and the intermediate mesh size of 50—70 mesh because the nucleation from quartz particles can be optimized so as not to form larger gap areas in this case [30,31]. The nucleation ratio is 60% in this case. The particle size becomes smaller as the mesh number increases. The size of grains also becomes smaller and uniform using silica particles [28]. For the HP cast growth using the fused quartz seed particles, the most important point is to avoid coexistence of large and small grains during growth, resulting in obtaining uniform distribution of only small grains.

The mixed coating of Si and SiO_2 powders affects the grain size and orientation to be more uniform as the ratio of SiO_2 increases, but it does not improve the minority carrier lifetime [28]. At the bottom of an ingot, the fractions of $\Sigma 3$ grain boundary and random grain boundary were 58% and 16%, respectively. This ratio is similar to that in the case of a Si_3N_4 coating and the fraction of random grain boundary is not so large because grains still nucleate from the Si_3N_4 coated bottom [28,32]. Thus, SiO_2 powders are not a good nucleation agent. The SiC-based coating reduces the mean grain size, but many dendrites and twins generate and the density of random grain boundaries becomes low [16]. The SiC-based coating is not a good nucleation agent, too.

The merits to use the non-Si particles or the mixed coating are to reduce the thickness of the red zone and increase the yield of materials. The demerits to use the non-Si nucleation agent as seed are the low initial fraction of random grain boundaries and the contamination from SiO_2 particles. The particles or O elements are dissolved into a Si melt and the O concentration in an ingot increases. Some impurities in the particles diffuse into the ingot and reduce the minority carrier lifetime in the bottom of the ingot. For these nucleation agents, the impurities from the crucible and Si_3N_4 coating material are still unavoidable problems which limit the solar cell performance.

5.3 Behavior and control of dislocations and dislocation clusters in ingots

5.3.1 Dislocations

Generally, the EPD in ingots grown by the HP cast method is on the order of 10^5 cm^{-2} in a high-dislocation-density area and on the order of 10^3/cm^2 in a low-dislocation-density area [4]. The EPD in ingots grown by the conventional cast method is on the order of 10^6 cm^{-2}, which is higher than that in the ingots grown by the HP cast method [5]. Generation of stress in an ingot is affected by many factors such as grain size, grain orientation, grain boundary character, growth behavior, shape of growing interface, impurities, precipitates, contact with crucible and so on. The grain size and distribution of grain size are the important factors in the HP cast method. The average grain size is very small at the bottom of an HP ingot and it increases in the lower part of the ingot during growth as shown in Fig. 5.2A. Near the bottom of the ingot, the stress is relaxed by many small grains and the movement of dislocations is inhibited by existence of many random grain boundaries which terminate dislocation clusters [2]. The dislocation density becomes lower comparing with that in the conventional ingot, especially in the lower parts of both ingots. The average grain size increases and remains almost constant in the middle and upper parts of the ingot.

Fig. 5.3 shows the comparison of the conventional and HP multi-crystalline Si ingots, such as (a) lifetime mappings of both ingots, (b) grain structures of wafers at the top and bottom of both ingots and (c) EPD mappings of wafers at the top and bottom of both ingots [16]. The HP ingot

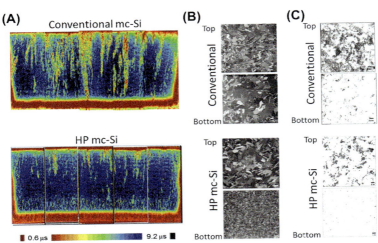

Fig. 5.3 Comparison between distributions of minority carrier lifetime, grains and EPD for ingots grown by the conventional and HP cast methods. (A) Longitudinal minority carrier lifetime mapping, (B) grain structures and (C) EPD mappings. For these measurements, the height of the ingots is 78 cm and the size of the wafers are 15.6 cm × 15.6 cm. Si seed particles (0.3–0.5 cm in size) were used for the HP cast method. *(Referred from Fig. 3 in C.W. Lan et al. J. Cryst. Growth, 468 (2017) 17.)*

was grown using small Si particles. As shown in Fig. 5.3A, the lifetime is more uniform and dislocation clusters are largely reduced for the HP ingot. As shown in Fig. 5.3B and C, the grains near the bottom of the HP ingot is smaller and more uniform, and the high-EPD areas (EPD > 10^5 cm^{-2}) are much smaller. The EPD is lowest in the bottom part of the HP ingot. However, the density of dislocation increases as the average grain size increases as shown in Fig. 5 3 (see Section 5.4.1). The total dislocation density in the HP ingot is usually lower than that in the ingot grown by the cast method [33]. Defects such as dislocations are localized in smaller grains surrounded by larger grains in the upper part of the ingot [4]. Such behavior of defects is similar to that of defects in the small grain region in the case of the dendritic cast method as shown in Fig. 4.17. The coexistence of large and small grains generates stress, concentrates the stress in the small grain regions and increases the density of dislocations and dislocation clusters. The stress tends to be relaxed by generation of dislocations and dislocation clusters in the small grain region and the small grains often contain many dislocations. Dislocation are also generated near small grains which appear by local confinement of small melts and generation of stress during solidification [34]. Multiplication of dislocations in the ingot occurs with the ingot height and strongly reduce the performance of solar cells. The dislocation density increases and the minority carrier lifetime decreases as the ingot grows taller owing to the multiplication of dislocations originated from the bottom of the ingot and the side wall of the crucible [2]. This behavior of the multiplication of dislocations is known by the similar distributions of the minority carrier lifetime in ingots as shown in Fig. 5.3. The increasing speed of dislocations with the height for the HP cast method is 3 times slower than that for the cast method [33]. Generation of dislocations due to thermal stress is inevitable for the ingot growth accompanied by contact with the crucible wall because Si expands by 11% during solidification [35]. Dislocations are usually generated near the random grain boundaries due to the local high stress or from the small angle grain boundaries during crystal growth as shown in Fig. 4.14 [36,37]. Random grain boundaries together with local high stress may act as origins of dislocations in multicrystalline ingots.

The shape of growing interface between crystal and melt is one of the important factors to reduce dislocations in an ingot. Generally, as small grains grow further from the crucible wall, the growing interface tends to become concave in the growth direction. In this case, much defects are introduced in the ingot by the thermal stress due to the large radial thermal gradient. For the convex growing interface in the growth direction, grains grow outward, dislocations move outwards and nucleation on the side wall of crucible is somewhat reduced, even though the same thermal stress exists. For the flat growing interface, the radial thermal gradient is smaller and the thermal stress is much reduced. Generation of dislocations also tends to be reduced for the flat interface [4,9].

5.3.2 Dislocation clusters

A smaller area fraction of densely clustered dislocations is observed in a HP ingot than a muticrystalline cast ingot [38]. Stokkan et al. [2] studied the termination of dislocation clusters during ingot growth using an industrially produced HP ingot. The dislocation clusters expand in diameter with the ingot height if they are not hindered by grain boundaries. As a result, dislocation clusters are few and small near the bottom of the ingot, and there are more dislocation clusters with every size toward the top of the ingot. Some small dislocation clusters tend to vertically and laterally become larger independent of the ingot height, and large clusters become more numerous toward the top. The main factor to terminate the dislocation clusters is the short distance between random grain boundaries or the small grain size because more rapid termination of dislocation clusters is observed in smaller grains in the bottom part of the HP ingot. Thus, the termination mechanism of dislocation clusters during growth can be explained by the interaction with random grain boundaries because they have amorphous nature operating as free surfaces. The longer length fraction or density of random grain boundary from seeds leads to the smaller area of high-EPD cluster in the HP ingot [3,15]. In the comparison between crystal defects such as dislocation clusters in the HP and mono-like ingots, the defect density is lower and more uniform in the ingot grown by the HP cast method than in the ingot grown by the mono-like cast method because random grain boundaries prevent the movement of dislocations in the HP ingot [39].

From Stokkan et al. [2], there is a correlation between the size of dislocation clusters and the size of grains in which the dislocation clusters exist. The dislocation clusters with small size ($3-10$ mm^2) tend to exist within a single crystalline grain with no twins and within one grain with twins or a twinned grain. The dislocation clusters with large size ($100-300$ mm^2) tend to exist within the twinned grain and within consecutive multiple grains. The dislocation cluster which appears within the twinned grain is confined inside the twinned grain, and it moves laterally with the twinned grain as the twin boundaries are shifted during growth. Sometimes, the twinned grain traverses the other grain during growth and forms a random grain boundary with an adjacent grain by which the movement of the dislocation cluster is prevented. Thus, the twins also play an important role to prevent the movement of dislocations.

From the 3-D visualization of dislocation clusters in a HP ingot by processing PL images, sub-grain boundaries with angular deviation of less than $10°$ cause generation of dislocation clusters, and the low dislocation density in the HP ingot is explained by the annihilation of dislocations [40] (see Sections 3.3.2 and 5.4.2).

5.4 Structure and defects in Si ingots using the HP method

5.4.1 Grain structures

At the initial stage of growth using the HP cast method, small randomly oriented spherical and equiaxed grains appear in an ingot grown from small seed particles [18,41]. The initial growth region is strongly related to the seed layer orientation and the crystallographic orientation along the growth direction spread in all directions except for the preferable orientations such as <101> and <111> [14]. Fig. 5.4 shows the vertical distribution of grains in an ingot grown from small seed particles, together with the inverse pole figure of the vertical wafer [41]. The initial mean grain size is small in the lower part of the ingot, but grains elongate and become columnar in the growth direction as the ingot height increases. The length of each columnar grain increases as the grain width at the initial growing interface increases [14]. In the lower part of the HP ingot, the mean grain size is small and rapidly increases during growth as shown in Fig. 5.2A [17–19]. The distributions of grain size and grain orientation tend to be more uniform than those in a conventional cast ingot [42]. In the middle and top parts of the HP ingot, the mean grain size is almost constant with the ingot height, but it increases for the industrial HP ingot [17]. The grain size distribution in the HP ingot is very homogeneous from the bottom to the top due to very small grains as shown in Fig. 5.2B. As the vertical temperature gradient becomes larger, the growth direction of grains becomes more parallel and

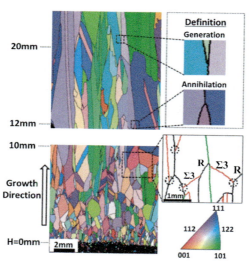

Fig. 5.4 Distribution of grains in an ingot grown from small seed particles, together with the inverse pole figure of the vertical wafer. At the initial stage of growth, small randomly oriented spherical and equiaxed grains appear in the ingot. As the ingot height increases, the grain size increases, and grains elongate and become columnar in the growth direction. Processes of the annihilation and generation interactions of grain boundaries are shown in the upper right figure. *(Referred from Fig. 1 in R.R. Prakash et al. Appl. Phys. Express 8 (2015) 035502.)*

uniform, and the grains become longer columnar [43]. The grains have more <112> orientations and random grain boundaries at a large vertical temperature gradient. The slowly moving rate of the vertical temperature gradient or the low growth rate of a cast ingot can reduce thermal stresses in the Si ingot during crystallization [44]. As the growth rate increases during directional crystallization, the grains become thinner and longer and are more oriented to the symmetry axis [45].

As shown in the lower right figure of Fig. 5.4, $\Sigma 3$ grain boundaries have already inclined nature to the growth front before an annihilation process as $\Sigma 3 + R = R$ (R: random grain boundary). Such interactions of $\Sigma 3$ grain boundaries with random grain boundaries contribute to the increase in grain size during growth because the annihilation processes are more dominant than the generation ones of $R = \Sigma 3 + R$. Thus, $\Sigma 3$ and random grain boundaries are dominant in large grains formed by the annihilation processes in the ingot. Random grain boundaries have also inclined nature after generation of $\Sigma 3$ grain boundaries ($R = \Sigma 3 + R$). The total number of grain boundary interactions decreases with the ingot height because the grain size increases with the ingot height as shown in Fig. 5.2. A $\Sigma 3$ grain boundary is generated from a random grain boundary, but the electrical activity of the random grain boundary is almost not affected by the $\Sigma 3$ generation at the steady state, which can be confirmed by EBIC [46].

Fig. 5.3 shows the comparison between distributions of minority carrier lifetime, grains and EPD for ingots grown be the conventional and HP cast methods [8]. Si seed particles (0.3—0.5 cm in size) were used for growth of the ingot by the HP cast method. Both longitudinal minority carrier lifetime mappings with a width of 78 cm are shown in Fig. 5.3A. The distribution of minority carrier lifetime for the ingot grown by the HP cast method is more uniform than that for the ingot grown by the conventional cast method. The distribution of minority carrier lifetime is largely improved near the top part of the HP ingot because propagation of dislocations is prevented by many random grain boundaries introduced by Si seed particles [11]. The grain structures at the bottom and top parts of the HP and conventional ingots are shown in Fig. 5.3B, and the EPD mappings of the HP and conventional ingots are shown in Fig. 5.3C. The size of these wafers is 15.6 cm $\times$ 15.6 cm. Near the bottom of the HP ingot, grains are smaller and more uniform and high-EPD ($>10^5$ cm^{-2}) areas are much smaller comparing with those in the ingot grown by the conventional cast method. The grain structure at the bottom region of the HP ingot is very different from that of the ingot grown by the conventional cast method, but both grain structures are not significantly different at the middle and upper parts of the ingots because the crystal growth is an essential process to go toward an energetically favorable state as the crystal grows [17]. At the top of each ingot, the average grain sizes are very similar regardless of the growth method due to the same reason [19]. Wafers toward the top of the ingot contain a higher density of dislocation clusters and a lower density of grain boundaries or larger grains [7].

Area fractions of grain orientations are much more homogeneously distributed in the bottom part of a HP ingot than in a conventional ingot because the surface area fraction of the {112} orientation is quite large in this part of the conventional ingot [17]. However, the {112} grains are also dominant in the lower part of the HP ingot because {112} grains have the smallest surface energy next to {111} grains [4,28]. Fig. 5.5 shows the area fractions of several grain orientations as a function of the total ingot height of 710 mm-high G1 ingots for the conventional type (a) and the HP type (b) [17]. These {112} and {111} grains are still dominant in the middle and top parts of both types of ingot as shown in Fig. 5.5 [17]. Fig. 1.6 shows the {111} and {112} grains are dominant in the upper part of the conventional ingot. Especially, even though the area fraction of the {111} orientation do not largely vary in the middle and top parts of the conventional ingot, the area fraction of the {111} orientation in the HP ingot gradually increases even in the middle and top parts of the ingot with the ingot height due to reduce the surface energy during the growth [17]. As shown in

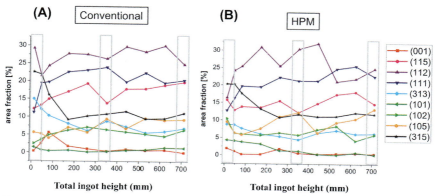

Fig. 5.5 Area fractions of several grain orientations as a function of the total ingot height of 710 mm-high G1 ingots for the conventional type (A) and the HP type (B). *(Referred from Fig. 7 in M. Trempa et al. J. Cryst. Growth 459 (2017) 67.)*

Fig. 5.5, the area fractions of all grain orientation are quite similar in both types of ingot in the middle and top parts of these ingots because the crystal growth is a thermodynamic process to go toward its stable state.

5.4.2 Generation and annihilation of grain boundaries

Grain boundaries interact with each other during ingot growth by the generation and annihilation processes and the number of each type of grain boundary changes with the ingot height. The generation processes increase the total number of grain boundaries and the annihilation processes decrease it. The processes of the annihilation and generation interactions of grain boundaries are shown in the upper right of Fig. 5.4. Both the dominant annihilation and generation interactions involves $\Sigma 3$ grain boundaries because of the significant lower energy of $\Sigma 3$ grain boundaries compared with those of other grain boundaries. Generally, $\Sigma 3$ grain boundaries increase and random grain boundaries decrease with the ingot height because $\Sigma 3$ grain boundary is more energetically favorable than random grain boundary [47], but their fractions basically depend on the initial amounts of these grain boundaries.

In an ingot grown by the conventional cast method, the fractions of $\Sigma 3$ grain boundary and random grain boundary were about 50% and 20% in the lower part of the ingot, respectively, as shown in Fig. 5.6A [17]. In a typical Si multicrystalline ingot, about 50 % of total grain boundaries are $\Sigma 3$ grain boundaries as shown in Fig. 1.6. In an ingot grown by the HP

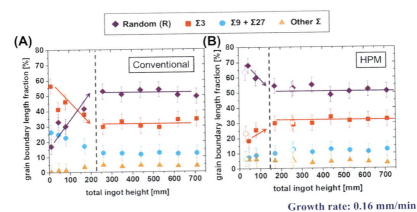

Fig. 5.6 Fractions of grain boundaries in ingots grown by the conventional and HP cast methods as a function of the ingot height. *(Referred from Fig. 8 in M. Trempa et al. J. Cryst. Growth 459 (2017) 67.)*

cast method using Si particles [4,17], as shown in Fig. 5.6B, the fractions of $\Sigma3$ grain boundary and random grain boundary were about 15% and 70% in the lower part of the ingot, respectively. However, at the middle and upper parts of the HP ingot, the fractions were kept constant as 30% and 50% for $\Sigma3$ grain boundary and random grain boundary, respectively. These fractions are quite similar to those at the middle and upper parts of the conventional cast ingot as shown in Fig. 5.6A [17—19]. Prakash et al. [41] studied grain boundary interactions in a Si multicrystalline ingot grown from small seed particles. The total number of the generation and annihilation interactions is very high at the initial stage of the growth, but it decreases with the ingot height. The fraction of annihilation increases and that of generation decreases with the ingot height. The overall fraction of annihilation is higher than that of generation. Thus, the total number of grain boundaries decreases with the ingot height even though their initial number is very high. The total number of grain boundaries is inversely proportional to the increase in grain size.

From the 3-D visualization of reconstructed ingot image for several sections by processing PL images as shown in Fig. 5.7, the generation and annihilation processes of dislocation clusters can be visualized by tracking of dislocation clusters in a HP ingot [40]. In this 3-D image, each dislocation cluster expand rapidly after generation and then shrink rapidly after reaching the maximum point, thus the annihilation processes strongly act to reduce the dislocation density in the HP ingot [40] (see Section 5.3.2).

The generation and annihilation processes of several types of grain boundary are useful to understand the fraction variation of each type of grain boundary during growth [15,17,18,41]. At the initial stage of the growth of HP ingot, the number of random grain boundaries is higher than that of $\Sigma3$ grain boundaries as shown in Fig. 5.6 [17]. From Prakash et al. [41], in the generation processes, the dominant interactions are $R = \Sigma3 + R$, $GB = GB + \Sigma3$ (GB: larger Σ grain boundary than 27 except for $\Sigma3$, $\Sigma9$ and $\Sigma27$) and $R = R + R$, and finally $\Sigma3$ grain boundaries increase. In the annihilation processes, the dominant interactions are $\Sigma3 - R = R$, $GB + \Sigma3 = GB$ and $R + R = R$, and finally $\Sigma3$ and random grain boundaries decrease. Especially, at the initial stage of the growth for HP ingot with lower $\Sigma3$ and higher R than those in a conventional cast ingot, $GB = GB + \Sigma3$ and $R + R = R$ are the dominant interactions, so $\Sigma3$ grain boundaries increase and random grain boundaries decrease in the HP ingot. At the final stage for the HP ingot, $R = \Sigma3 + R$ and $\Sigma3 + R = R$ are the dominant interactions, so both

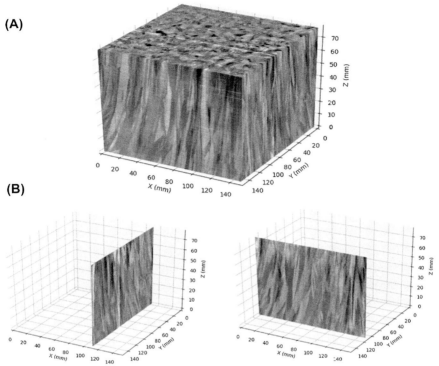

Fig. 5.7 3-D visualization of reconstructed HP ingot image (A) with several sections (B) by processing PL images. The generation and annihilation processes of dislocation clusters can be visualized. *(Referred from Fig. 3 in Y. Hayama et al. Sol. Energy Mater. Sol. Cells 189 (2019) 239.)*

fractions of Σ3 and random grain boundaries are almost constant. Form Trempa et al. [17], at the initial stage of the HP growth, $R = \Sigma 3 + R$, $R + R = R$ and $\Sigma 3 + R = R$ are the dominant interactions, so Σ3 grain boundaries are increase and random grain boundaries decrease in the HP ingot with lower initial Σ3. At the final stage of the conventional and HP growth, $R = \Sigma 3 + R$ and $\Sigma 3 + R = R$ are almost balanced, so both fractions of Σ3 and random grain boundaries are constant for both types of ingot as shown in Fig. 5.6. These fractions or the dominant interactions among them strongly depend on the initial relative amount of each type of grain boundary. That is to say, the frequency of each interaction depends on the number of available grain boundaries with each type. This point is very different between the conventional and HP cast ingots. At the initial stage of the growth of conventional ingot, the number of Σ3 grain boundaries is higher than that in the HP ingot. Therefore, $\Sigma 3 = \Sigma 3 + \Sigma 9$, $\Sigma 3 + \Sigma 3 = \Sigma 9$, and $\Sigma 3 + R = R$ are the dominant interactions [17], $\Sigma 3$ grain boundaries decrease and random grain boundaries increase in the conventional ingot as shown in Fig. 5.6 (a). These generation and annihilation processes based on Prakash et al. [41] and Trempa et al. [17] are summarized in Fig. 5.8.

The mechanism of the dominant generation process of $R = \Sigma 3 + R$ is explained as shown in Fig. 1.3 (see Section 1.2.2). At the growing interface, twin nucleation or formation of $\Sigma 3$ grain boundaries easily occurs in many grooves formed by random grain boundaries during ingot growth. Especially, at the initial stage of the growth of a HP ingot, there are many random grain boundaries in the ingot grown using Si or silica particles because of a large number of small grains [28]. Random grain boundaries tend to rapidly decrease in the growth direction in the lower part of the ingot because the mean grain size rapidly increases in this part. Random grain boundaries tend to remain almost constant in the middle and top parts of the ingot because the mean grain size is almost constant regardless of the ingot height in these parts. As the growth rate is higher, the fraction of random grain boundaries decreases faster because newly formed $\Sigma 3$ twin grain boundaries increase [10,16].

For the HP growth
Dominant generation (G) interactions
$R = \Sigma 3 + R$
$GB = GB + \Sigma 3$
$R = R + R$
$\Sigma 3$ increases.
Dominant annihilation (A) interactions
$\Sigma 3 + R = R$
$GB + \Sigma 3 = GB$
$R + R = R$
$\Sigma 3$ and R decreases.

Dominant factor for final fractions depends on the initial relative amount of each type of grain boundary.

At the initial stage of the HP growth
Dominant interactions
(G) $GB = GB + \Sigma 3$ or $R = \Sigma 3 + R$
(A) $R + R = R$
HP R >> Conv. R : $\Sigma 3$ increases and R decreases.

At the initial stage of the conv. growth
Dominant interactions
(G) $\Sigma 3 = \Sigma 3 + \Sigma 9$
(A) $\Sigma 3 + \Sigma 3 = \Sigma 9$, $\Sigma 3 + R = R$
Conv. $\Sigma 3$ >> HP $\Sigma 3$:
$\Sigma 3$ decreases and R increases.

At the final stage of the HP and conv. growth
Dominant interactions
(G) $R = \Sigma 3 + R$
(A) $\Sigma 3 + R = R$
$\Sigma 3$ and R are constant.

Fig. 5.8 Summarized generation and annihilation processes for the HP and conventional cast growth. *(Which is based on based on R.R. Prakash, K. Jiptner, J. Chen, Y. Miyamura, H. Harada, T. Sekiguchi, Appl. Phys. Exp. 8 (2015) 035502 and M. Trempa, I. Kupka, C. Kranert, T. Lehmann, C. Reimann, J. Friedrich, J. Cryst. Growth 459 (2017) 67.)*

Tang et al. [5] reported the slightly different results about distribution of grain boundaries in S2 ingots which is the name of high-quality ingots prepared in GCL. Their ingots also contain many small grains with uniform grain size and low density of dislocations. In the ingots, 58% of total grain boundaries are $\Sigma 3$ grain boundaries and the most of grain boundaries are high angle grain boundaries with a larger misorientation than $11°$ or large angle grain boundaries defined in Fig. 1.7. The fraction of $\Sigma 3$ grain boundaries is same as that of the conventional cast ingots. For the coherent grain boundaries, there is a trade-off between reduction of stress and reduction of dislocations or random grain boundaries in ingots. In the case of the S2 ingots, the coherent grain boundaries have few dislocations, and random grain boundaries with high angle can generate dislocations owing to reduce stress in Si multicrystalline ingots with high stress as shown in Figs. 4.14 and 4.15. The low growth rate (~ 0.1 mm min^{-1}) has no measurable influence on the fraction of random grain boundary [48]. As the growth rate increases, the fraction of random grain boundary becomes lower and that of $\Sigma 3$ grain boundary become higher [49]. However, the influence of the growth rate on the fraction of random grain boundary is quite small even though the growth rate is higher (near 0.5 mm min^{-1}) [48]. The density of random grain boundaries is an important factor to obtain high-quality multicrystalline ingots using the HP cast method. However, it is not possible to significantly increase the fraction of random grain boundary or reduce the mean grain size at 5 mm ingot height and above, which is known using different sizes of particle (0.3 and 3.8 μm), rod and chunk seeds [48].

5.4.3 Precipitations

Fe is present in the interstitial state (as Fe_i) of Si crystal or as metal silicide nano-precipitates such as $FeSi_2$. Autruffe et al. [50] studied Fe precipitation in a HP ingot in detail. The Fe_i concentration decreases close to an active grain boundary identified as random grain boundary. Fe is precipitated at the grain boundary during ingot cooling. Thus, regions with lower Fe_i concentration are formed close to the active grain boundary, which are referred as depleted regions or denuded zones. The onset of Fe precipitation requires supersaturation. The Fe supersaturation ratio estimated from the depleted region width is much higher than that for the standard multi-crystalline ingot. On the other hand, the density of precipitation sites in the HP ingot is lower than that in conventional ingot, due to the relatively low dislocation density. However, most of the Fe precipitation occurs in the intra-granular regions, probably on dislocations or dislocation clusters, and the contribution of grain boundary precipitation is low even though grain sizes are smaller in the HP ingot [51]. Existence of Fe precipitation at extended defects strongly affects the P gettering efficacy [52]. Nanoscale metallic precipitates containing Fe exist at recombination active dislocations. After P gettering, a higher Fe_i concentration is observed at dense dislocation clusters and grain boundaries, due to dissolving metal precipitates decorating defects [38,53]. Metallic precipitates containing Cu strongly affects the quality of Si crystals. All Fe and Cu precipitates seen in Fe + Cu contained crystals are co-located and appear generally smaller than those seen in Fe-only contained crystals [52]. The most harmful dislocations are connected to precipitates with ppm-range Cu and are not affected even by hydrogenation [51] (see Section 1.4.3). On the other hand, Si_3N_4 and SiC precipitations are more easily formed from seed particles for the HP cast method because they have more grain boundaries [51].

5.5 Electrical properties and solar cells

5.5.1 Electrical properties

The intra-grain lifetime is higher in the middle of a HP ingot and decreases gradually toward the top and bottom of the ingot because of the higher impurity concentration near the top and bottom parts [7]. The recombination velocity of a HP ingot is about 200 cm s^{-1} and that of a conventional ingot is about 1000 cm s^{-1} [8]. The area of recombination active defects or dislocations is small and uniform in the lower part of an ingot, but it becomes larger and almost kept constant in the middle and upper parts of the ingot. This area becomes lower than or almost same as that in the middle and upper parts of a HP ingot grown by the conventional cast method [17,42]. The recombination active defects exist higher within {111} oriented grains than within other oriented grains for both ingots [42].

Adamczyk et al. [54] studied the recombination activity of grain boundaries in a HP ingot. At the bottom of the ingot, the most of recombination active grain boundaries are the random and small-angle grain boundaries and the most of recombination less-active grain boundaries are the $\Sigma 3$ grain boundaries even though some $\Sigma 3$ grain boundaries are recombination active. The random and small-angle grain boundaries become almost recombination less-active after P gettering and hydrogenation. In the middle part of the ingot, the quality of the ingot known by the internal quantum efficiency (IQE) map becomes much better because of a lower intragrain impurity content. After the P gettering and hydrogenation, all small-angle grain boundaries remain recombination active, but only 6% and 2% of total segments of the random and $\Sigma 3$ grain boundaries remain recombination active, respectively. The recombination active segments are incoherent and induce stress in the lattice, especially the active segments of $\Sigma 3$ grain boundaries after the P gettering contain impurities such as Fe, Ni and Cu (see Section 2.2.3).

A much larger lifetime improvement is observed after P gettering of a HP ingot than a muticrystalline cast ingot because of the smaller area fraction of high dislocation density in the HP ingot [38]. Especially, the lifetime in the middle part of the HP ingot is largely improved and exceeded the ungettered lifetime by both P gettering and subsequent hydrogenation [53]. Sio et al. [7,55] compared the electric properties of p-type wafers grown by the cast and HP cast methods although a conventional cast ingot with high purity could be no longer obtained. The recombination activity of grain boundaries (recombination velocity: about 200 cm s^{-1}) of HP cast wafers is lower than that (recombination velocity: about 1000 cm s^{-1}) of conventional cast wafers [55]. Thus, as-grown grain boundaries and dislocations in the HP cast ingots tend not to be recombination active, but they become active after P gettering. Therefore, the average minority carrier lifetime for an as-grown intra-grain of a wafer grown by the cast method is considerably lower than that for a wafer grown by the HP cast method, but the difference is diminished after the P gettering. The reduction of the average lifetime after the P gettering is due to the activation of grain boundaries during the process. The high diffusion length or high minority carrier lifetime can be obtained for the intra-grain regions of wafers grown by the HP cast method after both P gettering and hydrogenation (see Section 2.4.1). The P gettering and hydrogenation can reduce the defect density and surface recombination velocities to levels lower than those in the as-grown state. Especially, a hydrogenation step can significantly improve the overall lifetime due to its ability to passivate grain boundaries. Thus, the high grain boundary density in a HP ingot has a possibility to be neutralized for solar cells by the efficacy of hydrogenation. The dislocation density in the middle part of a HP ingot is not still high, comparing to an ingot grown by the cast method, and dislocations in wafers cut from the middle part of the HP ingot are not very recombination active in the as-grown state (see Section 2.2.4). However, dislocation clusters in HP wafers are very recombination active even in the as-grown state, especially when they are located at the top and bottom of the ingot. The carrier recombination at such dislocation clusters is more detrimental to the overall cell performance than carrier recombination at grain boundaries and in intra-grain regions. These wafers with dislocation clusters cut from the bottom of the HP ingot are more negatively affected by the lower intra-grain lifetime due to impurities and the great density of recombination active grain boundaries.

Phang et al. [1] studied the electric properties of an n-type ingot grown by the HP cast method. In this case, B diffusion is used for preparation of a surface p-layer to make a p-n junction. In wafers with a high-density of grain boundaries from the bottom of the ingot, grain boundaries are already recombination active in as-grown state. After the B diffusion, the minority carrier lifetime significantly degrades even for wafers from the middle and top parts of the ingot. This degradation is occurred both at grain boundaries and in intra-grain regions. The subsequent P diffusion (825°C for 40 min) provides effective gettering and improves the intra-grain lifetime. Subsequent hydrogenation (rapid thermal annealing at 500°C for 10 s) can passivate the grain boundaries and significantly improve the lifetime, particularly in wafers with small grain size and high grain boundary density. The hydrogenation is not effective on dislocation clusters, and dislocations remain still recombination active even after the hydrogenation. The grain boundaries remain still recombination active only after the P diffusion, but they become recombination non-active after the hydrogenation due to passivation of the grain boundaries same as the p-type wafers. Thus, the average minority carrier lifetime is largely improved both at grain boundaries and in intra-grain regions from the bottom to the top after the B and P diffusions and hydrogenation. However, comparing to the lifetime of wafers using only P diffusions and hydrogenation, B diffusion has a large negative impact on the final lifetime of the n-type wafers even after subsequent P gettering and hydrogenation because of its higher diffusion temperatures (950 °C for 70 min) compared to P diffusion. The reported average and best lifetimes of n-type HP wafers are 1.8 and 2.0 ms after P gettering, respectively [1,8]. The reported average and best lifetimes of p-type HP wafers are 1.2 and 1.4 ms after P gettering and hydrogenation, respectively [7].

5.5.2 Solar cells

The initial grain structure with the low area fraction of recombination active defects such as dislocations strongly influences the solar cell efficiency. For the p-type Al-BSF solar cells, the average conversion efficiency prepared

High performance (HP) cast method   219

from ingots grown by the HP cast method is 17.4% and it becomes higher as 17.8% using the high-purity coating material and crucible [4]. The recent maximum and average solar cell efficiencies of a HP ingot are improved to 18.8% and 18.2%−18.5%, respectively, and the scattering width is 1.0% when it is 1.6% and 2.4% for the cast and mono-like solar cells [8,33,42]. These wafers are cut from the whole ingots except regions near the crucible wall and top surface. With an advanced cell structure such as the PERC structure, an average efficiency becomes higher to 19.6%−21.63% [3]. The reactive ion etching is used for texturing for these cells. The p-type champion cell efficiency is higher than 21.23% reported by Trina Solar Inc. and the n-type champion cell efficiency wafer is 21.9% reported by Fraunhofer ISE [3].

If the concentration of Fe$_i$ and Cu impurities remain above 10^{10} and 10^{13} cm^{-3}, respectively, HP multicrystalline Si wafers will limit the conversion efficiency above about 22.5% even using the PERC cell structure when dislocations are not reduced further, because the high lifetime in the wafers are still reduced by Fe$_i$ compared to CZ single wafers and the low-lifetime areas are limited by precipitates, most likely Cu [56]. Yang et al. [4] suggested that the improvement of the conversion efficiency of HP solar cells was due to the reduction of dislocation density in ingots with smaller grains through nucleation and grain control. The solar cell performance of an ingot grown using the non-Si nucleation agent is usually inferior to that using the Si particles because of the low fraction of random grain boundaries in the case of the non-Si agent [8]. The fraction of random grain boundaries is very important to determine the quality and defect structure of the ingot because the random grain boundaries interfere the movement of dislocations and dislocation clusters and reduce the dislocation density [2,15]. The fraction of random grain boundaries should be kept about 50%−60% for the HP cast ingots. In future, the reduction of metal impurities and dislocations is the most important subject to largely improve the conversion efficiency of HP solar cells.

5.6 Key points for improvement

For the HP cast method, many small grains and random grain boundaries are intentionally introduced in an ingot. Movement of dislocations is prevented by the many random grain boundaries. Thus, propagation of dislocations much delay and many dislocation clusters appear just in the top part of the ingot. The many random grain boundaries can be neutralized by the hydrogenation. These mechanisms can simply explain the efficacy of the HP cast method for solar cells. However, there are still some problems for this method.

The crystal structure is generally determined by minimization of the surface energy of grains and the grain boundaries energy. The initial crystallographic structure of an ingot grown by the HP cast method is not energetically stable because the structure varies to approach to the stable state as the ingot height increases. In the middle and upper parts of the ingot, the structure becomes quite similar to that of an ingot grown by the conventional cast method, because the crystal growth is an essential process to go toward an energetically favorable state [17]. Even in a G6 ingot with the height of 35 cm, the defect density becomes high near the top of the ingot because of reduction of random grain boundaries [8]. When the ingot grows taller, dislocation clusters even in the HP multi-crystalline Si ingot will become more problematic [3]. The dislocation clusters originated from precipitates and particles appear in the middle part of the HP ingot and spread in cascade toward the top of the ingot. A novel technology should be developed to keep the unstable initial state even in the middle and upper parts of the ingot.

There are still several other problems for the HP cast method. (1) The initial grain structure is undefined yet and the area fraction of recombination active defects is less controllable. The growth mechanism to determine the initial grain structure should be made clearer. The proper grain size and amount of grain boundaries also should be clear because grains with as small as the wafer thickness and a large amount of defects are expected to be harmful for the solar cell performance [33]. (2) To reduce the red zone near the bottom of an ingot, the purity of crucibles and coating materials such as Si_3N_4 and nucleation agent should be improved and the thickness of the seed layer should be decreased [3,12]. (3) A certain fraction of grain boundaries remains as recombination active ones even after P gettering and hydrogenation [7]. Dissolved impurities in ingots lead the minority carrier lifetime to be lower. The impurity control should be improved for the HP growth. (4) The Si multicrystalline wafers grown by the HP cast method have a larger share than single wafers in the present market of solar cells because of their high productivity. But they lose 1 % efficiency because of their inefficient texture structure. The low–cost effective texturing technologies should be developed. (5) The low–cost diamond–wire cutting is not effective for HP wafers because the wafer surface is too smooth to have good acid texturing in the current solar cell processing [57]. (6) The columnar grain growth is not yet achieved in the HP ingot [19]. (7) The remained stress in the ingot will be a final problem. These problems should be solved for widely expanding the HP method.

In future, a Si ingot grown by the HP method will be much more effective to obtain a large-scale Si multicrystalline ingot with uniform smaller grains by controlling the material, structure and arrangement of seed particles and keeping the grain size smaller all over the ingot. Although the grain size is smaller and the density of random grain boundary is higher comparing to the conventional cast ingot, the density of dislocation is still much higher and the solar cell performance is less than those of the CZ solar cells. As this problem will be also solved by controlling the seed particles and keeping the grain size smaller, the research about the dependence of the seed particles on the ingot structure and the annihilation mechanism of small grains during growth should be quite important for the future HP cast method.

References

[1] S.P. Phang, H.C. Sio, C.F. Yang, C.W. Lan, Y.M. Yang, A.W.H. Yu, B.S.L. Hsu, C.W.C. Hsu, D. Macdonald, Jpn. J. Appl. Phys. 56 (2017) 08MB10.

[2] G. Stokkan, Y. Hu, Ø. Mjøs, M. Juel, Sol. Energy Mater. Sol. Cells 130 (2014) 679.

[3] C.W. Lan, Growth of high-performance multi-crystalline silicon, in: D. Yang (Ed.), Chapter 10, Section Two Crystalline Silicon Growth, Hand Book of "Photovoltaic Silicon Material", Springer, Berlin Heidelberg, 2018, pp. 1–17 (On line).

[4] Y.M. Yang, A. Yu, B. Hsu, W.C. Hsu, A. Yang, C.W. Lan, Prog. Photovoltaics Res. Appl. 23 (2015) 340.

[5] X. Tang, L. Francis, L. Gong, F. Wang, J.-P. Raskin, D. Flandre, S. Zhang, D. You, L. Wu, B. Dai, J. Cryst. Growth 117 (2013) 225.

[6] C.W. Lan, C.F. Yang, A. Lan, M. Yang, A. Yu, H.P. Hsu, B. Hsu, C. Hsu, CrystEngComm 18 (2016) 1474.

[7] H.C. Sio, S.P. Phang, P. Zheng, Q. Wang, W. Chen, H. Jin, D. Macdonald, Jpn. J. Appl. Phys. 56 (2017) 08MB16.

[8] C.W. Lan, A. Lan, C.F. Yang, H.P. Hsu, M. Yang, A. Yu, B. Hsu, W.C. Hsu, A. Yang, J. Cryst. Growth 468 (2017) 17.

[9] Y. Azuma, N. Usami, K. Fujiwara, T. Ujihrara, K. Nakajima, J. Cryst. Growth 276 (2005) 393.

[10] Y.T. Wong, C. Hsu, C.W. Lan, J. Cryst. Growth 387 (2014) 10.

[11] D. Zhu, L. Ming, M. Huang, Z. Zhang, X. Huang, J. Cryst. Growth 386 (2014) 52.

[12] G. Zhong, Q. Yu, X. Huang, L. Liu, J. Cryst. Growth 402 (2014) 65.

[13] K.E. Ekström, G. Stokkan, A. Autruffe, R. Søndenå, H. Dalaker, L. Arnberg, M.D. Sabatino, J. Cryst. Growth 441 (2016) 95.

[14] G.W. Alam, E. Pihan, B. Marie, N. Mangelinck-Nöel, Phys. Status Solide C 14 (2017) 1700177.

[15] C. Reimann, M. Trempa, T. Lehmann, K. Rosshirt, J. Stenzenberger, J. Friedrich, J. Cryst. Growth 434 (2016) 88.

[16] M. Trempa, I. Kupka, C. Kranert, C. Reimann, J. Friedrich, Photo Interpret. 35 (2017) 36.

[17] M. Trempa, I. Kupka, C. Kranert, T. Lehmann, C. Reimann, J. Friedrich, J. Cryst. Growth 459 (2017) 67.

[18] R.R. Prakash, T. Sekiguchi, K. Jiptner, Y. Miyamura, J. Chen, H. Harada, K. Kalimoto, J. Cryst. Growth 401 (2014) 717.

[19] T. Strauch, M. Demant, P. Krenckel, R. Riepe, S. Rein, J. Cryst. Growth 454 (2016) 147.

[20] C. Huang, H. Zhang, S. Yuan, Y. Wu, X. Zhang, D. You, L. Wang, X. Yu, Y. Wan, D. Yang, Sol. Energy Mater. Sol. Cells 179 (2018) 312.

[21] S. Yuan, D. Hu, X. Yu, L. He, Q. Lei, H. Chen, X. Zhang, Y. Xu, D. Yang, Sol. Energy Mater. Sol. Cells 174 (2018) 202.

[22] T. Muramatsu, Y. Hayama, K. Kutsukake, K. Maeda, T. Matsumoto, H. Kudo, K. Fujiwara, N. Usami, J. Cryst. Growth 499 (2018) 62.

[23] G. Anandha babu, I. Takahashi, S. Matsushima, N. Usami, J. Cryst. Growth 441 (2016) 124.

[24] G. Anandha babu, I. Takahashi, S. Matsushima, N. Usami, J. Cryst. Growth 468 (2017) 620.

[25] T. Muramatsu, I. Takahashi, G. Anandha babu, N. Usami, Jpn. J. Appl. Phys. 56 (2017) 075502.

[26] I. Kupka, T. Lehmann, M. Trempa, C. Kranert, C. Reimann, J. Friedrich, J. Cryst. Growth 465 (2017) 18.

[27] H. Zhang, D. You, C. Huang, Y. Wu, Y. Xu, P. Wu, J. Cryst. Growth 435 (2016) 91.

[28] Y.T. Wong, C.T. Hsieh, A. Lan, C. Hsu, C.W. Lan, J. Cryst. Growth 404 (2014) 59.

[29] Z. Wu, G. Zhong, X. Zhou, Z. Zhang, Z. Wang, W. Chen, X. Huang, J. Cryst. Growth 441 (2016) 58.

[30] J. Ding, Y. Yu, W. Chen, X. Zhou, Z. Wu, G. Zhong, X. Huang, J. Cryst. Growth 211 (2000) 13.

[31] J. Ding, Y. Yu, W. Chen, X. Zhou, Z. Wu, G. Zhong, X. Huang, J. Cryst. Growth 454 (2016) 186.

[32] I. Brynjulfsen, L. Arnberg, J. Cryst. Growth 331 (2011) 64.

[33] X. Zhang, L. Gong, B. Wu, M. Zhou, B. Dai, Sol. Energy Mater. Sol. Cells 139 (2015) 27.

[34] K. Kutsukake, T. Abe, N. Usami, K. Fujiwara, I. Yonenaga, K.M.K. Nakajima, J. Appl. Phys. 110 (2011) 083530.

[35] K. Nakajima, R. Murai, K. Morishita, K. Kutsukake, N. Usami, J. Cryst. Growth 344 (2012) 6.

[36] I. Takahashi, N. Usami, K. Kutsukake, G. Stokkan, K. Morishita, K. Nakajima, J. Cryst. Growth 312 (2010) 897.

[37] K. Nakajima, K. Kutsukake, K. Fujiwara, N. Usami, S. Ono, I. Yamasaki, in: Proceedings of the 35th IEEE Photovoltaic Specialists Conference, 2010, p. 817.

[38] S. Castellanos, K.E. Ekstrøm, A. Autruffe, M.A. Jensen, A.E. Morishige, J. Hofstetter, P. Yen, B. Lai, G. Stokkan, C. del Cañizo, T. Buonassis, IEEE J. Photovol. 6 (2016) 632.

[39] C.C. Hsieh, Y.C. Wu, A. Lan, H.P. Hsu, C. Hsu, C.W. Lan, J. Cryst. Growth 419 (2015) 1.

[40] Y. Hayama, T. Matsumoto, T. Muramatsu, K. Kutsukake, H. Kudo, N. Usami, Sol. Energy Mater. Sol. Cells 189 (2019) 239.

[41] R.R. Prakash, K. Jiptner, J. Chen, Y. Miyamura, H. Harada, T. Sekiguchi, Appl. Phys. Exp. 8 (2015) 035502.

[42] T. Lehmann, C. Reimann, E. Meissner, J. Friedrich, Acta Mater. 106 (2016) 98.

[43] W. Chen, Q. Wang, D. Yang, L.D. Li, X.G. Yu, L. Wang, H. Jin, J. Cryst. Growth 467 (2017) 65.

[44] X. Yang, W. Ma, G. Lv, K. Wei, C. Zhang, S. Li, D. Chen, Metal. Mater. Trans. 2E (2015) 39.

[45] V. Pupăzan, A. Popescu, O.M. Bunoiu, D. Vizmam, Proc. Phys. Conf. TIM-11, AIP Conf. Proc. 1472 (2012) 210.

[46] X. Luo, R.R. Prakash, J. Chen, K. Jiptner, T. Sekiguchi, Superlattice. Microst. 99 (2016) 136.

[47] M. Kohyama, R. Yamamoto, M. Doyama, Phys. Status Solidi 138 (1986) 387.

[48] M. Trempa, C. Kranert, I. Kupka, C. Reimann, J. Friedrich, J. Cryst. Growth 514 (2019) 114.

[49] H.K. Lin, M.C. Wu, C.C. Chen, C.W. Lan, J. Cryst. Growth 439 (2016) 40.

[50] A. Autruffe, M. M'hamdi, F. Schindler, F.D. Heinz, K.E. Ekstrøm, M.C. Schubert, M.D. Sabatino, G. Stokkan, J. Appl. Phys. 122 (2017) 135103.

[51] G. Stokkan, M.D. Sabatino, R. Søndenå, M. Juel, A. Autruffe, K. Adamczyk, H.V. Skarstad, K.E. Ekstrøm, M.S. Wiig, C.C. You, H. Haug, M. M'hamdi, Phys. Status Solidi. 214 (2017) 1700319.

[52] D.P. Fenning, A.S. Zuschlag, M.I. Bertoni, B. Lai, G. Hahn, T. Buonassis, J. Appl. Phys. 113 (2013) 214504.

[53] K. Adamczyk, R. Søndenå, M. M'hamdi, A. Autruffe, G. Stokkan, M.D. Sabatino, Phys. Status Solidi C 13 (2016) 812.

[54] K. Adamczyk, R. Søndenå, G. Stokkan, E. Looney, M. Jensen, B. Lai, M. Rinio, M.D. Sabatino, J. Appl. Phys. 123 (2018) 055705.

[55] H.C. Sio, D. Macdonald, Sol. Energy Mater. Sol. Cells 144 (2016) 339.

[56] P.P. Altermatt, Z. Xiong, Q. He, W.W. Deng, F. Ye, Y. Yang, Y. Chen, Z.Q. Feng, P.J. Verlinden, A. Liu, D.H. Macdonald, T. Luka, D. Lausch, M. Turek, C. Hagendorf, H. Wagner-Mohnsen, J. Schön, W. Kwapil, F. Frühauf, O. Breitenstein, E.E. Looney, T. Buonassisi, D.B. Needleman, C.M. Jackson, A.R. Arehart, S.A. Ringel, K.R. McIntosh, M.D. Abbott, B.A. Sudbury, A. Zuschlag, C. Winter, D. Skorka, G. Hahn, D. Chung, B. Mitchell, P. Geelan-Small, T. Trupke, Sol. Energy 175 (2018) 68.

[57] Y.C. Wu, A. Lan, C.F. Yang, C.W. Hsu, C.M. Lu, A. Yang, C.W. Lan, Cryst. Growth Des. 16 (2016) 6641.

CHAPTER 6

Mono-like cast method

6.1 Concept and feature of the mono-like cast method

For the structure control of a Si multicrystalline ingot, the grain size is one of the important factors to reduce dislocations and increase uniformity in the ingot. As shown in Fig. 4.24, there is a trade-off between reduction of stress and enlargement of grain size. An ingot with a large grain has large stress and few random grain boundaries. In this case, the stress is mainly relaxed by introduction of dislocations in the large grain or introduction of many small grains near crucible wall. Many random grain boundaries cannot be introduced in the ingot with the large grain or the large grain cannot be rearranged to reduce the stress during crystal growth. For the growth of a Si ingot with high-quality large grains, an $<100>$ seed crystal (5 cm $\times$ 5 cm) was tried to be used on the bottom (5 cm $\times$ 5 cm) of crucible by the Bridgman method [1]. The ingot had an axially columnar grain structure with mainly $<110>$ orientation. To largely reduce the density of random grain boundaries and increase the performance of solar cells using a single surface orientation for efficient texture structure, the mono–like cast method was proposed [2–4]. The concept of this method is to intentionally grow a Si ingot with a quite large single grain in the center of the ingot or wholly grow a Si single ingot using one or several seed crystals arranged on the bottom of the crucible [2,3]. Another merit of this method is that the mono–like ingot can be cut using the low–cost and damage-free multiwire sawing with fixed diamond abrasive, which is still difficult to be used for the conventional and HP cast ingots. This method has some demerit such as metal contamination from the seed plates, many small grains grown from the crucible wall and dislocation generation from the seed junctions. The metal contamination should be prevented by increasing the purity of the seed crystals (see Section 6.5.1). The highly contaminated region in the ingot bottom closed to the seed crystals is usually called as "red zone". The small grains are prevented mainly using several types of large angle grain boundaries (see Section 6.3).

Crystal Growth of Si Ingots for Solar Cells Using Cast Furnaces
ISBN 978-0-12-819748-6
https://doi.org/10.1016/B978-0-12-819748-6.00006-2

Copyright © 2020 Elsevier Inc.
All rights reserved.

225

The dislocation generation are prevented by properly arrangement of seed orientations (see Section 6.4.4). These solutions are precisely explained in each below section.

Moreover, the feedstock set on the seed plates causes another peculiar problem of dislocation generation near the seed surface, Trempa et al. [5] studied the effect of the pressure from feedstock on the seed surface. The small pressure (some MPa) of Si feedstock particles applied on the surface of seed plates generates dislocations near the seed surface. The dislocations penetrate several hundred microns into the seed volume by high dislocation velocities up to 1 mm s^{-1}. The penetration depth increases with the surface pressure. The dislocations propagate into the Si ingot. The precise back-melting of the seed surface is required to remove the dislocations in the seed plates. The most promising solution is the usage of homogeneously arranged feedstock particles such as roundish fluidized bed reactor materials without sharp edges to reduce pressure.

The concept of the mono–like cast method is similar to the vertical Bridgmen method developed by P. W. Bridgman (1882–1961), in which a seed is set on the bottom of crucible and an ingot is unidirectionally grown from the seed [6]. The mono–like cast method basically uses the cast furnace and is called as the MONO^2, mono-like, mono-cast, seeded cast and quasi-single cast methods, which are also a seed-assisted cast method [3,6–13]. Several companies name their mono–like wafers such as S2 wafers from GCL, Maple wafers from JA Solar, U-grade wafers from SAS and Virtus wafers from Renasolar. Growth of ingots with such large grains was already demonstrated as 3-grain ingots by the CZ method, in which three {110} seeds were used [14]. The ingot is bounded by {111} planes and has twin grain boundaries. Azuma et al. [15] used Ge single seed plates with the <111> preferential orientation for directional solidification growth of the SiGe and Ge ingots and found that a large single grain with the preferred orientation of {110} was finally obtained without polycrystallization during growth. These pioneering results have given important implications for the mono–like cast method using Si single seed plates (see Section 1.3.1).

6.2 How to control to obtain a large single grain

6.2.1 Growth of mono-like ingots

In the mono-like cast method, one or multiple Si single-crystal seed plates prepared by the CZ method are paved on the bottom surface of a Si_3N_4

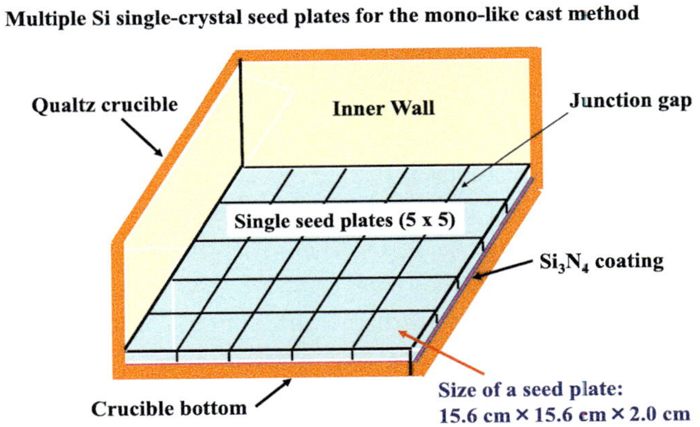

Multiple Si single-crystal seed plates for the mono-like cast method

Fig. 6.1 Multiple Si single-crystal seed plates (5 × 5) paved on the bottom surface of a Si_3N_4 coated crucible for the mono-like cast method.

coated crucible as shown in Fig. 6.1. Most of the seed plates are quadrangle, in some cases disk shaped plates are used to obtain mushroom-type interface [12,16]. The thickness of the Si crystal seed plates is usually 2 cm. The crucible is filled with Si feedstock on the seed plates. The furnace has usually two zone heaters which are the combination with top and bottom heaters or that with top and side heaters. Usually, to extract heat from the crucible bottom, some kinds of heat extraction methods are used for the growth. Jouini et al. use a heat extraction device which allows a flexible control of the heat extraction by an adjustable conductive contribution to the heat transfer using Ar gas layers on the water-cooled box [3]. Ar gas is flowed into the surface of the Si melt or ingot through a tube. After the upper part of the seed plates is melted, a mono-like ingot is grown from the surface of the remained seeds toward the top by directional solidification while controlling the temperature gradient in the furnace. A large vertical temperature gradient near the seed plates is beneficial for the seed preservation [17]. The temperature gradient at the growing interface is shown to be 0.174 K cm^{-1} on the solid side and 0.109 K cm^{-1} on the liquid side [2]. However, the large vertical temperature gradient has a possibility to increase the convexity of the seed–melt interface resulting in high thermal stress and dislocations [18]. The radial heat flux can be reduced by an insulated susceptor which decreases thermal energy from the heater and influences the horizontal temperature gradient in the corner of the seed, and such reduction of the radial heat flux is beneficial for the seed preservation [17]. Several kinds of insulation partition block are used to control the temperature distribution near the seed-melt or growing interfaces. The insulation partition block can be designed using the simulation of the temperature distribution in the furnace, and it is known that reducing the width or increasing the thickness of the partition block can reducing the convexity of the seed-melt interface to reduce thermal stress [19]. The temperature field is more uniform and the thermal stress is relatively lower in a growing Si ingot using a moving partition block than in an ingot using a fixed one, and the moving partition block is beneficial for the mono-like cast method to control the shape of the growing interface to more flat one or slightly convex one with low thermal stress [20] (see Section 3.1.1) Sometimes, a silica rod is used to dip into the melt to check the position of the growing interface during the melting process of the seed plates and crystal growth [21]. For the seeding process, the suitable time to spend in the melt is important to reduce the back diffusion of harmful metal impurities such as Fe [18].

6.2.2 Seed plates and gaps between seeds

This method uses several seed plates with a thickness of 1.5–2.0 cm set on the bottom of a crucible as shown in Fig. 6.1. Important points to obtain a high-quality ingot are the flatness of the seed support, the precision of seed placement, the low-impurity levels in crucible and Si_3N_4 coating and the low-stress level from heavy feedstock loads [22]. A junction gap always exists between adjacent seed plates as shown in Fig. 6.6. This process causes several serious problems for the quality of ingots. The first one is generation of dislocations in the initial seed plates. Even using dislocation-free seed plates, dislocations already exist near the surface of the seed plates during growth and annealing due to thermal shock and saw damage [23,24]. It can be observed by the Bragg diffraction images of them. The one more problem is generation of dislocations from junctions between adjacent seed plates. This generation depends on many factors such as the junction gap, orientation and arrangement of seed plates. To reduce such dislocations, the mechanism and technologies have been studied and developed for a long time.

At the initial stage of growth, junction gaps between seed plates are rapidly filled with a Si melt and single crystals grow on the seed side walls toward the gap center. At the bottom of the gaps, the temperature is 40 K below the melting point of Si and a rapid solidification can occur in the gaps [25]. As the seed plates, the <100> oriented seed is usually used for ease of alkaline texturing of solar cells. The <100> oriented seed is more favorable than the <111> oriented seed on the view point of twin formation [3]. For the <111> oriented seed, the lateral expansion of small grains from the crucible wall is partly suppressed comparing with the <100> oriented seed [26]. The seed plates have always a small misorientation between them which is a deviation angle of about $1°$. This small misfit ($<1°$) is enough to cause local stress which forms dislocations and small angle grain boundaries in the gap center [25].

Trempa et al. [25–27] studied the gaps between seeds in detail using the Laue scanner [28]. The gaps between seeds generate no visible distortion in an ingot even if the gap width is 1 cm. For the {100} and {111} oriented gap walls, regions between seed walls have a single structure without grain boundaries independently of the gap width. For the {110} oriented gap walls and 1 cm-wide gaps, regions between seed walls have a multicrystalline structure with a denser network of small angle grain boundaries. For wider gaps than 0.1 cm, especially for the <111> and <110> seed orientations, the quite large number of dislocation-loops and small angle grain boundaries are observed in crystals grown above the gaps and along the gap lines mainly due to the thermal shock at the side walls. The origin of these defects locates in the central region of each seed gap below the seed surface. From Ekstøm et al. [29], the area near seed junctions with low minority carrier lifetimes much more widely spreads as the junction gap increases from 0.05 to 0.91 cm. The large gaps considerably generate dislocations both above and below the seed interface because of the growth and defect-formation processes caused in the gap (see Section 6.4.1). In small gaps, the generation of dislocations mainly depends on the seed misorientation between adjacent seeds.

Growth for the {110} and {112} oriented gap walls is more sensitive to the slip systems than growth for the {100} and {111} oriented gap walls [25]. Especially, a multicrystalline structure often appears inside a large gap between as-cut {110} walls with small particles after sawing process. Dislocation formed at such side walls can move and multiply on the {111} glide plane and parallel to the <110> direction because the slip system {111} <110> is available perpendicular to the gap plane or parallel to the growth direction inside the gaps. The system {111} <110> means the dislocations move on the {111} plane and toward the <110> direction. For smaller gaps than 0.1 cm, there no growth dependence on the oriented gap walls. For wider gaps than 0.1 cm, multicrystals grow directly on the crucible bottom.

6.3 Growth and control of small grains appeared from crucible wall

The mono-like cast method is useful for reducing the number of grain boundaries by forming a quite large single-crystal-like grain or a mono-like crystal, but it has still a severe problem that polycrystalline growth from the crucible walls always occurs [3,7,8]. An ingot grows inside a quartz crucible while the ingot is contacting with the crucible wall. Many small grains with different orientations randomly grow from the crucible wall because the surface of the wall has many nucleation sites. Small grains nucleate most frequently around side corners of the ingot [13]. Such small grains expand toward one large grain or a mono-like area in the center of the crucible and continue to the top of the ingot during growth. Thus, the multicrystalline area with small grains appears along and around the inner surface of the crucible wall and encloses the mono-like area to the top. The small grain area or region becomes larger toward the center with the ingot height. Such a small grain area has inferior quality comparing with the mono-like area and decreases the yield of wafers cut from the ingot because of its un-uniformity. The small grain area has a much higher dislocation density than the center of the ingot [29]. Such parasitic small grains grown from the crucible wall tend to multiply twins which do not severely deteriorate the quality of the ingot. From results of Trempa et al. [26], for $<100>$ oriented seeds, the dominant orientation of the small grains is $<221>$ in the growth direction and the dominant twin is $\Sigma3$. The growth of the small grain region is dominated by a twinning formation mechanism on $\{111\}$ facet planes, which depends on the relative orientation of the $\{111\}$ planes to the growth direction or to the seed orientation. Jouini et al. reported the effect of the shape of solid–liquid growing interface on the multicrystallization [3]. The convex interface in the growth direction is better to prevent the extension of such small grains toward the mono-like area. According to the numerical simulation, the small grain area can be reduced by keeping the melt temperature near the crucible wall higher and reducing the growth rate along the crucible wall or the growth rate along the seed [30]. This concept is used for the growth of a mono-like ingot from a (100) disk-shaped single seed with 20 cm diameter paved on the center of the crucible bottom [31].

Several low Σ grain boundaries can be used to prevent propagation of grain boundaries or small grains grown from crucible wall. To obtain such low Σ grain boundaries, single crystal plates with selected orientations are arranged on the bottom of a crucible and used as seed crystals. Low Σ grain boundaries such as Σ3 and Σ9 were tried to be artificially manipulated by the tri-Si technology [32]. Due to the two Σ3 and one Σ9 grain boundaries, a structure stability appears to reduce cross slip in the grown ingot with {110} surface orientations [32]. Fig. 6.2A shows the normalized grain boundary energy as a function of the rotation angle or the misorientation (deg) for (100) Si twist grain boundaries with the <100> axis at 1473 K [33]. α means the dihedral angle (deg.) of a groove formed at each grain boundary. The grain boundary energy, is simply calculated by

$$\gamma_{gb} = 2\gamma_{sl}\cos(\alpha/2) \quad (6.1)$$

where γ_{sl} is the solid-liquid interface energy of the groove. The perfect Σ5 grain boundary has a second deepest energy cusp next to the Σ13 grain boundary. So, it causes no dislocation generation during growth to minimize its energy. Fig. 6.2B shows the grain boundary energy (J m^{-2}) as a function of the misorientation (deg) for Si [100] symmetric tilt grain boundaries with the <100> axis [34]. In Fig. 6.2B, the black dots show

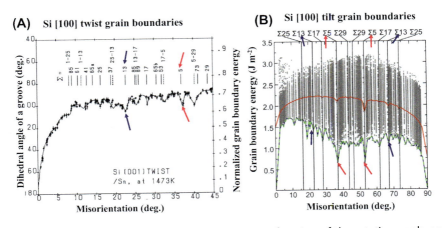

Fig. 6.2 (A) Normalized grain boundary energy as a function of the rotation angle or the misorientation (deg) for (100) Si twist grain boundaries with the <100> axis at 1473 K α means the dihedral angle (deg.) of a groove formed at each grain boundary. (B) Grain boundary energy (Jm^{-2}) as a function of the misorientation (deg) for Si [100] symmetric tilt grain boundaries with the <100> axis. The black dots show the grain boundary energy spectra of all the states for each misorientation and the green line shows the equilibrium ensemble-averaged grain boundary energy. The perfect Σ5 grain boundary has a deep energy cusp and it causes no dislocation generation during growth to minimize its energy. (A) Referred from Fig. 1 in A. Otsuki Interface Science **9** (2001) 293. (b). Referred from Fig. 4 in J. Han et al. Acta Mater, 104 (2016) 259.)

the grain boundary energy spectra of all the states for each misorientation, the green line shows the equilibrium ensemble-averaged grain boundary energy and the non-equilibrium ensemble-averaged grain boundary energy. The equilibrium ensemble-averaged grain boundary energy means the average of the lowest energy state and the non-equilibrium ensemble-averaged grain boundary energy means the average of the overall grain boundary energies of each grain boundary at each tilt angle.

Kutsukake et al. [11,13,35] demonstrated the effect of arrangement of {310}/{310} single crystal plates ({310} face-to-face contacted planes) on preventing the propagation of small grains grown from crucible wall as shown in Fig. 6.3. {310} Σ5 grain boundaries are effectively used in the work. The Σ5 grain boundary is artificially made by the twist operation of {310} planes with the rotation axis of <310> and the rotation angle of ±180°. The Σ5 grain boundary is formed by the coincidence site-lattice, has the second lowest Σ value after Σ3, and has the quite low energy and

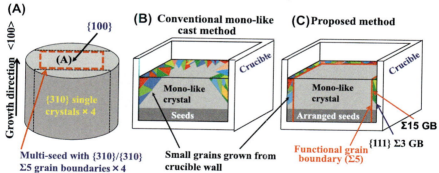

Fig. 6.3 Si single crystal plates with selected orientations to prevent propagation of grain boundaries. (A) Arrangement of {310} single crystals to make a multi-seed with {310}/{310} Σ5 grain boundaries by a face-to-face contact with each {310} surface. These combined crystals can be used as {100} seed crystals for the growth in the <100> direction. (B) Schematic illustration of an ingot grown on multi-seeds by the conventional mono-like cast method. Many small grains usually grow from the crucible wall into the mono-like crystal. (C) Schematic illustration of an ingot grown on arranged seeds to artificially introduce the {310} Σ5 grain boundaries. The {310}/{310} Σ5 grain boundaries can extend from the seed crystals in the growth direction. The small grain regions for the growth using such functional grain boundaries is smaller than those without using them. *(A) Referred from Fig. 1 in K. Kutsukake et al. Appl. Phys. Exp. 6 (2013) 025505. (B) Referred from Private Communication with K. Kutsukake, 2019.)*

low electric activity. The {310} Σ5 grain boundary has an energy of 0.42 or 0.26 Jm^{-2} [36]. The Σ5 grain boundary extends stably in the <100> growth direction because the {310} grain boundary plane is parallel to the <100> orientation. The {310} and {210} plates can be arranged to be parallel to the <100> rotation axis or growth direction and extend in the same growth direction [37]. As shown in Table 1, the {310} planes incline in the <100> direction with an angle of 36.87° around the <100> rotation axis. The {310}/{310} Σ5 grain boundary is formed by a face-to-face contact of two {310} single crystal planes. A {310} single crystal is arranged to make a face-to-face contact with another {310} single crystal (A) which is set at the center of the crucible and other three {310} single crystals are also arranged to make a face-to-face contact with the other sides of the {310} single crystal (A) set at the center as shown in Fig. 6.3A [11]. Finally, the {310} single crystal (A) set at the center is surrounded by the four {310} single crystals to make four {310}/{310} Σ5 grain boundaries called as functional grain boundaries. These combined crystals can be used as {100} seed crystals for the growth in the <100> direction.

For the conventional mono-like cast method, many small grains usually grow from the crucible wall into the mono-like crystal as shown in Fig. 6.3B. When the {310} Σ5 grain boundaries are artificially introduced using the arranged seeds, the {310}/{310} Σ5 grain boundaries can extend from the seed crystals parallel to the crucible wall in the <100> growth direction as shown in Figs. 6.3C and 6.4 [11,13,35].

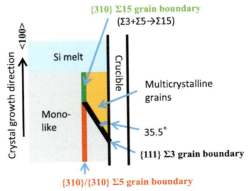

Fig. 6.4 Schematic illustration of the reaction between the {310}/{310} Σ5, {111}Σ3 and {310} Σ15 grain boundaries closed to the crucible wall. *(Referred from Fig. 4 in K. Kutsukake et al. Appl. Phys. Express 6 (2013) 025505.)*

Fig. 6.4 shows the schematic illustration of the reaction between the $\{310\}/\{310\}$ $\Sigma 5$, $\{111\}\Sigma 3$ and $\{310\}$ $\Sigma 15$ grain boundaries closed to the crucible wall [11]. $\{111\}\Sigma 3$ grain boundaries are often formed on the crucible side wall, and they are dominant for a Si multicrystalline ingot grown in a quartz crucible as shown in Fig. 1.6. The $\{111\}$ $\Sigma 3$ grain boundaries have tendency to move toward the center of an ingot because $\{111\}$ planes incline in the $<100>$ growth direction with an angle of $35.7°$ and extend along the $<110>$ direction as shown in Fig. 1.12. The Σ value of a grain boundary formed by interaction between grain boundaries composed by coincidence site-lattice can be simply represented by the product or quotient of Σ values of these grain boundaries as shown in Section 1.3.3.1 [38]. The grain boundary character of a grain boundary formed by interaction between the $\{310\}$ $\Sigma 5$ and $\{111\}$ $\Sigma 3$ grain boundaries is $\{310\}$ $\Sigma 15$ which has a misorientation angle of $48.2°$ about $<210>$ axis [11,39]. This can be known by $\Sigma 3 + \Sigma 5 = \Sigma 15$ from Eq. (1.22). The $\{310\}$ $\Sigma 15$ grain boundaries have stable configuration in the $<100>$ orientation like the $\{310\}/\{310\}$ $\Sigma 5$ grain boundaries. The $\Sigma 15$ grain boundary tends to extend in the growth direction not to expand their impact range or rather to decrease their area because they have not any special plane with low energy. Therefore, small grains grown from crucible wall have tendency not to propagate toward the center of an ingot by the arrangement of $\{310\}$ single crystal plates on the bottom of the crucible and are confined within narrow spaces close to the crucible wall as shown in Figs. 6.3C and 6.4. For the growth of rectangular ingots using such $\{310\}$ single crystal plates as seeds, small grains have tendency to frequently appear from the corners of an ingot at the initial stage of the growth [13]. The small grain regions for the growth using the functional grain boundaries is smaller than those without using them. The multicrystalline regions are confined within narrow spaces between the crucible wall and $\Sigma 15$ grain boundaries formed by interaction between the $\Sigma 5$ and $\Sigma 3$ grain boundaries. Therefore, a large single grain or mono-like crystal can be obtained which is surrounded by the $\Sigma 15$ grain boundary planes. Such effect of the $\Sigma 15$ grain boundaries does not depend on the shape of phase boundary or interface between solid and liquid, but it weakens around cross-points of the $\Sigma 15$ grain boundaries [13].

The perfect $\Sigma 5$ grain boundary has the quite deep energy cusp $(36.87°)$ as experimentally and theoretically shown in the grain boundary energy maps as a function of the rotation angle, θ or misorientation as shown in Fig. 6.2A and B [33,34]. Therefore, the $\Sigma 5$ grain boundary with a small angle deviation can generate dislocation to reduce its boundary energy. When the {310} $\Sigma 5$ grain boundary with a small angle deviation is artificially formed in an Si crystal grown from bi–crystal seeds, the deviation decreases as the distance from the seeds or the ingot height increases during growth and the grain boundary transforms into more stable structure similar to the perfect $\Sigma 5$ grain boundary [36]. In this case, the density of grain boundary dislocations decreases and the dislocation density near the grain boundary increases as the deviation decreases or the grain boundary energy decreases with the ingot height. Σ 5, 13, 17 and 25 grain boundaries have ordered configurations at 0 K [40], that is consistent with the experimentally observed energy cusps at these misorientations shown in Fig. 6.2A. From the simulation of grain boundary energy as a function of the rotation angle, the $\Sigma 5$ grain boundaries have the deepest energy minimum points among other Σ grain boundaries such as Σ 13, 17 and 25 grain boundaries except for the $\Sigma 3$ grain boundary shown in Fig. 6.2B [34] (see Sections 1.3.3.1 and 1.3.3.2). The simulated energy minimum points of Σ 13 grain boundary have not energy cusps because there are several metastable states around Σ 13 and the transformation occurs between the nearest states [41]. Wu et al. [42] introduced $\Sigma 5$ grain boundaries to reduce the small grain regions from the crucible wall and also introduced $20°$-tilt-angle grain boundaries to stop propagation of dislocations from the $\Sigma 5$ grain boundary region (see Section 6.4). However, these methods still cannot perfectly prevent the multiplication and appearance of dislocations in the upper part of the ingot.

The perfect $\Sigma 13$ grain boundary has the deepest energy cusp ($22.62°$) in the grain boundary energy map as a function of the misorientation as shown in Fig. 6.2A [33]. The simulated energy minimum points of Σ 13 have not energy cusps because there are several metastable states around them and the transformation occurs between the nearest states as shown in Fig. 6.2B [34]. Zhang et al. used {510} $\Sigma 13$ grain boundaries at the seed junctions to suppress sub-grain boundaries and dislocations [41]. For this seed arrangement, the seed crystals are cut from an $<100>$ oriented Si ingot along the {510} cutting planes and they are alternately rotated by $180°$ around the [510] axis. The sub-grain boundaries and dislocations can be effectively suppressed by $\Sigma 13$ grain boundaries which can compensate the extra grain boundary energy caused by small angle deviation. On the other hand, band-like poly-crystalline regions originate from the vertical $\Sigma 13$ grain boundary and $\Sigma 3$ twin boundaries also generate from the $\Sigma 13$ grain boundary because there are many metastable Σ relationships near its energy cusp. The $\Sigma13$ grain boundary can easily transform to other grain boundaries such as the $\Sigma37$ grain boundary which has an energy cusp of 18.92° and the vertical $\Sigma13$ grain boundary is mixed with $\Sigma37$ grain boundaries. The tilt angles of the mixed $\Sigma13$ and $\Sigma37$ grain boundaries are measured as 20.84° or 20.59° [42]. These effects of $\Sigma13$ grain boundaries should be compared with those of $\Sigma5$ grain boundaries.

6.4 Behavior of dislocations and precipitates in mono-like ingots

6.4.1 Generation and propagation of dislocations from Si seed plates on the bottom

For an ingot grown by the mono-like cast method using Si single plates as seeds on the crucible bottom, dislocations, dislocation clusters and small angle grain boundaries or sub-grain boundaries are the most dominant crystal defects. The dislocation density in the vicinity of seed plates are up to 10^4–10^5 cm^{-2} and it has a peak at the seed-grown crystal interface [22,43]. The dislocation density increases up to 10^5–10^6 cm^{-2} in dislocation clusters [22,43]. A high-density of dislocations are generated in the junction gap between seeds [29]. A sub-grain boundary with a low angle deviation appears in the center of the junction gap between seed plates with a width of 1.0 cm. The cause of the sub-grain boundary formation is the very small unavoidable mismatch of both Si lattices of the adjacent seed plates by < 1°, and the dislocation formation can be significantly reduced only if < 0.1° [44]. Generally, dislocations and sub-grain boundaries generate at the seed edges, in the gap center and at the interface between a seed and a solidified crystal by the thermal shock caused by the hot melt and the coalescence of the two growing interfaces [25]. The contact point of the seeds also forms a high-density of dislocations symmetrically around it because of large plastic deformation at this point by coalescence of crystals grown from both side surface of seeds. Most of the area surrounding junctions between seeds show low minority carrier lifetimes due to dislocation clusters [29]. Dislocations and sub-grain boundaries also appear from small grain regions near the crucible wall. Even in the middle part of a mono-like ingot with few initial metallic impurities, these dislocations and sub-grain boundaries easily propagate and multiply from the seed junctions, the surface of seeds and the crucible wall. Finally, they form dislocation clusters during growth and extend to the whole ingot. Dislocations in the middle part of the ingot are mainly located in clusters aligning above the seed junctions, clearly defined <111> directions [29]. Inclusions or precipitates in the ingot are also important origins of the generation of dislocations (see Section 3.3.3).

A discontinuity in EPD is observed at the interface between the unmelted seed and the crystal grown on it, and the EPD is much higher in the unmelted seed [45]. The cellular structure of dislocations can be observed by the Synchrotron radiation X-ray white beam topography and Rock curve imaging. The seed plates contain the cellular dislocation structure near their surface due to the thermo-mechanical stress during the heating and melting processes of the seeds, which is achieved in the seeds before crystallization [46]. Due to the cellular structure, an ingot grown on the seeds also have the cellular structure of dislocations. Such dislocation structure propagates and multiplies inward the ingot. However, the dislocation density in a seed just below the seed surface is much higher than that in a grown crystal just above the seed surface [29]. When saw damage remains on the surface of the seeds, the cellular dislocation structure is also produced near the surface above $900°C$ to minimize internal energy during annealing and dislocations propagate in various directions into the ingot [23]. A high-density of dislocations form on the bottom of the seeds with scratches due to saw damage set on a Si_3N_4 coating [23].

The dislocation generation strongly depends on the tilt angle between the seed plates and it is very easy at the tilt angle of $0°$ [42]. The seed junction formed at the tilt angle of $0°$ has still some misorientation of about $1°$ between both seeds and forms a non-perfect $\Sigma 1$ grain boundary or a small angle grain boundary [27]. It is quite different from artificially make a perfect $\Sigma 1$ grain boundary because of thermal stress near the grain boundary planes and difficulty of the perfect atomic-scale connection of both seed surface. Dislocations based on coincident structure with misoriented seeds normally generate at the seed junction and propagate along the growth direction in a crystal with larger shear stress. In order to confirm how dislocations generate from a seed junction, a Si crystal is grown on a special seed crystal with artificially manipulated grain boundaries as shown in the lower right of Fig. 6.5 [47]. The seed crystal is made by combining several single crystal plates with different orientations and a typical width of 4 mm. The structure of the seed crystal is designed so that all crystal plates regarded as grains are oriented to <110> in the growth direction. The rotation angles of each set of two adjacent crystal plates with {112} side face are 7.86 and $-7.93°$ around the <110> axis. The misorientation of the two adjacent crystal plates to generate a grain boundary is determined to be $15.7°$ by the X-ray analysis. Thus, the

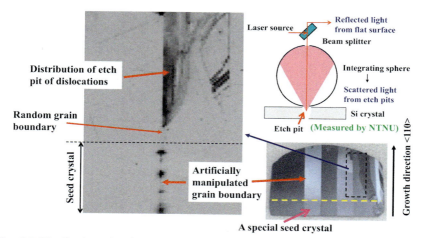

Fig. 6.5 Distribution of etch pit of dislocations in a Si crystal grown on a special seed crystal with artificially manipulated grain boundaries made by several single crystal sheets with different orientations. The yellow dashed line is the interface between the seed crystal and the grown crystal. The distribution of EPD was measured by the spatial distribution of scattered light from etch pits using the laser system. Dislocations appear from the random grain boundary and only one side of the random grain boundary. (Referred from Figs. 3 and 6 in I. Takahashi J. Cryst. Growth 312 (2010) 897.)

grain boundary is regarded to be a random grain boundary or a large-angle grain boundary. The yellow dashed line shows the interface between the combined seed crystal and the grown crystal. An array of dislocations appears around the grain boundary as growth proceeds, which is known by the crystal rotation along the growth axis using the x-ray rocking analysis [48]. The left of Fig. 6.5 shows the distribution of EPD of dislocations measured by the spatial distribution of scattered light from etch pits using the laser system (PVScan) [47,49]. In the measurement, a HeNe laser is irradiated on the sample surface during scan and the irradiated light is divided into the reflected light from the flat surface and the scattered light from etch pits. The EPD can be estimated from the intensity of the scattered light. The dislocations clearly generate from the random grain boundary during crystal growth, locally distribute near the random grain boundary, and appear on only one side of the random grain boundary. Then, the dislocations largely increase and spread in cascade along the growth direction inside only one grain with larger shear stress during crystal growth. Thus, the dislocations appear depending on the difference in crystallographic orientations in two grains on both sides of the grain boundary, which affects shear stress on slip plane in the grains because of anisotropy of the elastic modulus of Si. In the vicinity of the grain boundary, some grains have a lower dislocation density than 10^4 cm^{-2} and other grains have a higher dislocation density [47,50]. The reason why dislocations appear on only one side of the grain boundaries can be known by the distribution of shear stress near artificially manipulated grain boundaries as shown in Fig. 1.13. For bi-crystal Si ingots separated by near Σ9 and Σ27 grain boundaries, dislocations also appear on only one side of the central grain boundaries due to shear stress arising asymmetrically in {111} planes [51].

In the case of the <110> oriented crystal plates, the dislocations which are formed inside the seed gap and are aligned perpendicular to the growing interface and nearly parallel to the <110> growth direction can easily move and multiply on the preferred {111} <110> slip system which is also parallel to the growth direction [25,26]. During ingot growth, cascades of dislocations do not largely expand from the <110> growth direction, but the dislocation density in these cascades largely increase because of high concentration of cascade-like dislocations [52]. The dislocations concentrate near the {100}/{100} boundary planes formed from the junctions between the <110> oriented seed plates [52]. In the case of the <100> oriented crystal plates, dislocations formed inside the seed gap and multiplicated parallel to the growth direction also spread in cascade, but the multiplication and density are much reduced comparing with the <110> oriented crystal plates.

A new random grain boundary is formed in a small gap (250 μm) between the seeds [26]. However, random grain boundaries do not strongly act as origins of dislocations unlike in multicrystalline ingots because large Σ grain boundaries or large angle grain boundaries with a small angle deviation formed by the misorientation between seeds have not large energy enough to overcompensate the dislocation formation energy due to the flat energy curve as shown in Fig. 1.9. A random grain boundary or a Σ33 grain boundary prohibit the formation of dislocations [44]. Tsoutsouva et al. [53–55] studied crystal defects in a junction between seeds in detail. Sub-grain boundaries generate at the junction even through the misorientation between seeds is smaller than 0.02°. Low Σ such as non-perfect Σ3 grain boundaries with small misorientation formed by the seed arrangement easily generate dislocations from the seed junction to reduce the grain boundary energy [56]. At the initial stage of the growth for the mono-like cast method, the Si growth melt penetrates into joint boundaries between the adjacent seed plates toward the crucible bottom coated with Si_3N_4. Sub-grain boundaries are created at the center

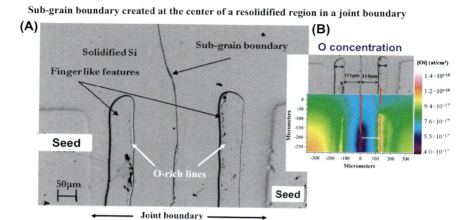

Fig. 6.6 (A) Sub-grain boundary created at the center of a re-solidified region in the joint boundary during growth owing to relative misorientation or stress between the seed plates, which was observed by an optical microscopy. (B) O-rich line region observed by μ-FTIR in the finger like features. *(Referred from Figs. 3 and 5 in M. G. Tsoutsouva et al. J. Cryst. Growth 401 (2014) 397.)*

of re-solidified regions in the joint boundaries during growth because the relative misorientation or stress exists between the seed plates as shown in Fig. 6.6A [54]. Dislocations generated in distorted zones near the sub-grain boundaries propagate away from the sub-grain boundaries toward the sides of the un-melted seed plates and inside of the ingot above the seeds. The misorientation due to presence of sub-grain boundaries rapidly increases with the ingot height as the sub-grain boundaries propagate toward along the growth direction, resulting in rapid multiplication of dislocations. The misorientation is 0.01° just above seeds, 0.04° at 1.5 mm higher and 0.5° at 50 mm above the bottom of the ingot. Due to relative misorientation (tilt or twist) between seeds, dislocations with a screw component generate at the bottom of an ingot in the area between seeds, and bunches of dislocations with a screw component localize above the top of the seeds. A relative misorientation between seed plates as small as 0.02° can produce sub-grain boundaries and cascades of dislocations. The finger like features in Fig. 6.6 are a memory of the contamination of the surface of the seeds during melting process, and they contain O-rich lines with O interstitials (O_i) observed by μ-FTIR which are also due to the contamination caused by the interaction between melted Si feedstock and oxidized Si_3N_4 coating as shown in Fig. 6.6B [54]. The O and C concentrations are also higher in sub-grain boundaries located inside junction gaps. Silicon oxynitride ($Si_xO_yN_z$) precipitates can be detected in the center of the O-rich zone (see Section 1.3.3.2).

To reduce cascade dislocations and sub-grain boundaries from seeds and seed gaps and reduce misalignment of the seeds as origin of these dislocations, large area single crystal seeds are tried to be used together with a leverage furnace, a high-purity crucible with high-purity coating materials, the seeds placed on a machined-flat surface and the feedstock loaded keeping sub-critical [22]. Ingots with low-dislocation density (10^2–10^3 cm^{-2}) and high lifetime (530 μs) can be obtained using large area seeding [22]. Generally, to reduce thermal shock, the large area seeds is heated to a temperature very closed to Mp of Si before contacting with the melt.

6.4.2 Formation of precipitates and propagation of dislocations during growth

Generally, Si_3N_4 particles precipitated from N-saturated Si melt often appear in ingots grown using a Si_3N_4 coating crucible [57] as explained in Section 3.4. SiC particles frequently appear around Si_3N_4 particles with needle-like, columnar and leaf-like shapes in Si multicrystalline ingots [57–59]. In a mono-like ingot, SiC precipitates appear in the bottom part of the ingot, the SiC, Si_3N_4 and SiCN precipitates appear in the middle part of the ingot and small angle grain boundaries appear after increase of the precipitates in the upper part of the ingot [60]. These precipitates act as origins of the small angle grain boundaries (see Section 3.3.4).

The C, O and N concentrations are high at the re-solidified regions in the joint boundaries between the plates, which is related to direct contact of the penetrating melt with Si_3N_4 coating and contamination of the side surface of the seed plates. The C concentration is also high around the sub-grain boundaries created at the center of re-solidified regions in the joint boundaries between the plates because rejection of C occurs ahead of the growing fronts due to segregation and C builds up ahead the upward-moving front [54]. In mono-like ingots, precipitates of $Si_xO_yN_z$ are observed at the re-solidified regions in the joint boundaries, especially in the O-rich line region because of the high concentrations of O and N as shown in Fig. 6.6B. Precipitates of silicon oxycarbide ($Si_xC_yO_z$) and SiC are also observed together with Si_3N_4, and $Si_xO_yN_z$ along sub-grain boundaries created at the center of re-solidified regions because the C concentration is high in the penetrating melt regions between the plates. The $Si_xC_yO_z$ precipitates in the sub-grain boundaries act as nucleation sites of Si_3N_4 particles [54].

The presence of precipitates in an ingot facilitates generation of dislocations by local stress owing to the difference between thermal expansion coefficients of precipitate and Si. The local stress can be relaxed by generation of dislocations. Thus, the local stress caused by precipitates is the origin of dislocations and subsequently small angle grain boundaries [60]. The precipitates of impurities related with the generation of dislocations mainly consist of Si, C, N and O. Fig. 6.7 shows (A) Si single plates set on the bottom of the crucible and the surface of the grown ingot and (B) the vertical cross section of the ingot, the distribution of resistivity and the distribution of minority carrier lifetime [61]. Dislocations nucleated from precipitates or particles spread in cascades from 1/3 or 1/2 of the height toward the top of the ingot during growth as shown in Fig. 6 7 [2]. The cascades of dislocations appear as a network of sub-grain boundaries or small angle grain boundaries [3]. The sub-grain boundaries are constituted by a set of dense, perfectly organized immobile dislocations with the distance of 90 nm, aligned along the growing direction and having an edge character, and they are strong obstacles for mobile dislocations [62]. They have a misorientation of about 0.2°, associated with a rotation around the [001]

Fig. 6.7 Propagation of dislocations in cascades from 1/3 or 1/2 of the height toward the top of the ingot during growth. (A) Si single plates set on the bottom of the crucible and the surface of the grown ingot. (B) Vertical cross section of the ingot, distribution of resistivity and distribution of minority carrier lifetime. *(Referred from OHP 6 in A. Jouini et al. in Crystalline Silicon for Solar Cells, CSSC-7, Fukuoka, Japan, October 22–25, 2013.)*

growth axis [62] (see Section 1.3.3.1). Some inclusions such as SiC and Si_3N_4 particles act as origins or nucleation sites of grains [57]. The super-cooling required for the nucleation of Si on SiC is very low near 1 K. The grains formed from these inclusions expand and rapidly enlarge themselves all the way to the top of an ingot [8]. The minority carrier lifetime is low in regions with higher dislocation density than 10^5 cm^{-2} because of metal precipitates entangled in many dislocations. In grains with low dislocation density (10^3-10^5 cm^{-2}), mid-band trap levels measured by the Deep Level Transient Spectroscopy (DLTS) are negligible in an as-grown ingot because any lifetime killers are completely precipitated in as-grown materials [2]. In grains with high dislocation density ($>10^6$ cm^{-2}), no traps are also detected by DLTS in an as-grown ingot because metal precipitates in the densely dislocated regions are strongly bound to the local structural defects and are not dissolved into the lattice [2].

6.4.3 Propagation and multiplication of dislocations in the upper part of ingots

Dislocations generated from the seed junctions rapidly increase and propagate in the top part of the ingot, and sometimes form cascade of dislocations. The dislocation density in the central single region is much lower than that in the surrounding polycrystalline region [63]. These dislocations are highly mobile above 900°C and travel several mm per hour at low stress levels [62]. The dislocation density increases as the ingot height increase even though few dislocations appear in the mono-like part near the bottom of the ingot. The background dislocation density evolved with the ingot height is clearly correlated with thermomechanical stresses generated during cooling [45]. Such dislocations drastically impact the reduction of the minority carrier lifetime [64]. Large amount of small angle brain boundaries (rotation angle less than 10°) or sub-grain boundaries which are formed from the aggregated dislocations also propagate and multiply in the upper part of a mono-like ingot [65]. The sub-grain length per unit area also increases drastically (over two orders of magnitude) from the bottom to the top [45]. The deviation of the sub-grain orientation on both sides of a sub-grain boundary increases along an ingot height within 3° [66].

Many dislocation clusters can be observed around small angle brain boundaries. The length density of small angle grain boundaries is about 2.0×10^3 mm/mm^2 in wafers cut from the top part of the ingot, but few small angle brain boundaries appear in wafers cut from the bottom part [65].

The area fraction of crystal defects in a wafer from the top part of the ingot is extremely larger than that in a wafer from the bottom part. Therefore, the conversion efficiency of solar cells using wafers from the top part of the ingot is much lower than that using wafers from the bottom part of the ingot. Especially, above the central region of the seed gaps, the dislocation density increases up to $5 \times 10^6 \, cm^{-2}$ with increasing ingot height comparing the dislocation density of $2 \times 10^5 \, cm^{-2}$ above the adjacent regions outside the seed gaps [25]. On the other hand, for an ingot growth using a <110> oriented seed with the {111} <110> slip system, the propagation of dislocations originating from the seed can be efficiently blocked by a horizontally arranged successive multi-twin boundaries in a $\Sigma 3$ <111> relation between them (see Sections 1.3.3 and 6.4.1).

6.4.4 Concept to control of dislocations using several grain boundaries

The generation, multiplication and propagation of dislocations in a mono-like ingot during growth should be suppressed and many trials are performed for this purpose. Basically, the modification of local structure in a Si multicrystalline ingot during directional solidification strongly depends on the initial grain boundary structure made by seed plates [48]. To prevent the generation and propagation of dislocations, the initial grain boundaries are designed by several complicated and well-examined arrangements of seed plates to intentionally form the non-perfect highly symmetrical grain boundaries with small angle deviations from each deep energy minimum and the large angle grain boundaries with low symmetry. The tilt grain boundaries with small misorientation are also used as the large angle grain boundaries.

At the low growth rate of 0.16 mm min^{-1}, Trempa et al. [27] reported that the non-perfect highly symmetrical grain boundaries ($\Sigma 3$ and $\Sigma 9$) could cause a grain boundary splitting, the dislocation and twin generation and a multicrystalline structure to minimize their energy. Actually, a non-perfect $\Sigma 3$ grain boundary prepared by artificial seeds with small tilt angles of 3.5 and 7.0° is varied to a perfect $\Sigma 3$ grain boundary accompanied by small grains during growth when the growth rate is as high as 1.0 mm mim^{-1} [67]. Some of these non-perfect highly symmetrical grain boundaries belong to large angle grain boundaries as a definition, but they behave like small angle grain boundaries due to their small angle deviation from the energy minimum as shown in Fig. 1.9. The non-perfect $\Sigma 9$ grain boundaries (angle deviation:$\theta = 2.5°$) cause less the splitting, dislocation generation and multicrystalline structure than the non-perfect $\Sigma 3$ grain boundaries ($\theta = 1.8-3.5°$). Such effect of the non-perfect $\Sigma 3$ grain boundaries increases as the deviation angle increases. The low symmetrical grain boundaries such as random grain boundaries with much high grain boundary energy could keep their structure and cause no dislocation generation or grain boundary splitting because the energy gain due to angle changing is not large enough to overcompensate the dislocation formation energy or splitting energy because the energy curve is almost flat as shown in Fig. 1.9. When the random grain boundaries are formed in seed junctions, few dislocations ($1 \times 10^4 \, cm^{-2}$) and twins appear at the seed junctions because the energy gain is not large enough to form dislocation and twins because their grain boundary energy is too far away from the energy minimum. A region surrounded by 4 random grain boundaries is nearly free of dislocations.

The concept to control of dislocations using such behavior of the non-perfect highly symmetrical grain boundaries and large angle grain boundaries or random grain boundaries can be explained as follows.

(1) Control of the generation of dislocations from the seed junctions

    (1) To prevent the generation of dislocations from the seed junctions, the near-perfect highly symmetrical grain boundaries with quite small angle deviations from each deep energy minimum can be used. These grain boundaries have not enough energy to generate dislocations due to their quite small angle deviations.

    (2) To prevent the generation of dislocations from the seed junctions, the large angle grain boundaries with low symmetry or random grain boundaries can be used. These grain boundaries have not enough energy to generate dislocations due to their almost flat boundary energy curve for the angle deviations as shown in Fig. 1.9.

(2) Control of the multiplication and propagation of dislocations from the crucible wall

    (1) To prevent the multiplication and propagation of dislocations from the crucible wall, the non-perfect highly symmetrical grain boundaries with small angle deviations from each deep energy minimum can be used. Dislocations coming to the grain boundaries can be trapped by the grain boundaries to minimize their grain boundary energy as shown in Fig. 1.9. These dislocations are difficult to move away from the grain boundaries.

246   Crystal Growth of Si Ingots for Solar Cells Using Cast Furnaces

(2) To prevent the multiplication and propagation of dislocations from the crucible wall, the large angle grain boundaries with low symmetry or random grain boundaries can be used. Dislocations coming to the grain boundaries can be stopped by the grain boundaries. The orientations of slip plane are largely changed on both sides of the grain boundaries. Therefore, it is difficult for dislocations to move across the grain boundaries during keeping the Burgers vector, b constant. However, dislocations cannot be trapped by them because the boundary energy gain for small angle deviations is not enough large to trap them.

6.4.5 Control of dislocations using the large-angle and non-perfect highly symmetrical grain boundaries

Hu et al. [66] intentionally induced grain boundaries by maintaining twist or rotated angular deviation of $10-45°$ between adjacent <100> oriented Si seed plates set on the bottom of a crucible. Such induced grain boundaries belong to the large angle grain boundaries ($\theta_R > 10°$) and are regarded as the CSL grain boundaries ($\Sigma = 5, 13, 17 -$) or random grain boundaries ($\Sigma > 27$), which depend on whether the seed plates are symmetrically or asymmetrically arranged. The asymmetrical arrangement is used in this case, but the symmetrical arrangements is more effective for such induced grain boundaries. The seed plates are cut with the intended angles to keep the rotation angle. The angular deviation is kept larger than the maximum deviation angle of the sub-grain boundaries. In an ingot grown by this method, sub-grain boundaries generated from junctions between the adjacent seed plates are suppressed near the intentionally induced grain boundaries which extend in the growth direction. These grain boundaries can suppress the multiplication of scattered dislocations or sub-grain boundaries by preventing the slip of dislocations. Thus, the formation and multiplication of sub-grain boundaries generated from the seed junctions are effectively suppressed using these grain boundaries.

Takahashi et al. [56] proposed a novel method named as seed manipulation for artificially controlled defect technique (SMART) to reduce dislocations in the center of an ingot as shown in Fig. 6.8. To trap impurities and relax stress, artificially designed seeds with several orientations are arranged to intentionally form dislocation cluster regions or functional defect regions which are enclosed by thin plates as seed partition and arranged along the crucible wall. Small angle grain boundaries (small angle GBs shown in Fig. 6.8) are introduced as origins of the dislocation

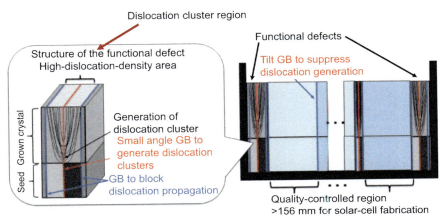

Fig. 6.8 Artificially controlled defect technique to reduce dislocations in the center of an ingot. Small angle grain boundaries (GBs) are introduced as origins of the dislocation clusters. Grain boundaries are intentionally arranged in the growth direction on the both sides of each dislocation cluster region. The dislocation cluster regions and tilt grain boundaries prevent a spread of dislocations grown from the crucible wall. *(Referred from Fig. 1 in I. Takahashi et al. Jpn. Appl. Phys. Express 8 (2015) 105501.)*

clusters because small angle grain boundaries have a larger capability to generate dislocations than other grain boundaries such as the tilt and random grain boundaries. Random grain boundaries are also intentionally arranged in the growth direction on both sides of each dislocation cluster region to prevent both generation and propagation of dislocations. The large angle grain boundaries such as the random grain boundaries prohibit the internal generation of dislocations and do not emit dislocations by itself [44]. The dislocation cluster regions and tilt grain boundaries effectively prevent a spread of dislocations grown from the crucible wall. The tilt grain boundaries used in the growth have grain boundary plains of {130} or {120} on each side and a growth direction of <100>, so they may have a title angle of 45° (large angle grain boundary) or 8.2° (small angle grain boundary) between them. Fig. 6.9A and B show orientation maps of the growth direction and the grain boundary plane, respectively. Fig. 6.9C shows an etch-pit image of the sample grown by the artificially controlled defect technique. The functional defect regions can generate dislocations in the periphery of the ingot along the crucible wall. Tilt grain boundaries can be effectively used to minimize the dislocation generation from the seed junctions and to block dislocations from the dislocation cluster regions as shown in Fig. 6.9. So, these tilt grain boundaries are large angle grain

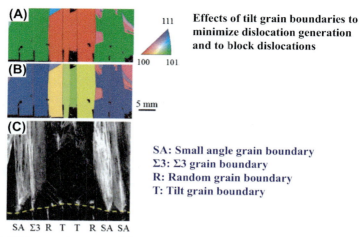

Fig. 6.9 (A) Orientation map of the growth direction. (B) Orientation map of the grain boundary plane. (C) Etch-pit image of the sample grown by the artificially controlled defect technique. Tilt grain boundaries effectively used to minimize the dislocation generation from the seed junctions and to block dislocations from the dislocation cluster regions. *(Referred from Fig. 2 in I. Takahashi et al. Jpn. Appl. Phys. Express 8 (2015) 105501.)*

boundaries. Thus, no dislocation cluster are generated from the grain boundary at the center of the ingot. To obtain low defect density, Wu et at [42]. found that the tilt angle away from CSL grain boundaries is the best choice and the generation of dislocations was strongly suppressed at the tilt angle of 10–30° even through the seed junctions have some gaps. A grain boundary formed from the tilt angle of 40° is Σ5 and grain boundaries formed from the tilt angles of 20° and 30° are low symmetric CSL grain boundaries or large angle grain boundaries which have a higher energy. To grow an (100) mono-like ingot for better texturing without color mismatch, Lan et al. [68] used (100) seed plates separated by 0.1 cm thick (100) thin plates as seed partition with a large tilt angle of ±20–30°. These tilt grain boundaries can be categorized as large angle grain boundaries. Therefore, the tilt grain boundaries on both sides of a partition area can more effectively prevent the propagation of dislocations generated in a seed junction than near-perfect Σ5 grain boundaries. Most of dislocations are concentrated near the sub-grain boundaries inside the partition plates. The junctions with such a large tilt angle have a little effect on the minority carrier lifetime.

Even if an ingot is grown by these methods, dislocations or dislocation clusters are newly generated from impurity inclusions at the upper part of the ingot. Regions with defects such as dislocations, dislocation clusters and sub-grain-boundaries increase from the bottom to the top of an ingot. The large angle grain boundaries are growing diagonally and annihilate each other or enlarge the area with the unfavorable second orientation [44]. The defect regions of the ingot are expected to be smaller enough to confirm the effect than those of an ingot grown by the regular mono-like cast method. The minority carrier lifetime and uniformity of the ingot are improved by these methods because of the lower density of defects or recombination centers [66]. Recently, a Si ingot with few dislocations generated from joint boundaries is reported by the artificially controlled defect technique using designed seeds with several orientations [69]. Moreover, the SMART can be used to confirm the effects of the misorientation of artificially designed joint seeds on dislocation generation [70] (see Section 1.3.3.2).

6.5 Quality of Si ingots using the mono-like cast method

6.5.1 Dislocation density, micro-twins and impurity concentration

The average residual strain in a mono-like cast ingot is 20%—25% higher than in a multicrystalline Si ingot grown under the same growth condition, but the local high-strained areas decrease in the mono-like cast ingot due to decrease of grain boundaries or dislocation clusters [71]. The dislocation density in a mono-like cast ingot is lower than that in the multicrystalline Si ingot even though the residual strain is higher [72]. Both strain and dislocations are mainly concentrated in the periphery of the mono-like cast ingot [73]. However, the average dislocation density is still high on the order of 10^4-10^5 cm^{-2} because the stress is caused by expansion due to the solidification of the Si melt. Generally, in such ingots grown by the mono-like cast method, the EPD is within 10^4-10^6 cm^{-2} in a high-dislocation-density area and on the order of 10^3 cm^{-2} in a low-dislocation-density area [2,10,74]. The dislocation density in the mono-like region is normally lower than that in the surrounding small-grain regions close to the crucible wall. However, dislocations appeared in the small-grain regions easily propagate and multiply toward the center and upper parts of the mono-like ingot to reduce stress remained in these parts. The dislocation density drastically increases from the bottom to the top of the ingot. The typical EPDs are $1-1.5 \times 10^4$ cm^{-2}, $4 \times 10^4-1.5 \times 10^5$ cm^{-2} and $3-6 \times 10^5$ cm^{-2} at the bottom, in the middle and at the top of the mono-like region, respectively [63]. By intentionally creating a convex growing interface during growth using a disk seed, the EPDs are reduced to 2×10^4 cm^{-2} and $3-6 \times 10^3$ cm^{-2} at the bottom and at the top, respectively [12]. This reducing effect of the convex growing interface on dislocations is similar to that of the NOC method which can create the convex growing interface in the growth direction during growth (see Section 7.4.3).

Lantreibecq et al. [45] studied micro-twins in a mono-like ingot. Sometimes, micro-twins propagate at the growing interface and are observed close to the bottom of an ingot (at 4.7 cm high), which exhibit small dimensions (<2 mm) and align in a direction. Their density and individual length strongly increase with ingot height, but they become moderate between height positions with a width of several mm. For the growth using {100} seeds, the micro-twins can be observed as traces <011> and <0-11> directions on the {100} surface because the four {111} planes have the same angle (54.73°) with the growth direction. They lie in only (1-11) and (111) planes. High C concentration favors the formation process of the micro-twins because the nucleation may be linked to C content.

Metallic impurities such as Fe_i largely diffuse from a quartz crucible and a Si_3N_4 coating into thick seed plates set on the bottom of the crucible during the heating and melting processes at high temperature. Then, these impurities diffuse into a Si ingot through the seed plates during ingot growth and make a thick bottom region with low quality and low minority carrier lifetime named as red zone [2,3,75]. The Fe_i concentration in the bottom region of the ingot is $5 \times 10^{12}-5 \times 10^{13}$ cm^{-3}, which is very high comparing with the Fe_i concentration of 1.4×10^{11} cm^{-3} in the top region [75,76]. The minority carrier lifetime degrades toward the crucible base because of the solid-state diffusion of metal impurities and the high O concentration from the crucible [76]. The length of the red zone is proportional to the height of the remaining seed [77]. On the other hand, Trempa et al. claimed that the main mechanism to form the red zone was not the Fe_i diffusion because the crucible had only locally contact with the lower surface of the seed plates due to the roughness of the inner crucible surface [75]. Using diffusion barriers to prevent the Fe_i diffusion from the quartz crucible and Si_3N_4 coating, the main contamination path is shown to be the gas phase diffusion from the furnace atmosphere through the Si feedstock downwards to the seed plates during heating and meting the feedstock [75]. The main Fe sources are the graphite furnace parts such as graphite materials, stainless camber and silica crucibles.

Thus, the bottom region of the ingot has lower minority carrier lifetime owing to high density of solid-state diffused metallic impurities dominated by Fe_i. The seed plate has relatively higher minority carrier lifetime than the first regrown part called by the red zone because the seed contains excess vacancies which reduce the influence of diffused Fe atoms by moving them from interstitial to substitutional positions [3]. For a wafer cut from an ingot grown by the mono-like cast method, Fe_i is accumulated in a high-dislocation density area in the wafer annealed at 600 °C and Fe_i is diffused uniformly over the wafer annealed at 800°C and 1000°C [78]. The ratio of Fe_i concentrations in the high- and low-dislocation density areas is about 9 in the wafer annealed at 600°C and it is almost 1 in the wafer annealed at 800°C and 1000°C [78]. The high density of dislocations act as trapping sites of Fe. The Fe_i concentration in the high-dislocation density area becomes much lower than that in the low-dislocation density area by slow cooling from 800 °C to 400 °C, which attributes to precipitation of Fe_i [78]. Therefore, the high-dislocation areas can be used for accumulation of many Fe_i atoms by annealing at 600 °C and for precipitation of them by slow cooling [78]. The SMART can be effectively used to introduce such high-dislocation areas in a large-scale ingot grown by the mono-like cast method (see Sections 1.4.4 and 6.4.5).

From Stoddard et al. [2], the O concentration in an ingot is on the order of 10^{18} cm^{-3} at the beginning of the growth due to the O segregation coefficient of 1.25 and on the order of 10^{17} cm^{-3} in the top of the ingot. The O concentration in an ingot is $1.4-3 \times 10^{17}$ cm^{-3} in the middle of the ingot, which is lower than that in an ingot grown by the CZ method [10,12,64]. To reduce the O concentration, a high gas flow rate and a firmly Si_3N_4 coating are effective [12]. The low O concentration is effective to reduce B-O defects in p-type mono-like ingots and increase the performance of solar cells. An ingot with a shiny and flat surface is obtained by reduction of C contamination using a high gas flow rate [12]. The C concentration with the segregation coefficient of 0.07 is on the order of 10^{16} cm^{-3} at the beginning of the growth and on the order of 10^{17} cm^{-3} in the top of the ingot [2]. The C concentration can be reduced to $1-3 \times 10^{16}$ cm^{-3} using a high gas flow rate of 80 L min^{-1} [12]. The N concentration with the segregation coefficient of 7×10^{-4} is on the order of 10^{15} cm^{-3} in the ingot [2].

6.5.2 Electrical properties

Phang et al. [79] studied the electric properties of n-type wafers grown by the mono-like cast method. After B diffusion, the average minority carrier lifetime degrades even for wafers cut from the middle part of an ingot. The subsequent P diffusion (825°C for 40 min) improves the lifetime and the subsequent hydrogenation (rapid thermal annealing at 500°C for 10 s) can more improve the lifetime. The hydrogenation is very effective even on passivating boundaries between seed plates. Thus, the average minority carrier lifetime is largely improved by the B and P diffusions and hydrogenation. However, comparing to the lifetime of wafers using only P diffusions and hydrogenation, B diffusion has a large negative impact on the final lifetime of the n-type wafers even after subsequent P gettering and hydrogenation because of its higher diffusion temperatures (950 °C for 70 min) compared to P diffusion (see Section 5.5.1).

The minority carrier lifetime in a mono–like part of an ingot is generally higher and more homogeneous than that in its small grain parts grown from the crucible wall. The average minority carrier lifetime in a mono–like part of an n-type ingot is usually higher than those of ingots grown by the conventional and HP cast methods because of the lower concentration of localized recombination centers and defects [10,79]. The minority carrier lifetime in the middle part of an n-type ingot is largely improved to 465 μs or 1.5 ms after gettering and passivation by reducing the C and O concentrations [12,64]. The highest average minority carrier lifetime of an n-type ingot is 3.3 ms [79]. The highest diffusion lengths at 0.1 sun is 1,400 μm, which is higher than the thickness of solar cells.

6.5.3 Solar cells

As the surface of mono–like wafers has uniform {100} crystal orientation, this point is beneficial to surface texturing of solar cells using alkaline solution. The cell performance improvement triggered by alkaline texturing becomes larger for the mono–like wafers with more uniform {100} crystal orientation, and the difference in cell conversion efficiency between alkaline and acidic texturing is up to an absolute value of 1.05% [4] (see Section 3.5.2). The average conversion efficiency of n-type solar cells is 18.5%—19.0% when it is 17.8%—18.2% and 19.0%—19.5% for the HP and CZ solar cells, respectively, and the scattering width has a somewhat wider value of 2.4% because of a long tail on the low efficiency side [80]. Using the advanced high efficiency process, the highest conversion efficiency of 21.5% is obtained for an n-type solar cell [64]. The average conversion efficiency of p-type solar cells is 18.1% [66]. Fig. 6.10 shows the conversion efficiency of n-type solar cells prepared

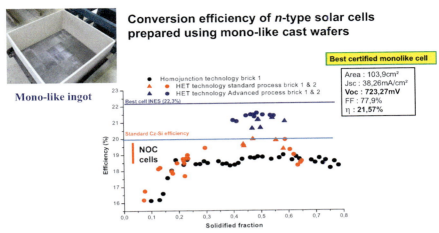

Fig. 6.10 Conversion efficiency of *n*-type solar cells prepared using mono-like cast wafers as a function of the solidification fraction. The wafer size is 15.6 × 15.6 cm². The solar cells have the PERT structure prepared in INES. The highest conversion efficiency of 21.57% can be obtained using the HIT technology advanced processes. The high-performance solar cells can be obtained only in the middle parts of the ingots. The scattering width of the NOC solar cells is much narrower than that of the mono-like solar cells using the same processes and structure as shown by the bar in the left corner of this figure. *(Referred from Figs. 6 and 8 in F. Jay et al. Sol. Energy Mater. Sol. Cells 130 (2014) 690.)*

using mono-like cast wafers as a function of the solidification fraction [66]. The wafer size is 15.6 × 15.6 cm². The solar cells have the passivated emitter, rear totally diffused (PERT) structure prepared in National Solar Energy Institute (INES) in France. The conversion efficiency strongly depends on the process technologies. The highest conversion efficiency of 21.57% can be obtained using advanced processes of the amorphous Si (α-Si:H/c-Si) heterojunction (HET) solar cell technology processes. The key to the technology is the very high surface passivation quality by the α-Si:H layers [81]. The high-performance solar cells can be obtained in the middle parts of the ingots, but the conversion efficiency of the solar cells prepared from the lower parts of the ingots is much lower. Such unhomogeneity of the ingots is the most serious problem for the mono-like cast method (see Section 7.5.2).

Generally, the performance of solar cells prepared using wafers cut from the middle part of a mono-like ingot is highest, but that of solar cells prepared using wafers cut from the top part of the ingot is lowest because of the highest dislocation density and impurity content [63,65]. The performance of solar cells prepared from the bottom part with the low dislocation density is not higher than that of solar cells prepared from the middle part because of large impurity diffusion through the seed plates [64]. Solar cells using wafers with small grains over more than 30% of the total area show a less conversion efficiency. The thickness of wafers cut from a mono–like ingot can be reduced from 200 μm to 100 μm because of their single crystallinity [82].

6.6 Key points for improvement

Mono–like cast Si wafers with low defect density exhibit the fracture stress as high as the CZ wafers, and after removing the surface damage, both wafers present a minimum threshold below that they will never fail [83]. Thus, the mono–like cast wafers have good features such as CZ wafers. However, the mono–like cast method has still several problems to be solved for industrialization as described below.

The thicker red zone appears near the bottom of ingot comparing with that of the conventional multicrystalline ingot. Although the use of a high-purity crucible can lead to a much shorter bottom red zone in the conventional multicrystalline ingot, but it has little effect on the bottom red zone in the mono–like ingot [77]. To reduce the red zone, Si_3N_4 coating materials and nucleation agent should be improved together with the purity of crucible.

Even in the mono–like cast method using artificially manipulated single plates to prepare $\Sigma 15$ grain boundaries formed by interaction between the $\Sigma 5$ and $\Sigma 3$ grain boundaries, many dislocations still appear around the upper corners of the ingot [13]. To prevent such propagation of dislocations and reduce the dislocation density, the thermal field should be controlled to maintain the growing interface flat or convex in the growth direction and reduce the thermal stress, impurities and precipitations in the ingot. The yield of the ingot to obtain high-efficiency solar cells is not still good even using the convex growing interface [12]. Recently, the propagation of dislocations is largely suppressed in a large-scale ingot prepared by INES using the similar concept to control of dislocations as explained in Section 6.4.5 and precisely controlling the thermal field [69].

Even using dislocation-free seed plates, dislocations already exist in the initial seed plates [24]. A dislocation cell structure is produced near the surface of the seed plates during growth and annealing due to thermal shock and remaining saw damage [23]. It is very difficult to reduce the thermal shock, but the seed surface can be deeply etched or polished to remove the saw damage. Seed crystals used for ingot growth cannot be recycled for the next ingot growth as perfect single crystals because of severe contamination and imperfection due to metallic impurities and crystal defects. These points will affect the cost of ingots.

The uniformity of the mono-like ingot is not good because of the coexistence of a mono-like region and multicrystalline regions with many small grains grown from the crucible wall. The solar cells prepared by mono-like wafers have high efficiency close to CZ wafers, but their distribution of conversion efficiency is still wider than that of solar cells prepared by HP cast wafers [80]. Especially, a long low-efficiency tail appears in the distribution of the conversion efficiency because wafers from the top part of an ingot still have high density of dislocations and small angle grain boundaries [65]. Such tail leads to the bad coat-effectiveness of the ingot grown by the mono-like cast method.

In future, a Si ingot grown by the mono-like cast method will be much more effective to obtain a large-scale Si single-like ingot with few dislocations by effectively controlling the generation of sub-grain boundaries from seed junctions and reducing the small grain regions near crucible inner wall. Although the structure is almost single-like and the dislocations is much fewer comparing to the HP cast ingot, the density of dislocation is still much higher than that of the CZ single ingot. As the largest merit of the mono-like cast method is its low-cost manufacturing the single-like Si ingot using a cast furnace comparing to the CZ method, the density of dislocation should be largely reduced for widely expanding the mono-like cast ingot by removing the remained stress in the ingot. This subject will be quite important for the future mono-like cast method because this is the main point for the mono-like cast ingot to replace the CZ dislocation-free ingot.

References

[1] T.F. Ciszek, G.H. Schwuttke, K.H. Yang, J. Cryst. Growth 46 (1979) 527.

[2] N. Stoddard, B. Wu, L. Witting, M. Wagener, Y. Park, G. Rozgonyi, R. Clark, Solid State Phenom 131—133 (2008) 1.

[3] A. Jouini, D. Ponthenier, H. Lignier, N. Enjalbert, B. Marie, B. Drevet, E. Pihan, C. Cayron, T. Lafford, D. Camel, Progress Photovoltaics 20 (2012) 735—746.

[4] G. Zhong, Q. Qinghua, Yu, X. Huang, L. Liu, Sol. Energy 111 (2015) 218.

[5] M. Trempa, M. Beier, C. Reimann, K. Roβhirth, J. Friedrich, C. Löbel, L. Sylla, T. Richter, J. Cryst. Growth 454 (2016) 6.

[6] H.J. Scheel, P. Capper, P. Rudolph, Crystal Growth Technology: Semiconductors and Dielectrics, John Wiley & Sons, 2010, pp. 177—178.

[7] B. Marie, S. Bailly, A. Jouini, D. Ponthenier, N. Plassat, L. Dubost, E. Pihan, N. Enjalbert, J. P. Garandet, D. Camel, The 26th European Photovoltaic Solar Energy Conference (26th EU PVSEC), Hamburg, September 5–9, 2011.

[8] B. Wu, R. Clark, J. Cryst. Growth 318 (2011) 200.

[9] W. Ma, G. Zhong, L. Sun, Q. Yu, X. Huang, L. Liu, Sol. Energy Mater. Sol. Cells 100 (2012) 231.

[10] X. Gu, K. Guo, L. Chen, D. Wang, D. Yang, Sol. Energy Mater. Sol. Cells 101 (2012) 95.

[11] K. Kutsukake, N. Usami, Y. Ohno, Y. Tokumoto, I. Yonenaga, Appl. Phys. Express 6 (2013) 025505.

[12] Y. Miyamura, H. Harada, K. Jiptner, S. Nakano, B. Gao, K. Kakimoto, K. Nakamura, Y. Ohshita, A. Ogura, S. Sugawara, T. Sekiguchi, Appl. Phys. Express 8 (2016) 062301.

[13] K. Kutsukake, N. Usami, Y. Ohno, Y. Tokumoto, I. Yonenaga, IEEE J. Photovoltaics 4 (2014) 84.

[14] G. Martinelli, R. Kibizov, Appl. Phys. Lett. 62 (1993) 3263.

[15] Y. Azuma, N. Usami, K. Fujiwara, T. Ujihara, K. Nakajima, J. Cryst. Growth 276 (2005) 393.

[16] Y. Miyamura, H. Harada, K. Jiptner, J. Chen, R.R. Prakash, J.Y. Li, T. Sekiguchi, T. Kojima, Y. Ohshita, A. Ogura, M. Fukuzawa, S. Nakano, B. Gao, K. Kakimoto, Solid State Phenom. 205–206 (2014) 89.

[17] C. Ding, M. Huang, G. Zhong, L. Ming, X. Huang, J. Cryst. Growth 387 (2014) 73.

[18] A. Black, J. Medina, A. Piñeiro, E. Dieguez, J. Cryst. Growth 353 (2012) 12.

[19] Q. Yu, L. Liu, W. Ma, G. Zhong, X. Huang, J. Cryst. Growth 358 (2012) 5.

[20] X. Qi, W. Zhao, L. Liu, Y. Yang, G. Zhong, X. Huang, J. Cryst. Growth 398 (2014) 5.

[21] V.A. Oliveira, B. Marie, C. Cayron, M. Marinova, M.G. Tsoutsouva, H.C. Sio, T.A. Lafford, J. Baruchel, G. Audoit, A. Grenier, T.N. Tran Thi, D. Camel, Acta Mater. 121 (2016) 24.

[22] N. Stoddard, B.G. Wendrock, A. Krause, D. Oriwol, M. Bertoni, T.U. Naerland, I. Witting, L. Sylla, Cryst. Growth 452 (2016) 272.

[23] T. Ervik, G. Stokkan, T. Buonassisi, Ø. Mjøs, O. Lohne, Acta Mater. 67 (2014) 199.

[24] M.G. Tsoutsouva, T. Riberi-Beridot, G. Regula, G. Reinhart, J. Baruchel, F. Guittonneau, L. Barrallier, N. Mangelinck-Noël, Acta Mater. 115 (2016) 210.

[25] M. Trempa, C. Reimann, J. Friedrich, G. Müller, A. Krause, L. Sylla, T. Richter, J. Cryst. Growth 405 (2014) 131.

[26] M. Trempa, C. Reimann, J. Friedrich, G. Müller, D. Oriwol, J. Cryst. Growth 351 (2012) 131.

[27] M. Trempa, C. Reimann, J. Friedrich, G. Müller, A. Krause, L. Sylla, T. Richter, Cryst. Res. Technol. 50 (2015) 124.

[28] T. Lehmann, M. Trempa, E. Meissner, M. Zschorsch, C. Reimann, J. Friedrich, Acta Mater. 69 (2014) 1.

[29] K.E. Ekstøm, G. Stokkan, R. Søndenå, H. Dalaker, T. Lehmann, L. Arnberg, M. Di Sabatino, Phys. Status Solidi 212 (2015) 2278.

[30] B. Gao, S. Nakano, H. Harada, Y. Miyamura, T. Sekiguchi, K. Kakimoto, J. Cryst. Growth 352 (2012) 47.

[31] Y. Miyamura, H. Harada, K. Jiptner, J. Chen, R.R. Prakash, S. Nakano, B. Gao, K. Kakimoto, T. Sekiguchi, J. Cryst. Growth 401 (2014) 133.

[32] A.L. Endrös, Sol. Energy Mater. Sol. Cells 72 (2002) 109.

[33] A. Otsuki, Interface Sci. 9 (2001) 293.

[34] J. Han, V. Vitek, D.J. Srolovitz, Acta Mater. 104 (2016) 259.

[35] K. Kutsukake, Growth of cryatalline silicon for solar cells: mono-like method, in: D. Yang (Ed.), Chapter 11, Section Two Crystalline Silicon Growth, Hand Book of "Photovoltaic Silicon Material", Springer, Berlin Heidelberg, 2018, pp. 1–20 (On line).

[36] K. Kutsukake, N. Usami, K. Fujiwara, Y. Nose, T. Sugawara, T. Shishido, K. Nakajima, Mater. Trans. 48 (2007) 143.

[37] K. Kutsukake, N. Usami, K. Fujiwara, Y. Nose, K. Nakajima, J. Appl. Phys. 101 (2007) 063509.

[38] D.G. Brandon, Acta Metall. 14 (1966) 1479.

[39] T. Jain, H.K. Lin, C.W. Lan, J. Cryst. Growth 485 (2018) 8.

[40] S. von Alfthan, K. Kaski, A.P. Sutton, Phys. Rev. B 74 (2006) 134101.

[41] F. Zhang, X. Yu, C. Liu, Z. Zhang, L. Huang, Q. Lei, D. Hu, D. Yang, Sol. Energy Mater. Sol. Cells 200 (2019), 109985.

[42] Y.C. Wu, A. Lan, C.F. Yang, C.W. Hsu, C.M. Lu, A. Yang, C.W. Lan, Cryst. Growth Des. 16 (2016) 6641.

[43] K. Jiptner, Y. Miyamura, H. Harada, B. Gao, K. Kakimoto, T. Sekiguchi, Prog. Photovoltaics Res. Appl. 24 (2016) 1513.

[44] M. Trempa, G. Müller, J. Friedrich, C. Reimann, Grain boundaries in multicrystalline silicon, in: D. Yang (Ed.), Crystalline Silicon Growth, Hand Book of "Photovoltaic Silicon Material", Springer, Berlin Heidelberg, 2018, pp. 1–48 (On line).

[45] A. Lantreibecq, M. Legros, N. Plassat, J.P. Monchoux, E. Pihan, J. Cryst. Growth 483 (2018) 183.

[46] V.A. Oliveira, M. Rocha, A. Lantreibecq, M.G. Tsoutsouva, T.N. Tran-Thi, J. Baruchel, D. Camel, J. Cryst. Growth 489 (2018) 42.

[47] I. Takahashi, N. Usami, K. Kutsukake, G. Stokkan, K. Morishita, K. Nakajima, J. Cryst. Growth 312 (2010) 897.

[48] N. Usami, K. Kutsukake, K. Fujiwara, K. Nakajima, J. Appl. Phys. 102 (2007) 103504.

[49] G. Stokkan, in: Proceedings of the 22nd European Photovoltaic Solar Energy Conference, 2007, p. 1282.

[50] G. Stokkan, J. Cryst. Growth 384 (2013) 107.

[51] A. Autruffe, V.S. Hagen, L. Arnberg, M.D. Sabatino, J. Cryst. Growth 411 (2015) 12.

[52] F. Zhang, X. Yu, D. Hu, S. Yuan, L. He, R. Hu, D. Yang, Sol. Energy Mater. Sol. Cells 193 (2019) 214.

[53] M.G. Tsoutsouva, V.A. Oliveira, D. Camel, J. Baruchel, B. Marie, T.A. Lafford, Acta Mater. 88 (2015) 112.

[54] M.G. Tsoutsouva, V.A. Oliveira, D. Camel, T.N. Tran Thi, J. Baruchel, B. Marie, T.A. Lafford, J. Cryst. Growth 401 (2014) 397.

[55] M.G. Tsoutsouva, V.A. Oliveira, J. Baruchel, D. Camel, B. Marie, T.A. Lafford, J. Appl. Crystallogr. 48 (2015) 645.

[56] I. Takahashi, S. Joonwichien, T. Iwata, N. Usami, Jpn. Appl. Phys. Express 8 (2015) 105501.

[57] K. Nakajima, K. Morishita, R. Murai, N. Usami, J. Cryst. Growth 389 (2014) 112.

[58] A.K. Søilanda, E.J. Øvrelid, T.A. Engh, O. Lohne, J.K. Tuset, Ø. Gjerstad, Mater. Sci. Semicond. Process. 7 (2004) 39.

[59] C. Reimann, M. Trempa, J. Friedrich, G. Müller, J. Cryst. Growth 312 (2010) 1510.

[60] T. Tachibana, T. Sameshima, T. Kojima, K. Arafune, K. Kakimoto, Y. Miyamura, H. Harada, T. Sekiguchi, Y. Ohshita, A. Ogura, J. Appl. Phys. 111 (2012) 074505.

[61] A. Jouini, F. Jay, V. Amaral, E. Pihan, Y. Veschetti, Crystalline Silicon for Solar Cells, CSSC-7, October 22–25, 2013. Fukuoka, Japan.

[62] A. Lantreibecq, J.P. Monchoux, E. Pihan, B. Marie, M. Legros, Mater. Today: Proceedings 5 (2018) 14732.

[63] Y. Zhang, Z. Li, Q. Meng, Z. Hu, L. Liu, Sol. Energy Mater. Sol. Cells 132 (2015) 1.

[64] F. Jay, D. Muñoz, T. Desrues, E. Pihan, V.A. Oliveira, N. Enjalbert, A. Jouini, Sol. Energy Mater. Sol. Cells 130 (2014) 690.

[65] L. Gong, F. Wang, Q. Cai, D. You, B. Dai, Sol. Energy Mater. Sol. Cells 120 (2014) 289.

[66] D. Hu, S. Yuan, L. He, H. Chen, Y. Wan, X. Yu, D. Yang, Sol. Energy Mater. Sol. Cells 140 (2015) 121.

[67] M. Kitamura, N. Usami, T. Sugawara, K. Kutsukake, K. Fujiwara, Y. Nose, T. Shishido, K. Nakajima, J. Cryst. Growth 280 (2005) 419.

[68] C.Y. Lan, Y.C. Wu, A. Lan, C.F. Yang, C. Hsu, C.M. Lu, A. Yang, C.W. Lan, J. Cryst. Growth 475 (2017) 136.

[69] A. Jouini, CSSC-10, April 8—11, 2018. Sendai, Japan.

[70] T. Iwata, I. Takahashi, N. Usami, Jpn. J. Appl. Phys. 56 (2017) 075501.

[71] K. Jiptner, M. Fukuzawa, Y. Miyamura, H. Harada, T. Sekiguchi, Jpn. J. Appl. Phys. 52 (2013) 065501.

[72] S. Nakano, B. Gao, K. Jiptner, H. Harada, Y. Miyamura, T. Sekiguchi, M. Fukuzawa, K. Kakimoto, J. Cryst. Growth 474 (2017) 130.

[73] K. Jiptner, M. Fukuzawa, Y. Miyamura, H. Harada, K. Kakimoto, T. Sekiguchi, Phys. Status Solidi C 10 (2013) 141.

[74] A. Jouini, G. Fortin, E. Pihan, V. Amaral, D. Camel, F. Jay, D. Munoz, The 27th European Photovoltaic Solar Energy Conference (27th EU PVSEC), Frankfurt, Germany, September 24—28, 2012.

[75] M. Trempa, C. Reimann, J. Friedrich, G. Müller, L. Sylla, A. Krause, T. Richter, J. Cryst. Growth 429 (2015) 56.

[76] I. Guerrero, V. Parra, T. Carballo, A. Black, M. Miranda, D. Cancillo, B. Moralejo, J. Jiménez, J.F. Lelièvre, C. del Cañizo, Prog. Photovoltaics Res. Appl. 22 (2014) 923.

[77] G. Zhong, Q. Yu, X. Huang, L. Liu, J. Cryst. Growth 402 (2014) 65.

[78] Y. Hayama, I. Takahashi, N. Usami, J. Cryst. Growth 468 (2017) 610.

[79] S.P. Phang, H.C. Sio, C.F. Yang, C.W. Lan, Y.M. Yang, A.W.H. Yu, B.S.L. Hsu, C.W.C. Hsu, D. Macdonald, Jpn. J. Appl. Phys. 56 (2017) 08MB10.

[80] X. Zhang, L. Gong, B. Wu, M. Zhou, B. Dai, Sol. Energy Mater. Sol. Cells 139 (2015) 27.

[81] F. Jay, D. Muñoz, N. Enjalbert, G. D. Alonzo, J. Stendera, S. Dubois, D. Ponthenier, A. Jouini, P. J. Ribeyron, ResearchGate, Conference Paper, September, 2012.

[82] A. Jouini, F. Coustier, E. Flahaut, D. Sarti, "100 mm Mono-Like Silicon Wafers for High Solar Cells Efficiency", in: "100 mm Mono-Like Silicon Wafers for High Solar Cells Efficiency", M-A-27, 2nd, Intern. Conf. on Cryst. Silicon Photovol. SiliconPV, Leuven Belgium, April 2012, pp. 3—5.

[83] J. Barredo, V. Parra, I. Guerrero, A. Fraile, L. Hermanns, Prog. Photovoltaics Res. Appl. 22 (2014) 1204.

CHAPTER 7

Growth of Si ingots using cast furnaces by the NOC method

7.1 Development of the NOC method

7.1.1 Trigger for the development

A multicrystalline Si ingot is usually grown inside a quartz crucible during contacting the crucible wall from the top to the bottom by directional growth. For the ingot growth using such a crucible, the stress is mainly caused by compressive force from the crucible because a Si melt expands by 11% during solidification to grow the ingot and the crucible wall made of silica has insufficient flexibility to reduce the stress. In the case of ingot growth using a Si_3N_4 coated quartz crucible, however, an ingot can be more easily released from the crucible wall because of the lower thermal expansion coefficient of Si_3N_4 than that of Si [1]. To grow a high-quality Si ingot using a quartz crucible, it is necessary to prepare the Si ingot while controlling the stress because the driving force to generate dislocations is stress in the ingot. In conventional Si ingots grown by the cast method, the dislocation density is usually between 10^5 and 10^6 cm^{-2}. Therefore, the reduction of stress in the ingots has been one of the most important subjects for the cast growth.

In the initial attempt for the growth of a Si ingot without contact to a crucible wall, the floating cast method was proposed [2,3]. In this method, nucleation occurs on the surface of a Si melt using a floating Si seed crystal and an ingot grows inside the Si melt from the floating seed near the melt surface. The grain size of the ingot is remarkably larger than that of an ingot grown by the cast method. The most of grain boundaries in the ingot is $\Sigma 3$ grain boundary. This suggests that the structure of the ingot can be largely improved by controlling the nucleation site using a seed crystal. However, this attempt is unsuccessful in terms of reducing stress and taking out the grown ingot from the Si melt because the floating ingot finally grows in contact with the crucible wall even though the ingot is initially floating in

Crystal Growth of Si Ingots for Solar Cells Using Cast Furnaces
ISBN 978-0-12-819748-6
https://doi.org/10.1016/B978-0-12-819748-6.00007-4

Copyright © 2020 Elsevier Inc.
All rights reserved.

the Si melt [4,5]. The solidified ingot covers the surface of the Si melt, it grows inside the Si melt from the top to the bottom of the crucible and the remained Si melt is finally crystallized under the surface-covered ingot. The stress becomes much larger when the finally grown Si ingot breaks the surface-covered ingot out due to expansion by solidification. The large destructive power of this stress can be known by a fact that the grown ingot breaks through the bottom of a crucible with a thinner bottom of 0.25 cm thickness when a remained Si melt is solidified under a surface-covered crystal [6].

The NOC method was proposed to realize a Si ingot without any contact to the crucible wall and to reduce the expansion and contact stress in the Si ingot grown inside a conventional silica crucible. In this method, a Si melt has a large low-temperature region in its upper central part and the natural crystal growth occurs inside it. The diameter of the ingot is determined by the size of the low-temperature region in the melt (see Section 7.1.3). The growth rate is determined by the horizontally and vertically expanding rate of the low-temperature region. To realize such a large low-temperature region, an optimum temperature distribution must be created in the Si melt inside the hot zone by designing the furnace.

7.1.2 Comparison with the kyropoulos method

The NOC method is similar to the Kyropoulos method [5,7−9] only on the view point of growing an ingot inside a melt. However, the concept is completely different from each other on the view point whether the low-temperature region is intentionally established in a Si melt or not. The furnace of the Kyropoulos method has a cold spiral tube above the surface of a melt to promote growth of a crystal deeper inside the melt in vertical direction [5]. The Kyropolous method is mainly used for the growth of sapphire [7]. Sapphire contracts by 18% during solidification, so there are no problems for contacting the crucible wall. No nucleation occurs from the crucible wall because Mo can be used as the crucible material. The thermal expansion of sapphire is two orders of magnitude lower than that of Si, which lowers the buoyancy force during growth.

For growth of a Si ingot inside a Si melt, several trials have been performed using the Kyropoulos method [8,9]. Ravishankar attempted to grow Si ingots using the Kyropoulos method, but the growing ingots was in contact with the crucible wall during growth and it took a same shape of the crucible because a low-temperature region was not established in the Si melt [8]. His furnace had a water-cooled shaft under the crucible. Nouri

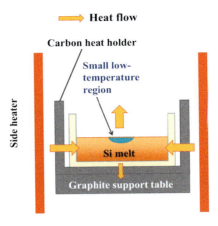

Kyropoulos method

Fig. 7.1 Schematic illustration of the typical heat flow of the Kyropoulos method. The heat is mainly removed from the surface and bottom of the Si melt through the seed holder and the crucible bottom. *(Referred from Fig. 1 in K. Nakajima et al. J. Cryst. Growth 499 (2018) 55.)*

et al. [9] also tried to grow Si ingots using the Kyropoulos method based on the 3- dimensional simulation. They grew a symmetric Si crystal (about 8 cm diameter) inside a Si melt using a seed without contact to the crucible wall and without any upward pulling. However, their ingots were still very small because the low-temperature region was quite small [9]. Fig. 7.1 shows a schematic illustration of the typical heat flow of the Kyropoulos method, which has a relatively long side heater [7]. The heat is mainly removed from the surface and bottom of the Si melt, and the latent heat is mainly removed through the cold tube as a seed holder and from the ingot surface in Ar gas atmosphere [8,9]. Thus, the Si melt is wholly cooled during growth under such configuration of heat flow and the large low-temperature region cannot be established by the Kyropoulos method.

In the growth of a Si ingot, Si expands during solidification, so there are serious problems for contacting the crucible wall. For the ingot growth using a conventional silica crucible, it is more difficult to prevent growth from the crucible wall because the wall surface is coated with a Si_3N_4 layer which strongly acts as nucleation sites. Therefore, for the growth of a Si ingot without contacting with the crucible wall, it is very important to create a large low temperature region in the Si melt by designing the furnace. Thus, the proposed NOC method is defined as a growth method to intentionally establish a distinct low-temperature region in a Si melt.

7.1.3 Principle of the NOC method

Fig. 7.2 shows the principle of the NOC method [10]. In this method, a Si melt has an intentionally established large low-temperature region in its upper central part, in which the melt temperature is kept lower than the melting point of Si (Mp = 1414°C). The low-temperature region allows natural crystal growth inside it. Nucleation occurs on the surface of the Si melt using a seed crystal and an ingot grows along the surface of the Si melt and inside the melt without contact with the crucible wall. Then, the growing ingot is slowly pulled upward while the ingot grows inside the low-temperature region. The diameter of the ingot is freely determined by the size of the low-temperature region in the melt. The NOC method can be used to prepare a Si single ingot with a diameter ratio larger than 0.9 [11], where the diameter ratio is the maximum diameter of the ingot divided by the crucible diameter. A Si ingot with a diameter almost as large as the crucible diameter can be grown by the NOC method.

In principle, the NOC method has several novel characteristics originated from its key feature that a large low-temperature region exists in a Si melt and an ingot can be grown inside a Si melt without contact with a crucible wall. (1) A large ingot with a large diameter and a large diameter ratio can be grown using a small crucible [11]. Si single ingots with a diameter as large as 80%−90% of the crucible diameter can be grown

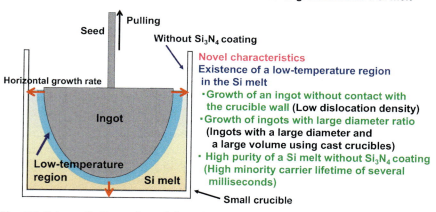

Fig. 7.2 Schematic illustration of the NOC method with its novel characteristics that originate from its key feature that ingots can be grown inside Si melt without contact to crucible wall owing to a low-temperature region. *(Referred from Fig. 1 in K. Nakajima J. Cryst. Growth 468 (2017) 705.)*

[12,13]. A Si ingot with the largest diameter of 45 cm is realized using a 50 cm diameter crucible [11]. The NOC method has a potential to obtain a larger single ingot using a larger crucible. (2) The EPD of dislocations in the center of an ingot is on the order between 10^2 and 10^4 cm^{-2} without necking technique [14]. A low dislocation density can be realized owing to the convex growing interface in the Si melt [15]. (3) The millisecond lifetime is obtained after gettering because the high–purity ingot is grown using a crucible without a Si$_3$N$_4$ coating [16]. (4) An ingot with a low O concentration can be obtained because the convection is suppressed to be very small in the Si melt using a very low co-rotation rate of the seed and crucible in the same direction [11]. (5) An ingot with a large volume can be realized because of the buoyancy effect from the Si melt. (6) The yield of the p-type regular solar cells with a high conversion efficiency is much larger than that of the HP cast method [10]. Using the cast furnace, p-type solar cells with the conversion efficiency and yield as high as those of CZ solar cells can be realized by the NOC method [10]. (7) The defect formation mechanism of the NOC method may be very different from that of the CZ method because pair annihilation between vacancies and interstitial atoms occurs during long diffusion in the hot ingot from the growing interface to the melt surface position.

These merits can be applied to both the CZ and cast furnaces with a large low-temperature region which are used for the NOC method. Using the cast furnace, a Si single ingot with a large volume can be effectively grown using a thick seed. Using the CZ furnace with a bottom heater, a Si single ingot with a diameter larger than 45 cm can be grown using a 50-cm diameter crucible. In this case, four wafers can be obtained for regular-size solar cells from the cross section of the ingot. One of final goals of the NOC method is to realize a uniform large Si single ingot with sufficient quality to obtain solar cells with the conversion efficiency and yield as high as those of the CZ solar cells using the cast furnace.

7.1.4 Furnace design to grow a Si ingot inside a Si melt

The furnace design is very important to intentionally establish a distinct low-temperature region in a Si melt. Fig. 7.3 shows the designed furnaces with two and three zone heaters [11]. In Fig. 7.3, Tc1, Tc2 and Tc3 are the thermocouples for the first, second and third heaters, respectively. The temperature is directly monitored by the thermocouple (Tc2) set near the side of the crucible wall or below the susceptor and it is determined from the value on the monitor of the heater used. The Ar pressure of 0.2 MPa is

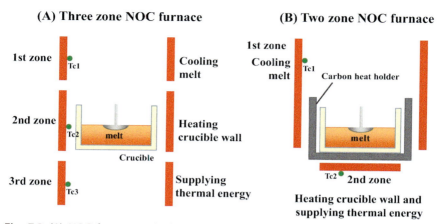

Fig. 7.3 (A) NOC furnace with three zone heaters, (B) NOC furnace with two zone heaters which has a carbon heat holder to maintain the heat of the crucible wall. These NOC furnaces have three roles such as cooling the Si melt, heating the crucible wall and supplying thermal energy to the Si melt assigned for each zone-heater. *(Referred from Fig. 1 in K. Nakajima et al. J. Electron. Materials, 45 (2016) 2837.)*

usually used, and the Ar flows from the top of the chamber to the melt surface and out from the side or bottom of the chamber.

In the furnace with three zone heaters, as shown in Fig. 7.3A, each heater has each role to obtain a large low-temperature region in a Si melt. The first heater is mainly used to cool the Si melt for the growth of an ingot by reducing the heater temperature. The second heater is mainly used to heat the crucible wall and maintain its heat to keep the melt temperature and the crucible wall at a high temperature. The third heater is used to supply more thermal energy from the crucible bottom to the Si melt to keep the melt temperature high, preferably higher near the periphery of the crucible than that in its center. Thus, the third heater plays an important role to create a large low-temperature region in the Si melt. The furnace with three zone heaters is essentially required and usually used for the NOC method as shown in Fig. 7.3A.

A simpler furnace with two zone heaters (NOC Furnace 600: KDN) is advantageous for practical use as shown in Fig. 7.3B. In this case, the first zone (side) heater is a cylindrical carbon heater and the second zone (bottom) heater is a disk-shaped carbon heater, which is set horizontally and below the first zone heater. The inner diameter of the first heater is 89 cm and the diameter of the second heater is 65 cm. The thicknesses of the carbon heaters are 1.8 and 1.5 cm for the first and second zone heaters, respectively. The maximum power of the first and second heaters is 140 and 60 kW, respectively. The first zone heater has a role to supply the thermal energy to the Si melt and keep the melt temperature high. The second zone heater has two roles such as heating the crucible wall and supplying thermal energy to the Si melt. A low-temperature region is established mainly by heating the side of a crucible using both side and bottom heaters. To satisfy the three roles using only the furnace with two zone heaters, a carbon heat holder is applied to keep the temperature near the crucible wall much higher than the melting point of Si as shown in Fig. 7.3B. The carbon heat holder can contain the entire quartz crucible. The holder has an outer diameter of 55 cm, an inner diameter of 35, 42 and 52 cm and a height of 21 cm. These inner diameters are used for crucibles with diameters of 33, 40 and 50 cm, respectively. The wall of the holder is made to be very thick (5.5—10.0 cm) to flow the thermal energy through the holder wall from the second heater to the top of the wall and prevent the cooling of the crucible side-wall. Owing to the existence of the carbon heat holder, both the Si melt and the crucible wall are simultaneously cooled by reducing the temperature of the second heater. The establishment of a large low-temperature region is cleverly performed by the combination of the first and second zone heaters and the carbon heat holder.

For the ingot growth using the furnace with two zone heaters, the first zone heater is used in the same manner as that in the furnace with three zone heaters. The Si melt temperature is mainly reduced by cooling the first heater. When the cooling of the Si melt by the first zone heater is insufficient to largely expand the growing ingot inside the Si melt, the temperature of the second bottom heater is reduced to increase the temperature reduction of the Si melt. In this case, both the Si melt and the crucible wall are cooled by reducing the temperature of the second zone heater. The total ΔT (K) associated with the growth can be written as

$$\Delta T = \Delta T_1 + \Delta T_2 \tag{7.1}$$

where ΔT is the total temperature reduction of a Si melt defined as the difference between the starting temperature, at which the initial crystal is first observed to start growing from the seed crystal, and the final temperature at the end of the cooling. The starting temperature can be determined by direct observation of the melt surface near the seed. ΔT_1 (K) is defined as the difference between the starting and final temperatures of the first heater during cooling the melt for the ingot growth, and ΔT_2 (K) is defined as the difference between the initial and final temperatures of the second heater during cooling the melt. The temperature reduction of the melt is determined by measuring the monitor value at the heater system. The measured temperature reduction does not always exactly correspond to the real amount of supercooling used for growth because of thermal inertia in the Si melt. However, it can be used as an indicator to relatively compare each temperature reduction among the NOC growth of ingots, and the approximate amount of the real temperature reduction can be known from the measured temperature reduction. The timing to finish the growth process can be determined by comparing the estimated volume of the growing ingot with the charged volume of the melt.

Each furnace has a rotating mechanism and a graphite susceptor on which a silica crucible is set. Each furnace should have two windows through which the melt surface near the seed and crucible wall can be directly observed. A seed crystal is attached to a rotating axis, which can move up and down so that the seed can come in contact with the melt surface. The ingot is continuously rotated at a rate of 0.1–1.5 rpm, and the crucible is continuously co-rotated in the same direction at a rate of 0.05–1.0 rpm during the ingot growth. The Si ingot grows naturally from the seed crystal inside the Si melt, then continues to grow while being slowly pulled upward, during ensuring that the crystal growth restricts inside the Si melt. The growth rate can be precisely controlled by varying the cooling rate of the Si melt or the expanding rate of the low–temperature region. At the final stage of growth, the ingot is pulled upward at a rate of 10 mm min^{-1} to separate it from the remaining melt. The timing to pulling the ingot upward is directly determined by the in–situ observation of the growing interface near the crucible wall.

7.2 Establishment of the low-temperature region in a Si melt

7.2.1 Concept to establish the low-temperature region in a Si melt

The NOC method is comparable to the CZ method [10], which has generally been used for the growth of a Si single ingot. In the CZ method, an ingot is grown above the surface of a Si melt, the growing interface is generally concave in the growth direction, and a low-temperature region is not intentionally established in the Si melt. In the NOC method, an ingot is grown inside a Si melt and the growing interface is convex in the growth direction, and a large low-temperature region is required in the Si melt.

The Si melt surrounding the low-temperature region prevents the contact between an ingot and the hotter crucible wall. Since the temperature of the crucible wall is higher than the melting point of Si, there is not serious nucleation from the crucible wall.

To realize such NOC growth, the large and deep low-temperature region should be established in the upper central part of a Si melt. The melt temperature near the bottom of crucible should be kept higher than that near the melt surface. The melt temperature near the inner side wall of crucible should be kept much higher than that in the central part of the Si melt. However, it is very difficult to create the large low-temperature region by cooling the surface of the Si melt using strong blow of Ar gas because the cooling part is limited only near the melt surface. A new concept is required to effectively create the large low-temperature region inside the Si melt.

A new concept for intentionally establishing a distinct low-temperature region is proposed on the view of controlling the heat flow in the designed furnace [17]. The concept is fundamentally based upon preventing the heat flow from the bottom heater into the Si melt, especially on the central part of the crucible bottom, but it is not based on locally cooling the Si melt. The innovative and ingenious idea for the heat-flow control is based on using an insulator plate on the crucible bottom [17]. The heat flow into the melt is basically controlled using insulating plates set under the crucible bottom. The cast furnace with two zone heaters can be used which was design as a simple furnace for practical use. The two zone heaters and a carbon heat holder are arranged as shown in Fig. 7.4B. Fig. 7.4 shows a schematic illustration of each concept of the heat flow toward the Si melt inside the quartz crucible for the conventional CZ and NOC furnaces. Each crucible is set on the graphite support table for both methods. For the conventional setup as shown in Fig. 7.4A, the Si melt is heated by the heat flow from the side and bottom heaters. Sometimes, instead of the bottom heater, a long side heater is used [18]. The small low-temperature region is normally obtained only near the center of the melt surface. The size of an ingot grown inside the region is quite small. For the NOC setup using an insulating plate set under the center of crucible bottom as shown in Fig. 7.4B, the Si melt is similarly heated by the heat flow from the side heater, but the heat flow from the bottom heater is extremely limited by the insulating plate, especially near the center of the crucible bottom. On the contrary, the periphery of the Si melt is strongly heated by the heat flow from the bottom heater through a periphery plate with a high thermal

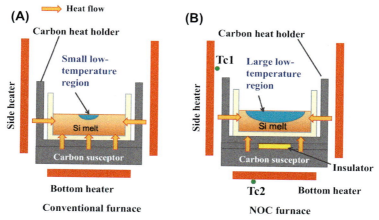

Fig. 7.4 Schematic illustration of the concept of the heat flow toward the Si melt. (A) Conventional setup of the furnace. The Si melt is heated by the heat flow from the side heater and the bottom heater. The low-temperature region is quite small. (B) The NOC setup using an insulating plate set under the center of the crucible bottom. The Si melt is heated by the heat flow from the side heater, but the heat flow from the bottom heater is extremely limited by the insulating plate. The low-temperature region can become quite large. *(Referred from Fig. 2 in K. Nakajima et al. J. Cryst. Growth 499 (2018) 55.)*

conductivity. The large low-temperature region can be clearly established in the central upper part of the Si melt using the proper size of insulating plate. An ingot with a large size can be freely grown by expanding low-temperature region into the Si melt.

7.2.2 Method to establish the low-temperature region in a Si melt

A disk-shaped graphite plate is used as the insulating plate, which has a low thermal conductivity of 0.2–0.4 W m^{-1} K^{-1} at Mp as shown in Fig. 7.5A [17]. The graphite plate is surrounded by another enclosure ring-shaped graphite plate with a high thermal conductivity of 40–120 W m^{-1} K^{-1} at Mp as shown in Fig. 7.5B. These graphite plates with low and high thermal conductivities are combined to one plate as shown in Fig. 7.5C. The combined plate is made to completely cover the bottom surface of crucibles. The heat flow is largely blocked by the insulator and the main heat flow is generated from the outer periphery of the disk-shaped graphite plate. The main heat flow is largely strengthened because most of heat from the bottom heater gets together and flows into the periphery of the Si melt

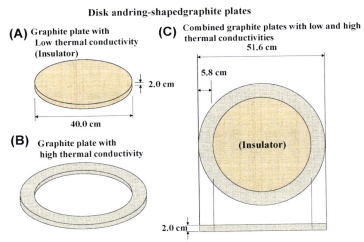

Fig. 7.5 (A) Disk-shaped graphite plate used as the insulating plate which has a low thermal conductivity of 0.35 W m^{-1} K^{-1} at Mp. (B) Ring-shaped graphite plate set around the disk-shaped graphite plate, which has a high thermal conductivities of 40 W m^{-1} K^{-1} at Mp. (C) Combined graphite plates with low and high thermal conductivities. The enclosure ring-shaped graphite plate is located outside of the disk-shaped graphite plate with low thermal conductivity. The thickness of the disk-shaped graphite plate used for the ingot growth is 0.5–3.0 cm. *(Referred from Fig. 4 in K. Nakajima et al. J. Cryst. Growth 499 (2018) 55.)*

only through the narrow space outside the insulator. The disk-shaped graphite plates with several thicknesses can be used in the range of 0.5–3.0 cm. For the ingot growth using a 50-cm diameter crucible, the enclosure graphite plate with a high thermal conductivity is located outside of the disk-shaped graphite plate with low thermal conductivity and with the thickness of 2.0 cm and the diameter of 40.0 cm.

7.2.3 Relationship between ΔT and the low-temperature region

In the NOC method, the melt temperature near the crucible wall must be kept higher than that in the low-temperature region to prevent crystal growth from the wall. The diameter of the ingot is determined by the size of the low-temperature region in the Si melt, which is mainly determined by ΔT during crystal growth and the temperature gradient in the Si melt when the crucible size and melt volume are constant. Fig. 7.6 shows a schematic illustration of the relationship between ΔT and ΔT_d (K) for the NOC furnace with the insulating plate [11,17]. Here, T_{a1} is the initial melt

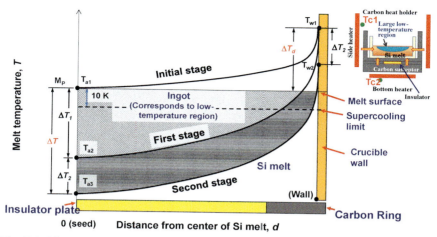

Fig. 7.6 Schematic illustration of the relationship between ΔT (K) and ΔT_d (K) for the furnace with the insulating plate. T_{a1} is the initial melt temperature below the seed which corresponds to Mp. T_{a2} is the melt temperature after cooling by ΔT_1. T_{w1} is the initial melt temperature close to the crucible wall. In the initial stage, the melt temperature is equal to Mp near the center of the Si melt and has a steep slope near the crucible wall. In the first stage, the Si melt was cooled from T_{a1} to T_{a2} ($\Delta T_1 = T_{a1} - T_{a2}$), and the melt temperature near the wall remained at T_{w1}. In the second stage, the Si melt was cooled from T_{a2} to T_{a3} ($\Delta T_2 = T_{a2} - T_{a3}$), and the melt temperature near the wall was reduced from T_{w1} to T_{w2}. T_{w2} is the melt temperature close to the wall after cooling by ΔT_2 ($\Delta T_2 \cong T_{w1} - T_{w2}$). ΔT is estimated as the sum of ΔT_1 and ΔT_2 ($\Delta T = \Delta T_1 + \Delta T_2$). ΔT_d is defined as the temperature difference between T_{w1} and T_{a1} ($\Delta T_d = T_{w1} - T_{a1} = T_{w1} - Mp$). The thermocouple Tc1 is located inside the first zone heater and the thermocouple Tc2 is located at the bottom of the second zone heater. (Referred from Fig. 5 in K. Nakajima et al. J. Cryst. Growth 499 (2018) 55.)

temperature directly below the seed crystal set on the central axis and it corresponds to the starting growth temperature from the seed. T_{a2} is the melt temperature after cooling by ΔT_1. T_{a1}, T_{a2} and T_{a3} are monitored using a thermocouple Tc1 located inside the first zone heater. The location of the thermocouples, Tc1 and Tc2 are shown in Figs. 7.4 and 7.6. Tc2 is located at the bottom of the second zone heater. The exact timing when T_{a1} is almost equal to the melting point of Si can be known by direct observation of the seed touching at which the initial crystal is first observed to start growing from the seed. T_{w1} is the initial melt temperature close to the crucible wall. In the initial stage before cooling, the melt temperature is equal to Mp near the center of the Si melt and has a steep slope near the crucible wall. The starting growth temperature from the seed also corresponds to Mp. In the first stage after cooling, the Si melt is cooled from T_{a1} to T_{a2} ($\Delta T_1 = T_{a1} - T_{a2}$) using only the first zone heater, and the melt temperature near the wall remains around T_{w1}, because a large ΔT can be still kept even though the melt temperature at the center of Si melt decreases by cooling the first zone heater, thanks to large ΔT_d caused by the insulating plate. A low-temperature region is created in the Si melt by this cooling and an ingot simultaneously grows inside the region. The size and shape of the ingot almost correspond to those of the low-temperature region. In the second stage, the Si melt is cooled from T_{a2} to T_{a3} ($\Delta T_2 = T_{a2} - T_{a3}$) using only the bottom heater, and the melt temperature near the wall is also reduced from T_{w1} to T_{w2} owing to the carbon heat holder. T_{w2} is the melt temperature close to the wall after cooling by ΔT_2 ($\Delta T_2 \cong T_{w1} - T_{w2}$). The total temperature reduction, ΔT is estimated as the sum of ΔT_1 and ΔT_2 ($\Delta T = \Delta T_1 + \Delta T_2$). ΔT_d is defined as the temperature difference between T_{w1} and T_{a1} ($\Delta T_d = T_{w1} - T_{a1} = T_{w1} - Mp$).

The temperature difference between the initial temperatures near the center of the Si melt and near the crucible wall, ΔT_d is the most important parameter for obtaining a large ingot inside the crucible. The melt temperature along the surface of the Si melt has a steep slope toward the crucible wall, and it becomes more steeper near the crucible wall as ΔT_d increases. ΔT_d is a key parameter determining the temperature gradient in the Si melt. When ΔT_d is sufficiently large to obtain a large ΔT_2, we can utilize the large ΔT_2 or ΔT before ΔT_2 reaches the supercooling limit of 10 K below Mp [19]. The maximum diameter of an ingot is determined by the final size of the low-temperature region which is mainly determined by ΔT during crystal growth and the temperature slope in the Si melt. ΔT is effective as a criterion for controlling the size and shape of ingot. The size of the insulating plate strongly affects ΔT and ΔT_d which determine the temperature slope toward the crucible wall. The large temperature slope can keep the melt temperature near the crucible wall higher than that in the low-temperature region to prevent nucleation on the crucible wall. The growth rate is basically determined by the expansion rate of the low-temperature region which can be precisely controlled by the temperature reduction and cooling rate of the Si melt (see Section 7.4.1).

7.2.4 Effect of ΔT_d on the low-temperature region and the ingot size

In the first cooling stage using the first zone heater, if T_{w1} is equal to T_{a1} ($\Delta T_d = 0$ K), crystals simultaneously grow from the seed crystal and the crucible wall during cooling, and a very small low-temperature region is established in the Si melt. A small temperature reduction of less than 10 K is available for the ingot growth in the Si melt ($\Delta T \leq 10$ K) in this case. Only a very small ingot grows from the seed. If ΔT becomes larger than 10 K in this case, many dendrite crystals suddenly appear from the crucible wall because the melt temperature near the crucible wall exceeds the supercooling limit of 10 K [20]. When ΔT_d is much larger than 10 K, the ingot continues to grow only from the seed crystal owing to the large low-temperature region. Dendrite crystals do not appear from the crucible wall in this case. The low-temperature region is sufficiently large to obtain a large ingot before dendrite crystals appear because ΔT_d is enough large to keep the wall temperature higher than the melt temperature inside the low-temperature region during cooling and growth. Only when ΔT becomes larger than $\Delta T_d + 10$ K, dendrite crystals appear from the crucible wall because the melt temperature near the crucible wall exceeds the supercooling limit of 10 K even though ΔT_d is large.

The furnace with two zone heaters (NOC Furnace 600: Dai-ich Kiden Corporation Limit.) are used to grow ingots by the NOC method [11]. To reduce the melt temperature near the center of the Si melt, it is effective to cool the melt only using the first zone heater. Fig. 7.7A shows an n-type small ingot (NOC-1) grown in a small low-temperature region because a hotter crucible wall could not be established in this case. As shown in Fig. 7.7B, crystals simultaneously grow from the seed crystal and crucible wall during cooling. The maximum value of ΔT is only 10.7 K in this growth. When ΔT reaches 10 K, many dendrite crystals start to grow from the crystals grown on the crucible wall because the melt temperature near the crucible wall is slightly over the supercooling limit of 10 K [19] as shown in Fig. 7.7B. When the low-temperature region is sufficiently large to obtain a large ΔT in the Si melt by maintaining the crucible wall at a higher temperature, the grown ingot (NOC-2) is much larger as shown in Fig. 7.7C. The maximum value of ΔT is 30.6 K in this growth. Even at such a high ΔT, no crystal grows from the crucible wall owing to the sufficiently high melt temperature near the crucible wall as shown in Fig. 7.7D. Thus, in this case, a large low-temperature region can be clearly established in the Si melt simply by reducing the temperature of the first zone heater.

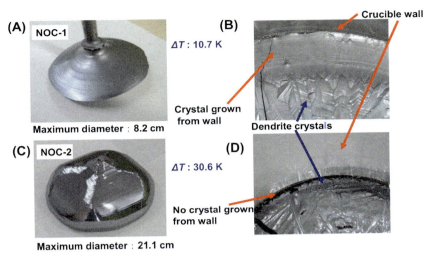

Fig. 7.7 (A) n-Type ingot (NOC-1) grown in a small low-temperature region in a Si melt owing to a small ΔT because the crucible wall was not much hotter than the low-temperature region and ΔT_d was small. (B) Crystal grown from the crucible wall during crystal growth from a seed crystal. ΔT was reached only 10.7 K. (C) n- Type ingot (NOC-2) grown in a relatively large low-temperature region created by a large ΔT because the crucible wall was much hotter than the low-temperature region and ΔT_d was large. (D) No crystal grew from the crucible wall during crystal growth from the seed. ΔT was reached 30.6 K. *(Referred from Fig. 3 in K. Nakajima et al. J. Electron. Materials, 45 (2016) 2837.)*

7.2.5 Effect of the insulating plate on the growth of Si ingots

The effects of the diameter and thickness of the insulating plate on the diameter and diameter ratio of grown ingots can be know from many ingots were grown using the different sizes of crucible and insulator. An ingot (NOC-3) is grown using a crucible with 33 cm diameter and an insulator of 20 cm diameter. The cooling rate is 0.2 K min^{-1} and the pulling rate is 0.2–0.4 mm min^{-1}. Fig. 7.8 shows the ingot of NOC-3, and the diameter and diameter ratio of the ingot are 14.7 cm and 0.44 when ΔT is 12.5 K, respectively [17]. The growth parameters and the crystal size of the ingot is shown in Table 7.1. The ingot has a square shape with 11.3 × 11.4 cm as shown in Fig. 7.8A. The height of the ingot is only 2.6 cm because of the relatively small ΔT. As shown in Fig. 7.8B, the bottom of the ingot is convex in the growth direction and the shape of the lower half part of the ingot almost corresponds to the shape of the

An ingot grown using a small insulator with a diameter of 20 cm

(A) Upper surface of the ingot   **(B)** Side shape of the ingot

 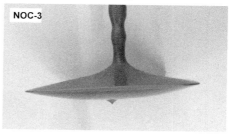

Diameter of the crucible: 33 cm,   Height of the ingot: 2.6 cm
Diameter of the ingot: 14.7 cm,   Mass of the ingot: 356 g
Diameter ratio: 0.44

Fig. 7.8 Effect of the insulator diameter on the diameter and diameter ratio of grown ingots. (A) Ingot of NOC-3 grown using an insulating plate with a diameter of 20 cm. The diameter and diameter ratio of the ingot are 14.7 cm and 0.44, respectively. The ingot has a square shape with 11.3 × 11.4 cm. (B) Side view to see the ingot height of only 2.6 cm. The bottom shape of the ingot is convex in the growth direction. *(Referred from Fig. 7 in K. Nakajima et al. J. Cryst. Growth 499 (2018) 55.)*

low-temperature region in the Si melt. This means that the isothermal line of the solid/liquid interface has a convex shape in the growth direction because the low-temperature region is expanding toward the bottom of the Si melt during radiating latent heat outside the region to the melt.

The diameter and diameter ratio of a grown ingot are strongly affected by the insulator size. The mass of the ingot also increases with ΔT and the insulator diameter. The convex growing interface is much easily obtained using a lower pulling rate of 0.1 mm min^{-1} and a lower cooling rate of 0.1 K min^{-1} at the end of the growth [10]. Dendrite crystals does not appear from the crucible wall even through ΔT is larger than 10 K which is the supercooling limit of generation of dendrite crystals [19]. This is because the melt temperature near the crucible wall is kept enough high not to go over the supercooling limit of 10 K owing to the insulator and carbon heat holder.

The heat flow from the bottom heater into the Si melt inside the crucible is largely affected by the size of the insulating plate which prevents the direct heat flow from the heater. The growth of Si bulk crystals inside the low-temperature region is also largely affected by the size of the insulating plate. Fig. 7.9 shows the maximum diameter of Si ingots as a function of the diameter of insulators [17]. The ingots are grown using many crucibles with different sizes. The maximum diameter of the ingots

Table 7.1 Growth parameters and crystal sizes for grown ingots.

	ΔT (K)	ΔT₁ (K)	ΔT₂ (K)	Crucible diameter (cm)	Cristal mass (g)	Cristal diameter (cm)	Cristal diameter ratio	Cristal length (cm)	Insulator diameter (cm)	Melt height (cm)
NOC-1	1.7	1.7	0	17	6.5	2.5	0.14	0.4	0	5.0
NOC-2	6.8	6.8	0	30	560.0	10.0	0.33	5.0	15.0	4.0
NOC-3	12.5	12.5	0	33	356.0	14.7	0.44	2.6	20.0	5.0
NOC-4	53.5	44.6	8.9	40	8,050	32.2	0.80	8.3	35.0	7.0
NOC-5	48.1	38.1	10.0	50	13,300	35.0	0.70	9.0	40.0	7.0
NOC-6	69.3	55.1	14.2	50	22,700	45.0	0.90	13.0	40.0	7.0

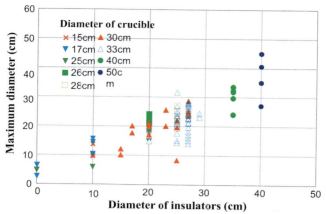

Fig. 7.9 Maximum diameter of Si ingots as a function of the diameter of insulators. The maximum diameter of the ingots clearly tends to become larger as the diameter of insulators increases. The diameter of crucible also strongly affects the maximum diameter of the ingots. *(Referred from Fig. 9 in K. Nakajima et al. J. Cryst. Growth 499 (2018) 55.)*

clearly tends to become larger as the diameter of insulators becomes larger because the large low-temperature region can be freely created by the large insulating plate. The diameter of crucible also strongly affects the maximum diameter of the ingots. The thickness of insulator strongly affects the heat flow from the bottom heater. Fig. 7.10 shows the diameter ratio as a

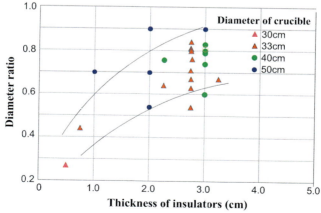

Fig. 7.10 Diameter ratio as a function of the thickness of insulators. The diameter ratio tends to increase as the thickness of insulators increases regardless of the crucible diameter. The thick insulator can effectively weaken the heat flow from the bottom heater into the melt. The thickness of 3 cm is enough to prevent the heat flow. *(Referred from Fig. 11 in K. Nakajima et al. J. Cryst. Growth 499 (2018) 55.)*

function of the thickness of insulator [17]. The diameter ratio tends to increase as the thickness of insulator increases regardless of the crucible diameter. This means that the thick insulator can effectively weaken the heat flow from the bottom heater. The effect of the thickness of insulators saturates near 3 cm which is enough thick to prevent the heat flow. This method is effective to obtain a wider ingot because of its large diameter ratio. As known from the experimental results about the relationship between the largest diameter ratio of 0.9—0.95 for ingots and the (insulator diameter/crucible diameter) ratio, the largest diameter ratio of ingots almost keeps constant and does not depend on this (insulator diameter/crucible diameter) ratio [17]. As the crucible diameter increases, the larger (insulator diameter/crucible diameter) ratio is required to obtain the same largest diameter ratio of ingots. This means that a larger insulator diameter is required as the crucible diameter increases. The effect of the size of the insulating plate on controlling the heat flow largely influences the shape of the growing interface and ingot.

The effects of the insulating plate on establishment of the low-temperature region in a Si melt can be confirmed by simulation (PHENIX: Cham Ltd.) of the temperature distribution in the Si melt. Fig. 7.11 shows the simulated temperature distributions in the Si melt without and

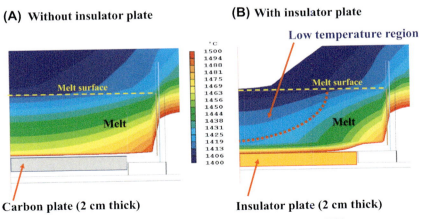

Heater power: Side heater = 59.5 kW, Bottom heater = 96.7 kW.

Fig. 7.11 Simulated temperature distributions in the Si melt (A) without and (B) with an insulator plate. The diameter and thickness of the carbon and insulator plates are set to be 40 cm and 2 cm, respectively. Each melt temperature is set to be near 1400 °C. Temperature differences between the surface and bottom in the Si melt are 66.0 and 27.0 °C without and with the insulator plate, respectively. The establishment of the large low-temperature region can be clearly shown with the insulator plate.

Table 7.2 Temperatures in the melt without and with an insulator plate 2-cm thick insulator plate.

	Without insulator	With insulator
Surface of the melt (°C)	1403.0	1402.9
Bottom of the melt (°C)	1469.0	1429.9
Temperature difference (°C)	66.0	27.0

*Heater power: Side heater = 59.5 kW, Bottom heater = 96.7 kW.

with a disk-shaped insulator plate set under the crucible bottom. Fig. 7.11 HERE In this simulation, the diameter and thickness of the carbon and insulator plates are set to be 40 cm and 2 cm, respectively. The thermal conductivities of the carbon and insulator plates are 100 W m^{-1} K^{-1} and 0.2 W m^{-1} K^{-1} at Mp, respectively. The inner diameter of the quartz crucible is 50 cm, and the wall and bottom thicknesses of the crucible are 0.5 cm and 1.0 cm, respectively. The height of the Si melt is 10 cm. The inner diameter, wall thickness and height of the cylinder-shaped carbon heater holder are 53 cm, 5 cm and 15 cm, respectively. The diameter of the hood which is set to cover the melt surface and prevent falling-down SiO particles is 1 m. The Ar pressure is 70 kPa. The heater powers of the side and the bottom heaters are 59.5 ad 96.7 kW, respectively. Table 7.2 shows the melt temperatures without and with an insulator plate simulated on the surface and bottom in the Si melt. Each melt temperature is set to be near 1400°C. Temperature differences between the surface and bottom in the Si melt are 66.0 and 27.0°C without and with the insulator plate, respectively. The temperature difference clearly becomes smaller using the insulator plate. From these results, the establishment of the large low-temperature region can be clearly confirmed in the case with the insulator plate as shown in Fig. 7.11B.

7.3 Growth of Si ingots using Si_3N_4 coated crucibles by the NOC method

A Si_3N_4 coated silica crucible with 30 cm-diameter is used to grow a Si ingot to prevent the incorporation dissolved O into the melt and the ingot during growth. The low-temperature region in the coated crucible is wider than that for the uncoated crucible because heat transfer through the crucible wall with Si_3N_4 coating is more strongly suppressed. Therefore, for the uncoated crucible, a larger temperature reduction is required to obtain an ingot with same diameter as an ingot grown using the coated crucible

[12]. The large temperature reduction and the long total growth time are used to obtain a larger ingot by expanding the low-temperature region. ΔT is set between 4.0 and 90.0 K. The diameter of the ingot rapidly increases almost linearly with ΔT. The total mass of the ingot is proportional to the product of ΔT and the total growth time, t (min) because the driving force of ingot growth is proportional to ΔT and its volume during pulling the ingot upward is almost proportional to t. The solidification ratio of the ingots increases almost linearly with the product of ΔT and t. The largest solidification ratio obtained is 95% for a crucible diameter of 30 cm.

In the NOC method, nucleation occurs at the center of melt surface using a seed crystal, and there are no other nucleation sites in this area. Si_3N_4 particles appeared on the melt surface strongly act as nucleation sites and cause generation of small grains. A single ingot can be obtained only when Si_3N_4 particles are effectively removed from the melt surface. Pizzini et al. reported that under an appropriate growth condition, Si crystals grown using Si_3N_4-coated silica crucibles acquired electrical properties very similar to those of Si crystals grown by the CZ method [21].

Impurities that degrade solar cell performance have small segregation coefficients except for O [22]. The concentration of impurities such as Fe that diffuse from the crucible wall into an ingot is expected to be much smaller than that in an ingot grown by the conventional cast method because the Si ingot is not in contact with the crucible wall. However, elemental Fe, which dissolves in a Si melt, is included in each ingot at the moving interface by segregation and Fe diffuses to the ingot surface from the melt. The concentration of Fe can be measured by ICP-MS method. The detection limit of this method is 0.5×10^{-9} g g^{-1} for the concentration of Fe. The concentration of Fe is $1.5-3.5 \times 10^{-9}$ g per 1 g crystal within the ring-shaped region in the distribution of minority carrier lifetime as shown in Fig. 7 20D, in which the minority carrier diffusion length is low. The concentration of Fe is $0.8-1.2 \times 10^{-9}$ g g^{-1} at the center of the ingot. The reason why the concentration of Fe is higher near the surface than at the center of the ingot may be that the peripheral part of the ingot solidifies later than the center part because the growth interface is convex to the growth direction.

The concentrations of O and C in ingots can be measured on longitudinal and horizontal cross sections of the ingots by FTIR. The O concentration in an ingot is almost the same as those in the remaining melt because the segregation coefficient of O in Si is 1.25 which is close to 1. The O concentrations are $0.5-1.6 \times 10^{17}$ cm^{-3} for ingots grown using crucibles coated with Si_3N_4 particles [6,23] and $2.5-8.3 \times 10^{17}$ cm^{-3} for ingots grown using crucibles without Si_3N_4 coating [11,12,24]. The higher O concentration of the latter is because of the reaction between the crucible wall and the Si melt and the incorporation of dissolved O into the ingots. The O concentration of the ingots grown in the Si_3N_4 coated crucibles is almost equal to that for the dendritic casting method. The O concentration of ingots grown by the conventional cast method is usually $1.0-8.0 \times 10^{17}$ cm^{-3} [23] and that by the CZ method is usually 1×10^{18} cm^{-3} [25]. As O in a Si melt has an equilibrium segregation coefficient of $k_0 \cong 1$, the O concentration at the bottom of an ingot is slightly lower than that at the top of the ingot for the NOC method using a coated crucible. This trend is opposite for ingots grown by the conventional casting method [26]. The distribution of O concentration in ingots grown by the CZ method shows that the O concentration is higher at the center than at the periphery of the ingots when radial direction flow in the Si melt is not strong [27]. In the NOC method, different O concentration distribution may be expected because the ingot is convex in the growth direction and the convection in the melt is very small, as shown in Figs. 7.23 and 7.24. A low O concentration is important because void nucleation is suppressed by increasing void surface energy [28].

The C concentration is very low due to little C contamination in the NOC growth system. The C concentration is less than 3×10^{16} cm^{-3} for the normally used methods except for the conventional cast method. The C concentration of the p- and n-type ingots grown using crucibles without Si_3N_4 coating by the NOC method, is about $2.4 \times 10^{15}-1 \times 10^{16}$ cm^{-3} [11]. When a crucible with Si_3N_4 coating is used for the ingot growth, precipitated Si_3N_4 particles formed in the N saturated Si melt seriously affect on the quality of an ingot because these particles act as nucleation sites to form small grains in the ingot [6,12,13,29].

Si ingots with low stress can be realized by the NOC method because stress is quite low due to non–contacting with the crucible wall. To evaluate EPD in cross sections of an ingot, the surface of cross sections is mechanically polished to a mirror finish and then chemically etched with Sopori solution [30]. The EPD is normally calculated by counting the number of etch pits in an area of 1×1 cm^2 through an optical microscope. The dislocation density in the high-quality area in the ingots is on the order of 10^3 cm^{-2} without necking technique. When a seed crystal is touched to the surface of the Si melt, the process generates many local dislocations in the ingot on the order of 10^5-10^6 cm^{-2} because of thermal shock due to rapid contact.

7.4 Growth of Si single ingots using the NOC method

The prices of Si multicrystalline and single wafers are not largely different. The Si multicrystalline wafers have a larger share than single wafers because of their high productivity. But they lose 1% efficiency because of their inefficient texture structure. The next target of the cast method should be to develop uniform large Si single ingots with sufficient quality to obtain solar cells with high conversion efficiency and high yield using a cast furnace. The NOC method has a potential to realize this target.

7.4.1 Growth rate of single ingots using the NOC method

The NOC method has the possibility of attaining a high growth rate using a high cooling rate even though an ingot grows inside a crucible similarly to the cast method [31–34]. The growth rate of the CZ method is determined by the pulling rate of an ingot above the surface of a Si melt [35–37]. The horizontal and vertical growth rates of the NOC method are determined by the expansion rate of the low-temperature region in a Si melt, which is mainly determined by the cooling rate of the Si melt and the temperature gradient in the Si melt [38]. A high-speed growth is required to obtain an ingot with a large diameter for practical use. The cooling rate required to obtain such a large ingot should be determined, and the surface or horizontal growth rate should be determined in the $<110>$ and $<100>$ directions inside a Si melt.

The horizontal growth rate of a Si single ingot can be experimentally determined by the growth of an n-type ingot using a silica crucible without a Si_3N_4 coating [38]. The circular crucible made of high-purity electric fused quartz is used for the growth. The main impurities in the crucible material are Al, Ca, and Fe, with impurity levels of 7.9, 0.6, and 0.17 ppm, respectively. The impurity level is measured by ICP-MS [16]. To suppress the reaction between the crucible wall and the Si melt by reducing convection currents, the Ar pressure is set to slightly below 0.1 MPa, even though a lower Ar pressure is important for reducing the O concentration in Si crystals in the CZ method [39]. The growth length on the top surface of the ingot can be determined by making marks in it. Fig. 7.12A shows an n-type ingot with a size of 20.3×20.3 cm^2, a diagonal length of 25.0 cm and a thickness of 4.0 cm [38]. The ingot is grown using a crucible with 28 cm diameter. The thickness of the Si melt is 5.0 cm and the cooling rate is 0.3 K min^{-1}. The melt temperature remains constant without cooling while the ingot is intermittently and slightly pulled upward after each

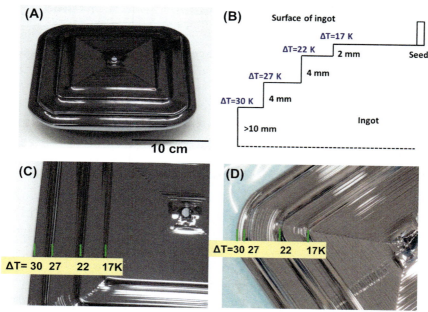

Fig. 7.12 (A) n-Type ingot with a square-like shape of size 20.3 × 20.3 cm². The top surface has four steps corresponding to the growth process. (B) Growth process of this ingot for the purpose of making marks to determine the growth rate. (C) Top surface of the ingot in the <110> direction. (D) Top surface of the ingot in the <100> direction. *(Referred from Fig. 2 in K. Nakajima et al. J. Crys. Growth 405 (2014) 44.)*

expansion of the crystal along the surface of the melt. The growth process of this ingot is shown in Fig. 7.12B [38]. To make marks to determine the growth rate, the ingot is grown during cooling the melt by $\Delta T = 30$ K while the ingot is pulled upward by 0.3, 0.4, 0.4 and 0.2 cm at $\Delta T = 17$, 22, 27 and 30 K with a pulling rate of 2 mm min^{-1}, respectively. The top surface of the ingot has four steps, which are located at distances of 5.2, 6.8, 9.0 and 10.0 cm in the <110> direction and at distances of 7.1, 8.0, 11.6 and 12.3 cm in the <100> direction from the center of the seed. The growth times for these growth lengths are 57, 73, 90 and 100 min, respectively. At $\Delta T = 30$ K, for finishing the ingot growth, the ingot is continuously pulled upward without cooling the melt. The ingot has a square-like shape because the growth rate in the <100> direction is larger than that in the <110> direction. Fig. 7.12C and D show the top surface of the ingot in the <110> and <100> directions, respectively [38]. The top surface has four steps corresponding to the growth process shown in

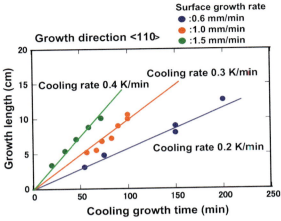

Fig. 7.13 Surface or horizontal growth length in the <110> direction as a function of the growth time. The growth rate was determined from the slope. *(Referred from Fig. 3 in K. Nakajima et al. J. Crys. Growth 405 (2014) 44.)*

Fig. 7.12B. The side face in the <110> direction is flat but the <100> corner has a fan-shaped face [13]. The side-face widths for the exact rectangle shape are 10.3 and 10.7 cm for this ingot, which correspond to the step at $\Delta T = 17$ K shown in Fig. 7.12D.

The growth rate can be determined from the growth length and growth time [38]. Fig. 7.13 shows the growth length as a function of the growth time in the <110> direction [38]. The cooling rates are 0.2, 0.3 and 0.4 K min^{-1}. The growth length almost linearly increases with the growth time. The growth rate given by the slope increases with the cooling rate. From these results, the growth rate is determined to be 0.6, 1.0 and 1.5 mm min^{-1} for the cooling rates of 0.2, 0.3 and 0.4 K min^{-1}, respectively. The growth rate in the <100> direction is also determined in the same manner to be 0.8, 1.3 and 1.9 mm min^{-1} for the cooling rates of 0.2, 0.3 and 0.4 K min^{-1}, respectively [38]. The growth rate in the <100> direction is about 1.3 times higher than that in the <110> direction. However, the growth rate in the <100> direction is not $\sqrt{2}$ times higher than that in the <110> direction because the <100> corners have circular fan-shaped faces as shown in Fig. 7.12D. The linear fits are used to determine the initial growth rates in each direction.

The horizontal growth rates in the <110> and <100> directions increase with the cooling rate. The horizontal growth rates at the cooling rate of 0.4 K min^{-1} are more than twice those at the cooling rate of

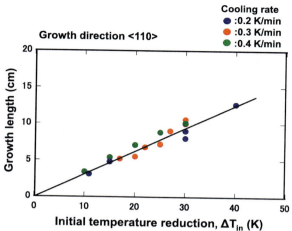

Fig. 7.14 Relationship between the grown length and ΔT_{in} in the <110> direction for cooling rates of 0.2, 0.3 and 0.4 K min^{-1}. The growth length almost linearly depended on ΔT_{in} regardless of the cooling rate. (Referred from Fig. 5 in K. Nakajima et al. J. Crys. Growth 405 (2014) 44.)

0.2 K min^{-1}. A high growth rate of about 2 mm min^{-1} is obtained in the <100> direction for the cooling rate of 0.4 K min^{-1} [38]. This growth rate is much higher than that of the cast method and is as high as that of the CZ method. The relationship between the growth length and ΔT is determined in the <110> direction for the cooling rates of 0.2, 0.3 and 0.4 K min^{-1} as shown in Fig. 7.14 [36]. The growth length almost linearly depends on ΔT regardless of the cooling rate, i.e., the growth length is simply determined by only ΔT. Therefore, we can use a higher cooling rate to obtain the same size of ingot and reduce the growth time.

The vertical growth rate is very important because it determines the maximum pulling rate at which an ingot can be pulled upward during keeping the ingot growth inside a Si melt, i.e., it is the actual growth rate of the ingot by the NOC method. During growth of the ingot along the surface of a Si melt, the ingot also grows inside the melt in the vertical direction. Just when the ingot grows in contact with the bottom of the crucible, the ingot swings in the melt, which can be observed through the windows of the furnace chamber. The vertical growth rate can be determined by direct observation of the first swing of the ingot at the initial stage of growth, and it can be estimated by the depth of the melt and the interval time between the start time of the surface growth from the seed crystal and

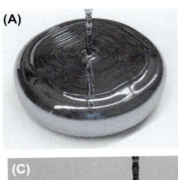
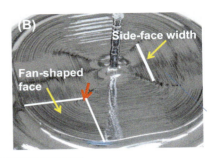
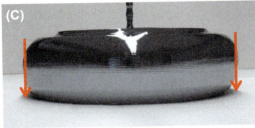

Diameter: 23 cm
Side-face width: 4.8–5.1 cm
Height: 7 cm

Fig. 7.15 (A) n-Type ingot with a circular shape. (B) Top surface of the ingot with short side-face widths and fan-shaped faces in the <100> direction. A fan-shaped face appeared at the point shown by the red arrow. (C) Ingot with a diameter that remained constant during pulling with a high cooling rate of 0.4 K min^{-1} because the horizontal growth rate increased with increasing cooling rate and the melt temperature markedly decreased. *(Referred from Fig. 8 in K. Nakajima et al. J. Crys. Growth 405 (2014) 44.)*

the final time when the ingot swings. The vertical growth rate tends to increase with the depth of the Si melt, and it is determined to be 0.3–0.6 mm min^{-1} at the melt depth of 5 cm [35]. The vertical growth rate is not particularly large because the melt depth used for the growth is limited to only 5 cm at the cooling rate of 0.2 K min^{-1}.

For the practical use of an ingot for solar cells, it is important to keep the initial size of the ingot constant during crystal growth. Fig. 7.15A shows an n-type Si single ingot grown using a crucible with 33 cm diameter, which has a diagonal length of 23 cm, a mass of 6440 g (solidification ratio: 64%) and a thickness of 7.0 cm [38]. The ingot is grown while cooling the melt by $\Delta T = 30$ K without pulling. The top surface of the ingot has a step located at distances of 8.0 cm in the <110> direction and 8.5 cm in the <100> direction from the center of the seed crystal. The growth time for these growth lengths is 58 min. At $\Delta T = 30$ K, the ingot is grown while pulling it upward with a pulling rate of 0.4 mm min^{-1}, and then the melt

286 Crystal Growth of Si Ingots for Solar Cells Using Cast Furnaces

cools by $\Delta T = 68.3$ K. The ingot is circular because the side-face width is very short (4.8−5.1 cm) as shown in Fig. 7.15B [38]. A fan-shaped face appears in the <100> direction at the point shown by the red arrow; thus, the top surface of the ingot has circular corners in the <100> direction. As shown in Fig. 7.15C, the diameter of the ingot remains constant during pulling with a high cooling rate of 0.4 K min^{-1} because the horizontal growth rate increases with increasing cooling rate and the melt temperature markedly decreases [38]. The ingot is sufficiently large to obtain a 15.6×15.6 cm^2 wafer even when using the crucible with 33 cm diameter. The same diagonal length of the top surface of ingots can be obtained for cooling rates between 0.2 and 0.4 K min^{-1}. A larger cooling rate can be used to obtain ingots with the same diameter.

It is desirable to clarify the maximum growth rate for the NOC method. The growth rate can be expressed as [38]

$$
\begin{aligned}
V(\text{cmmin}^{-1}) &= \mathrm{d}r_t/\mathrm{d}t \\
&= (\mathrm{d}\Delta T_{in}/\mathrm{d}t)/(\mathrm{d}\Delta T_{in}/\mathrm{d}r) \\
&= (\mathrm{d}\Delta T_{in}/\mathrm{d}t)/G_T^s.
\end{aligned}
\tag{7.2}
$$

Here, V (cm min^{-1}) is the horizontal growth rate, r (cm) is the grown length at the initial temperature reduction along the surface of the ingot, ΔT_{in}, and t (min) is the growth time required to obtain the growth length. $\mathrm{d}\Delta T_{in}/\mathrm{d}t$ is the cooling rate, which is expressed in terms of the initial temperature reduction, and $\mathrm{d}\Delta T_{in}/\mathrm{d}r$ is the temperature gradient near the surface of the melt on the growth interface of the expanding ingot along the surface of the Si melt, G_T^s (K cm^{-1}).

In the NOC method, the initial temperature reduction is defined as the difference between the starting temperature, at which the initial crystal is first observed to start growing from the seed crystal, and the final temperature, at which the pulling growth starts. Therefore, the initial temperature reduction is the temperature reduction that is used to induce crystal growth along the surface of the Si melt from the seed crystal. The low-temperature region in the central part of the Si melt becomes larger across the surface of the Si melt as the initial temperature reduction increases, which results in an increase of the temperature difference near the surface of the Si melt between the center and periphery of the low-temperature region. The temperature gradient across the crucible near the surface of the Si melt is almost constant when the low-temperature region is expanding because the side-face width (10−12 cm) is almost constant for $\Delta T_{in} = 30$ K regardless the cooling rate. The side-face width, d (cm), of a square-shaped crystal can be expressed as Eq. (8.1). The growth rate is almost proportional to the cooling rate [13].

In the case that G_T^s is not constant, the growth rate strongly depends on the temperature gradient in the Si melt as indicated by Eq. (7.2). When the temperature gradient is larger, it is more difficult to significantly expand the low-temperature region in the Si melt. The growth rate is also lower for a larger temperature gradient in the Si melt. To obtain an ingot with a diameter as large as the crucible diameter, a temperature gradient should not be large in the Si melt to easily obtain a large low-temperature region.

7.4.2 Growth of single ingots with a large diameter and a diameter ratio

A Si ingot with a large diameter is important for low-cost solar cells because the ingot with a larger diameter than 45 cm has a cross section from which four wafers required for regular-size (15.6×15.6 cm^2) solar cells can be obtained. To grow a single ingot with a large diameter and a diameter ratio by the NOC method, a large distinct low-temperature region should be established in a Si melt. For the growth using the furnace with two zone heaters, the carbon heat holder is quite important to maintain the heat of the crucible wall as shown in Fig. 7.3B. Sometimes, a cylinder-type carbon heat holder is used for the ingot growth. Si ingots with different diameters and diameter ratios are grown using different ΔT to confirm the possibility of growing ingots with a diameter as large as the crucible diameter, which is one of the merits of the NOC method. The Ar pressure is set at $50-96$ kPa in the growth.

As shown in Fig. 7.3, to determine ΔT, thermocouples are placed near each heater and the temperature of each heater is continuously read from a monitor on the heater system using the thermocouples during crystal growth. ΔT is determined as the sum of the temperature reductions of the heaters using Eq. (7.1). ΔT can be used as an indicator to compare each temperature reduction for each ingot growth inside the Si melt with an almost constant temperature gradient because ΔT almost corresponds to the supercooling in the center of the Si melt as shown in Fig. 7.6. Thus, the temperature reduction of the Si melt is approximately determined from ΔT measured at the heater system and the supercooling in the Si melt can be expressed by ΔT.

The maximum diameters of 28, 33.5 and 45.0 cm are obtained when ΔT is 46 K for a 33–cm–diameter crucible, 55.6 K for a 40-cm-diameter

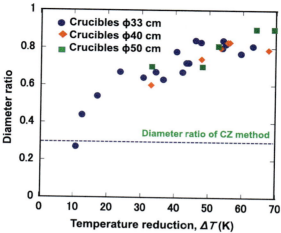

Fig. 7.16 Diameter ratio of n-type ingots as a function of ΔT. The diameter ratio increases with ΔT regardless of the crucible diameter. A diameter ratio of 0.9 can be obtained by expanding the low-temperature region in the Si melt. (Referred from Fig. 6 in K. Nakajima et al. J. Electron. Materials, 45 (2016) 2837.)

crucible and 69.3 K for a 50-cm-diameter crucible, respectively. The maximum diameter of the ingots increases with ΔT. The maximum diameter of the ingot grown using a 40- or 50-cm-diameter crucible is larger than that of the ingot grown using a 33-cm-diameter crucible even at the same ΔT. The diameter ratio can be determined as a function of ΔT as shown in Fig. 7.16 [11]. The diameter ratio increases with ΔT regardless of the crucible diameter. This tendency is very effective for obtaining an ingot with a large diameter because a larger diameter can be grown only using a larger crucible at the same ΔT. The diameter ratio becomes saturated when ΔT is large because the fringe of an ingot becomes closer to the hotter crucible wall as the diameter ratio increases. A diameter ratio of 0.90 can be obtained by largely expanding the low-temperature region in the Si melt. The maximum diameter ratio is 0.84, 0.83 and 0.90 when ΔT is 46 K for a 33-cm-diameter crucible, ΔT is 55.6 K for a 40-cm-diameter crucible and when ΔT is 69.3 K for a 50-cm-diameter crucible, respectively. The driving force of the ingot growth is proportional to ΔT for the system with two zone heaters, similarly to the system with three zone heater system [6,12]. The maximum diameter and diameter ratio of an ingot is mainly determined by the temperature reduction of the first zone heater and not strongly affected by that of the second zone heater.

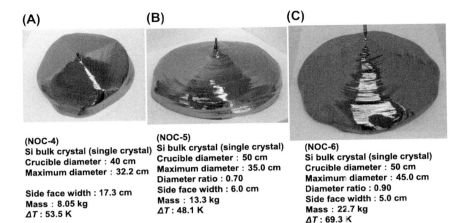

(NOC-4)
Si bulk crystal (single crystal)
Crucible diameter : 40 cm
Maximum diameter : 32.2 cm

Side face width : 17.3 cm
Mass : 8.05 kg
ΔT : 53.5 K

(NOC-5)
Si bulk crystal (single crystal)
Crucible diameter : 50 cm
Maximum diameter : 35.0 cm
Diameter ratio : 0.70
Side face width : 6.0 cm
Mass : 13.3 kg
ΔT : 48.1 K

(NOC-6)
Si bulk crystal (single crystal)
Crucible diameter : 50 cm
Maximum diameter : 45.0 cm
Diameter ratio : 0.90
Side face width : 5.0 cm
Mass : 22.7 kg
ΔT : 69.3 K

Fig. 7.17 *n*-Type bulk crystals (NOC-4, NOC-5 and NOC-6) grown using crucibles of (A) 40 and, (B) and (C) 50 cm diameter. The diameters of the ingots (A), (B) and (C) were 32.2, 35.0 and 45.0 cm, respectively. The diameter ratios were 0.80, 0.70 and 0.90, respectively. Using a Si melt with a depth of 7 cm, the lengths were 9.5, 9.0 and 13.0 cm, respectively. *(Referred from Fig. 5 in K. Nakajima et al. J. Electron. Materials, 45 (2016) 2837.)*

Fig. 7.17 shows *n*-type single ingots (NOC-4, NOC-5 and NOC-6), in which ingot (A) was grown using a crucible of 40 cm diameter and ingots (B) and (C) were grown using crucibles of 50 cm diameter [11]. The diameters of the ingots (A), (B) and (C) are 32.2, 35.0 and 45.0 cm, respectively, and the diameter ratios are 0.80, 0.70 and 0.90, respectively. Using Si melts with a depth of 7 cm, the lengths of ingots (A), (B) and (C) are 9.5, 9.0 and 13.0 cm, respectively. The maximum values of ΔT are 53.5, 48.1 and 69.3 K, respectively. For the growth of these ingots, the temperature of the second zone heater is cooled, but no crystal growth is observed from each crucible wall. A four-cornered pattern with a side-face width of 17.3 cm appears on the top surface of the ingot (NOC-4) in Fig. 7.17A. This value of the side-face width, which is important in providing a low temperature gradient near the growing interface [13], is the largest recorded value. Fig. 7.18 shows the largest single ingot with a diameter of 45.0 cm and a diameter ratio of 0.90, from which four wafers required for the regular-size solar cells can be obtained in cross-sections of the lower part of the ingot.

To obtain an ingot with a large diameter ratio, ΔT should be kept larger than 40 K by ensuring that ΔT_d is larger than 30 K as shown in Figs. 7.6 and 7.16. A large ΔT can be established for a large ΔT_d, and a large

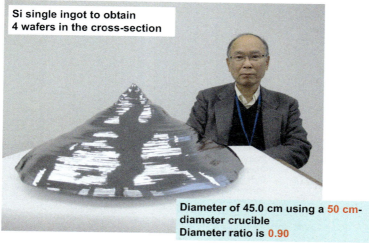

Fig. 7.18 Largest single ingot with a diameter of 45.0 cm and a diameter ratio of 0.90, from which four wafers can be obtained in its cross-section.

low-temperature region can be created for a large ΔT. A large ΔT_d must be established to obtain an ingot with a large diameter ratio by expanding the low-temperature region to close to the wall. To obtain a large ΔT_d, a steep temperature gradient near the crucible wall should be realized by keeping the melt temperature near the crucible wall high. The temperature gradient in the low-temperature region is required to be kept low over a large area. Using a large crucible, it is much easier to establish these temperature conditions.

7.4.3 Shape of the growing interface of Si single ingots grown by the NOC method

In the NOC method, the crystal diameter is basically determined by the size of the low-temperature region. An ingot is grown inside a Si melt and the growing interface is convex in the growth direction. In the CZ method, an ingot is grown above the surface of a Si melt and the growing interface is generally concave in the growth direction [37]. The shape of the solid/liquid interface is expected to strongly affect the quality of these ingots. The shape and quality of Si single ingots grown inside Si melts should be clarified, particularly on the view point of the effects of the convex growing interface on the distributions of dopants, dislocations, and O [16]. The crystallographic and electronic qualities of the resulting materials should be also evaluated [16].

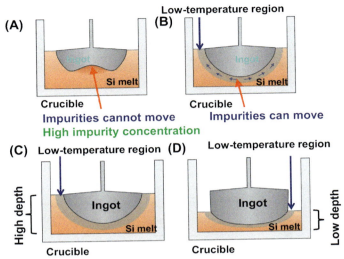

Fig. 7.19 (A) Concave growing interface, in which impurities can easily accumulate in a Si melt below the center of the growing interface. (B) Convex growing interface, in which impurities in the Si melt more easily move away from the center of the growing interface along its slope. (C) Growing interface with small curvature in a large-depth Si melt. (D) Growing interface with large curvature in a small-depth Si melt. *(Referred from Fig. 1 in K. Nakajima et al. Jpn. J. Appl. Phys. 54 (2015) 015504.)*

A convex growing interface in the growth direction of a Si ingot is compared with a concave growing interface shown in Fig. 7.19 [16]. When the solid–liquid interface is concave, impurities such as dopants and O can easily accumulate in the Si melt below the center of the growing interface as shown in Fig. 7.19A, and their concentrations become high at the center of the ingot. The boundary between the center with a high impurity concentration and the surrounding area with a low impurity concentration appears to be clear. For a convex growing interface, these impurities in the Si melt more easily move away from the center of the growing interface along its slope as shown in Fig. 7.19B. These impurities are expected to form a swirl pattern in the growing ingot because of the movement of the impurities and the rotation of the ingot. The curvature radius of the convex growing interface increases as the depth of the Si melt decreases during crystal growth as shown in Fig. 7.19C and D. The distribution of impurities in the ingot is expected to change as the curvature radius of the convex bottom of the ingot increases during crystal growth. The distribution of impurities for the convex growing interface is predicted to be more uniform than that for a concave growing interface. To effectively use the

NOC method, the crystallographic and electronic qualities of the ingot should be clarified, particularly on the view point of the effect of the convex bottom of the ingot on the distribution of impurities in the ingot. The effect of the convex bottom of the ingot on the distribution of dislocations and the effect of reducing dislocation density on the lifetime should be also clarified.

7.4.4 Dislocations in Si single ingots grown by the NOC method

Generally, low dislocation density ($<10^4$ cm^{-2}) can be obtained for a Si ingot grown inside a Si melt [9]. Fig. 7.20A and B show a *p*-type ingot grown using a 30 cm–diameter crucible without Si_3N_4 coating and its cross section cut 1.4 cm below its top surface, respectively [12]. ΔT for the growth is 30 K. The ingot has a mass of 4.8 kg, a diameter of 22 cm and a height of 7 cm. The top surface clearly shows a four-cornered pattern. No grain boundaries are observed in the cross section, and the surface

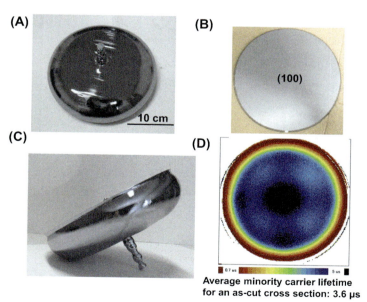

Fig. 7.20 (A) *p*-Type ingot grown using a crucible with 30 cm diameter. A four-cornered pattern is observed on the top surface of the ingot. (B) Cross section of the ingot. No grain boundaries are observed in the cross section. (C) Bottom of the ingot that is somewhat convex or almost flat in the growth direction. (D) Distribution of the minority carrier lifetime in the same as-cut cross section. The distribution of the high minority carrier lifetime is almost uniform except for the growing pattern. *(Referred from Fig. 4 in K. Nakajima J. Cryst. Growth 372 (2013) 121.)*

orientation of the cross section is (100). When an ingot has a concave bottom, a region with a low lifetime is concentrated in the center of the bottom cross section. When an ingot has a convex bottom, impurities and O in the Si melt more easily move from the center of the growing interface to the periphery [16]. Fig. 7.20C shows the bottom of the ingot, which is slightly convex in the growth direction. Fig. 7.20D shows the distribution of the minority carrier lifetime measured by μ-PCD in the same as-cut cross section. The cross section has a ring-shaped region in which the minority carrier lifetime is low, because the defect and impurity densities are higher at the periphery of the ingot. A high minority carrier lifetime is observed near the center of the cross section. Fig. 7.21 shows the EPD of dislocations in the same cross section shown in Fig. 7.20B and D [12]. The EPD is 1.9×10^3 cm^{-2} over most of the cross section but slightly high (6.8×10^3 cm^{-2}) at the periphery of the cross section because of the convex growing interface of the ingot [12]. The seeding process generates many local dislocations in an ingot with densities of 10^5–10^6 cm^{-2} because of thermal shock due to rapid contact. Dislocations in an ingot move to the periphery from the center during crystal growth and the pair annihilation of dislocations occurs near the center of the ingot during crystal growth. As a result, the dislocation density can be generally reduced to the order of 10^2–10^4 cm^{-2} without necking technique [12,13,40].

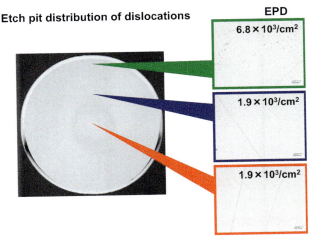

Fig. 7.21 Etch pit distribution of dislocations in the same cross section shown in Fig. 7.20B and D. The dislocation densities are 1.9×10^3 cm^{-2} across almost the entire cross section and 6.8×10^3 cm^{-2} at the periphery of the cross section. *(Referred from Fig. 5 in K. Nakajima J. Cryst. Growth 372 (2013) 121.)*

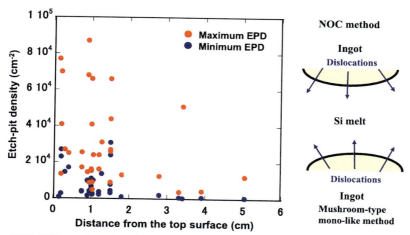

Fig. 7.22 EPD as a function of the distance from the top surface of the ingots. The EPD near the top surface is very high, but the EPD rapidly decreases as the distance from the top surface of ingots increase. (From Kazuo Nakajima et al., Growth of Si single ingots with large diameter (~45 cmφ) and high yield of high conversion efficiency (≥ 18 %) cells using a small crucible (~50 cmφ) by noncontact crucible method, 2016 IEEE 43rd Photovoltaic Specialists Conference (PVSC), © 2016 IEEE.)

To determine the effect of the convex bottom on the distribution of dislocations in an n-type ingot, the distribution of the EPD of dislocations was measured on cross sections of the ingot [16]. On the cross section just below the top surface, the dislocation density is on the order of 10^3 cm^{-2} over most of the cross-sectional area but much higher at the periphery of the cross section. On the bottom cross section, the dislocation density slightly decreases on the order of 10^3 cm^{-2} over most of the cross-sectional area but is much higher at the periphery. This means that dislocations in the ingot move to the periphery from the center during growth owing to the convex growing interface. This behavior of dislocations can be confirmed by Fig. 7.22, which shows the EPD as a function of the distance from the top surface of ingots [41]. The maximum and minimum EPDs are evaluated in a cross section of each ingot. The maximum and minimum EPDs usually appear near the periphery and center of ingots, respectively. The EPD near the top surface is high, but the EPD rapidly decreases as the distance from the top surface of ingots increases owing to the convex growing interface toward the melt [16]. This phenomenon is very similar to the effect of dislocation reduction by the mushroom-type mono-like cast method [42]. Dislocations tend to perpendicularly extend to the growing interface owing to minimize their length or energy (see Section 1.3.4.1). Finally, the minimum EPD decreases to 300 cm^{-2}. Thus, the dislocation density in the center of the ingots can be reduced to the order between 10^2 and 10^4 cm^{-2} without necking technique. A dislocation density of less than 10^3 cm^{-2} is very important for obtaining a large minority carrier diffusion length [43].

Dislocations remained in the seed crystal are concentrated at the center of an initial ingot directly below the grown seed crystal. However, even on the top cross section just below the top surface, the measured dislocation density is on the order of 10^3 cm^{-2} but much higher at the periphery of the cross section. On the bottom cross section, the dislocation density slightly decreases on the order of 10^3 cm^{-2} over most of the cross-sectional area. This distribution of dislocations in the ingot with a convex bottom can be explained by the movement of dislocations. (1) Dislocations directly below the grown seed easily move to the inside and periphery of the ingot during the crystal growth at a high temperature. (2) Dislocations easily moves to the surface area of the ingot during the crystal growth because of the convex growing interface. (3) The density of dislocations is also decreased by their pair annihilation near the center of the ingot during crystal growth. The convex growing interface accelerates such movement of dislocations and contributes to the low dislocation density at the center of the ingot.

7.4.5 Impurities in Si single ingots grown by the NOC method

In the NOC growth, the O concentration in the ingot is mainly determined by the reaction between crucible wall and Si melt and the SiO evaporation at the melt surface. In this case, the strength of the convection in the melt and the free surface area of the melt are important factors. In the CZ method, the O concentration in an ingot is mainly determined by Ar gas flow rate and furnace pressure [44]. Normally, the higher Ar gas flow rate makes the diffusion layer thinner, enhances SiO evaporation and decreases the O concentration in the ingot. In the case when the O-rich Si melt from the crucible bottom flows underneath the growing interface without releasing a large amount of SiO at the melt surface, the O concentration increases [44]. The lower furnace pressure makes the lower partial pressure of SiO in the Ar gas, enhances SiO evaporation and decreases the O concentration in the ingot.

It should be clarified whether the distribution of the O concentration in an ingot grown using the NOC method is affected by the convex bottom of an ingot. different from that grown by the CZ method because the resulting ingot is convex in the growth direction. The O concentration in ingots grown by the NOC method is relatively lower than that in ingots grown by the CZ method [6,11,12,29]. The O concentration in the ingot can be measured from longitudinal and horizontal cross sections of the ingot by the FTIR spectroscopy. O dissolves in a Si melt from the silica crucible and continuously evaporates from the surface of the Si melt. When the convection of the melt is suppressed, the O concentration is expected to be lowest near the melt surface and increase toward the bottom of the melt. Fig. 7.23A shows a schematic illustration of the distribution of the O concentration in an ingot grown by the NOC method [16]. The thickness of the Si melt is 5 cm. The ingot is pulled upward at a rate of 0.2 mm min^{-1} while cooling the melt by 39.9 K (= ΔT of 79.9 K – ΔT_{in}). The diagonal length of the top surface of the ingot is 28.6 cm and the ratio of the diagonal length to the crucible diameter is 0.95. The thickness of the ingot is 9.5 cm. Fig. 7.23B, C and D show the distributions of the O concentration in the ingot as a function of the distance from the ingot center directly below the seed crystal along the surface of the ingot, in the vertical direction to the bottom of the ingot, and in the slanting direction to

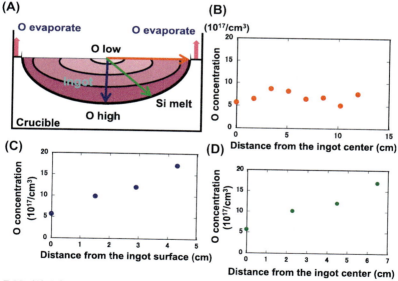

Fig. 7.23 (A) Schematic illustration of distribution of the oxygen concentration in an ingot. (B) Oxygen concentration along the surface of the ingot. (C) Oxygen concentration in the vertical direction from the surface to the bottom of the ingot. (D) Oxygen concentration in the slanting direction from the surface to the bottom of the ingot. The oxygen concentration near the seed was low and increased along the surface of the ingot and in the vertical and slanting directions. *(Referred from Fig. 6 in K. Nakajima et al. Jpn. J. Appl. Phys. 54 (2015) 015504.)*

the bottom of the ingot, respectively [16]. The O concentration slightly increases along the surface of the ingot and markedly increases in the vertical and slanting directions from the center to the bottom of the ingot during crystal growth. Fig. 7.23A schematically shows the concentric distribution of the O concentration on the seed axis. This distribution of the O concentration is due to the convex growing interface.

As known by the distribution of the O concentration in Fig. 7.23A, the O concentration near the seed is low in the initial stage of growth. It gradually increases along the ingot surface and rapidly increases in the vertical and slanting directions. This distribution of the O concentration in the ingot with a convex bottom can be explained by the O concentration in the Si melt. (1) The O concentration in the Si melt increases with the depth of the melt because the evaporation of O atoms more easily occurs near the surface of the melt. (2) The O concentration in the Si melt increases with the area of the ingot surface covering the melt surface because of the suppression of the evaporation of O atoms from the limited melt surface area. (3) The O concentration near the Si melt surface increases as the ingot expands because the large amount of O elements moves up along the slope of the growing interface from deep in the Si melt owing to convexity. The convex growing interface seems to accelerate the movement of O elements and contribute to the relatively low and uniform O concentration in the ingot. From these effects, the concentric distribution of the O concentration on the seed axis appears, as shown in Fig. 7.23A. The O concentration in the center of the ingot is about 5×10^{17} cm^{-3}.

The convection of the melt is an important factor in accelerating the evaporation of O atoms from the surface of the melt, similarly to in the CZ method [18]. However, in the NOC method, the convection is possibly suppressed, resulting in generation of the large low–temperature region in a Si melt. The ingot and crucible are continuously co–rotated in the same direction at low rates of 1.5 and 0.5 rpm during crystal growth, respectively. Therefore, the convection is not sufficiently large to affect the distribution of the O concentration in the Si melt. The small convection can also suppress the reaction between the Si melt and the crucible wall.

The O concentration in the Si melt strongly depends on the free surface area of the Si melt which is not covered by the ingot, because the evaporation of O atoms is suppressed when the free surface area is limited. Fig. 7.24 shows the O concentration of ingots ($\times 10^{17}$ cm^{-3}) grown using the NOC furnace as a function of the diameter ratio [15]. The O concentration near a diameter ratio of 0 is determined by that at the bottom of

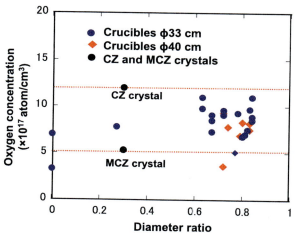

Fig. 7.24 Oxygen concentration in ingots grown using 33-, 40- and 50 cm-diameter crucibles without a Si$_3$N$_4$ coating as a function of the diameter ratio. The oxygen concentration of the ingots grown using the new and large furnace did not strongly depend on the diameter ratio. The oxygen concentration of the ingots has a tendency to become lower as the crucible diameter becomes larger. The minimum value was 3.6 × 10^{17} cm^{-3} for an ingot grown using a 40-cm-diameter crucible for a diameter ratio of more than 0.7. (Referred from Fig. 9 in K. Nakajima et al. J. Electron. Materials, 45 (2016) 2837.)

the grown seed crystal. The O concentrations of the ingots grown using the NOC furnace are lower than 1×10^{18} cm^{-3} and they do not strongly depend on the diameter ratio, even when the diameter ratio is larger than 0.6. The O concentration of the ingots grown using 40 and 50-cm-diameter crucibles has a tendency to be lower than that of the ingots grown using 33-cm-diameter crucibles because the ratio between the reactive area of the crucible wall and the total Si melt becomes lower as the crucible diameter increases. For a diameter ratio of more than 0.7, the lowest O concentration is 3.6×10^{17} cm^{-3} owing to gas flow control. It is comparable to that of an ingot grown by the magnetic-field-applied Czochralski (MCZ) method. This means that the convection in the Si melt and the reaction between the crucible wall and the Si melt are small in the NOC furnace. The convection is markedly suppressed by both the carbon heat holder and the low co-rotating rate of the seed and crucible (1 or 2 rpm). The carbon heat holder increases the uniformity of the temperature distribution in the Si melt in the furnace system with two zone heaters. The low rotation rate of seed or ingot increases the thickness of a boundary layer, $δ_T$ near the growing interface, in which heat is transported only by diffusion and the temperature oscillation amplitude is small as shown by the following equation [45],

$$\delta_T \propto w^{-1/2} \tag{7.3}$$

where w is the rotation rate of ingot in radians per second. The small amplitude prevents the convection which causes an unsteady flow. The oscillation amplitude is determined by the ingot rotation rate and not by the crucible rotation rate [45]. The obtained values of the O concentration are between the well-known values for ingots grown by the CZ and MCZ methods as shown in Fig. 7.24. These O concentrations of the CZ and MCZ data points are measured using crystals on the market at the same time by the FTIR method [11].

For a p-type Si ingot (B–doped), the O atoms form −B–O complexes which cause the light induced degradation. To prevent this degradation, the O concentration should be lower than 3×10^{17} cm^{-3} [46]. The main points to reduce the O concentration are a large crucible, a large amount of melt, small convection, large Ar gas flow, an evaporation method and a liquid–inert coating. For the NOC method, most of these points can be adopted except for a liquid–inert coating. Recently, a liquid–inert crucible is reported to reduce the O concentration in the Si melt and ingot [47−50]. This technology will be also effective for the NOC method (see Section 7.6).

In the NOC method, the growing interface of an ingot has a convex shape in the growth direction because of the large low-temperature region toward a deep Si melt and the small convection in the melt. The center of the ingot grows toward a deeper part of the melt, which has a higher N concentration than the periphery of the ingot. However, the N concentration in the ingot grown using a crucible without a Si$_3$N$_4$ coating is below the detection limit of the FTIR measurement of 1×10^{14} cm^{-3}.

7.5 Electrical properties and solar cells of Si single ingots

7.5.1 Resistivity and minority carrier lifetime of Si single ingots

The average resistivity decreases from the top to the bottom in an n–type NOC ingot [16]. Sometimes, swirl patterns caused by the local distribution of the resistivity appear in an ingot grown by the NOC method. The same swirl patterns are often observed in CZ wafers. The origins of swirl formation are proposed as O, C and re-melting of an ingot during growth

[45]. Non-uniform resistivity and O distributions suggest the formation of recombination-active oxide precipitates associated with swirl defects [51,52]. The oxide precipitates act as internal gettering sites for metal impurities and have deep energy levels within the band gap [53]. Thus, O impurities in the Si melt disperse to form a swirl pattern in the growing ingot because they easily move away from the center of the growing interface along its slope. The rotation of the ingot will enhance the swirl pattern. As known from Fig. 7.23, a convex growing interface clearly affects the distribution of O in these cross sections. However, the TR process at 1080 °C for 30s followed by PDG can dissolve almost all the swirl defects and result in the NOC wafers almost free of swirl defects and homogenizing the lifetime [53]. The origin of the low-lifetime swirl patten in the NOC wafers is the same oxide related defects as that in CZ wafers [53]. The distribution of defects limiting the charge carrier lifetime for NOC wafers is more homogeneous than that for CZ wafers because the morphology and/ or impurity decoration of oxide precipitates in the NOC wafers is different from that in the CZ wafers [51].

The SPV and μ-PCD methods can measure the minority carrier diffusion length in the ingots grown by the NOC method to evaluate the quality of ingots. Before the SPV and μ-PCD measurements, the sample surface is mechanically polished and chemically etched with hydrogen fluoride (HF) and nitric acid (HNO_3) solution. Then, the sample surface is electrically passivated with quinhydrone and ethanol solution [54]. In a multicrystalline ingot, a region with high diffusion length is always obtained near the center of the ingot. The maximum diffusion length is 464–610 μm and the average diffusion length is 140–173 μm. The ingot has a ring-shaped region in which the diffusion length is low. The diffusion length in the ring-shaped region is markedly improved by a P gettering process. Therefore, the average diffusion length in an entire wafer can be improved from 140 μm to 185 μm by the gettering process. The distribution of the minority carrier lifetime (47–118 μs) is similar to that of the diffusion length [29]. The highest minority carrier lifetime of p-type multicrystalline ingots grown by the NOC method is 117.6 μs measured by μ-PCD. This value is much higher than that of normal p-type ingots which are 10–30 μs [55].

The minority carrier lifetimes of an n-type single ingot are as high as 205 μs and 1.8 ms at an injection level of 1×10^{15} cm^{-3} before and after P gettering, respectively [40]. The impurity distribution in the ingot is consistent with that in the normal freezing case. Not all impurity species

Table 7.3 Minority carrier lifetime (μs) of wafers selected from ingots at 10^{15} cm^{-3} injection level.

	Piece	NOC-7	NOC-8	NOC-4
Top	25	725	742	1344
	26	1086	1165	1219
	101	583	379	562
Bottom	102	486	508	862

Furnace cleaning was performed between growth of NOC-8 and NOC-4

Fe contamination was reduced

From Table 2 in K. Nakajima et al. J. Crystal Growth, 468, 705 (2017)

From Table 7.2 in K. Nakajima et al. J. Crystal Growth, 468, 705 (2017).

respond satisfactorily to gettering, as some species diffuse too slowly. The greater lifetime improvement of the ingot is attributed to a reduced background concentration of difficult-to-getter impurities [40]. Generally, the minority carrier lifetime and efficiency of solar cells increase as the dislocation density or dislocation area decreases [56]. The dislocation density lower than 10^4 cm^{-2} is required to support minority carrier lifetime higher than 1 ms [57].

Table 7.3 lists the minority carrier lifetime of three n-type ingots (NOC-4, NOC-7 and NOC-8) grown before and after furnace cleaning using the NOC furnace. The minority carrier lifetime is measured after gettering by QSSPC in Massachusetts Institute of Technology (MIT). The furnace cleaning is performed between the growth of NOC-8 and NOC-4. The minority carrier lifetime at the injection level of 1×10^{15} cm^{-3} is largely improved even at the bottom of the ingot (NOC-4) after the furnace cleaning because Fe contamination is reduced. The minority carrier lifetime at the top of the ingots is higher than that at their bottom because of the presence of O precipitates in the latter part [53]. The minority carrier lifetime of the ingot (NOC-4) grown after furnace cleaning is much higher than that of the ingots (NOC-7 and NOC-8) grown before furnace cleaning all over the ingot. The millisecond lifetime is obtained after P gettering to control metal point defect concentrations [42]. The highest minority carrier lifetime is 1344 μs, measured after extended PDG.

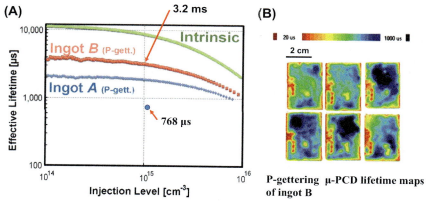

Fig. 7.25 (A) Minority carrier lifetimes of the n-type single ingots A and B as a function of the injection level after P gettering. They are as high as 1.8 ms and 3.2 ms at an injection level of 1×10^{15} cm^{-3}, respectively. (B) P-gettering μ-PCD lifetime maps of ingot B. Large areas with high minority carrier lifetimes (denoted as dark blue and black color) can be observed. *(Referred from Figs. 3 and 4 in S. Castellanos et al. Energy Procedia 92 (2016) 779.)*

Through the stringent impurity–control procedure at the growth stage to prevent in-diffusion of impurities to the melt as shown in Table 7.2 and the defect-engineering approach to get better P-gettering time-temperature profiles, the highest effective minority carrier lifetimes of a P-doped n-type ingot are improved to be as high as 3.2 ms and 768 μs at an injection level of 1×10^{15} cm^{-3} after and before the optimized extended P gettering, respectively, which are measured be QSSPC in MIT [16,42]. Fig. 7.25A shows the effective minority carrier lifetimes of n-type single ingots A and B as a function of the injection level after PDG [42]. Fig. 7.25B shows μ-PCD lifetime maps of ingot B after PDG [40,42]. In the maps shown in Fig. 7.25B, large areas with high minority carrier lifetimes (denoted as dark blue and black color) can be observed. The minority carrier lifetimes of ingots A and B are as high as 1.8 ms and 3.2 ms at an injection level of 1×10^{15} cm^{-3}, respectively. The lifetime-limiting defects are most likely getterable metal impurities [40]. The concentration from potential slow diffusers such as chromium (Cr) must be lower than 5×10^9 cm^{-3} [42]. The time-temperature profile is shown to enhance the extraction of interstitial Fe [58].

O precipitates still present at the bottom of several NOC ingots. Jensen et al. [52] studied the effect of the TR process for the ingots. Even though the ingots contain some amount of O impurities, the relative recombination strength of the swirl region is largely improved and the minority carries lifetime is also largely improved using the TR process following by the PDG process. Such TR process can enable swirl-free wafers throughout the ingots and higher lifetime. However, the minority carries lifetime is degraded after TR only. As-grown wafers cut from the top of NOC ingots have usually lifetimes of about 1 ms. After the TR process following the extended P gettering, the lifetimes are effectively improved to about 2 ms which is two times higher. The spatial homogeneity of the top ingot wafers observed in the as-grown state is not disrupted by any of the high temperature processes. Some getterable impurities can sufficiently move at the annealing temperature (see Sections 1.4.2 and 2.4.3).

7.5.2 Performance of solar cells prepared by the NOC method

The NOC method has a novel characteristic that an ingot can be grown inside a Si melt without contact with a crucible wall. Therefore, it is one of the effective growth methods to obtain Si high-quality single ingots with a large diameter and a large volume using the cast furnace and to obtain p-type solar cells with the high conversion efficiency and yield [10,11,41]. A p-type ingot with a convex growth interface in the growth direction is most suitable to prepare such solar cells.

The minority carrier lifetime of an n-type wafer is generally higher than that of a p-type wafer [59]. n-Type solar cells with the PERT structure were prepared using NOC wafers with a size of 15.6×15.6 cm^2 by INES in France. Fig. 7.26 shows the solar cell performance of these wafers cut from an n-type ingot grown after the furnace cleaning [60]. Using the same solar cell structure and process to obtain the conversion efficiency of 20% for n-type CZ wafers, the highest conversion efficiency of 19.8% and the average conversion efficiency of 19.3% are obtained for the NOC wafers. The top part of the ingot has a slightly higher conversion efficiency than the bottom part because of the large segregation coefficient (0.35) of P dopant and some O-related defects. The scattering width is only 1.2%, which is much narrower than that of the mono-like solar cells using the same processes and structure as shown by the bar in the left corner of Fig. 6.10 [56] (see Section 6.5.3). The scattering width is defined as the diffence between the highest and lowest conversion efficiencies. The highest conversion efficiency is almost the same as the conversion efficiency of the CZ solar cells.

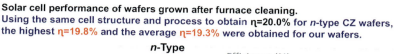

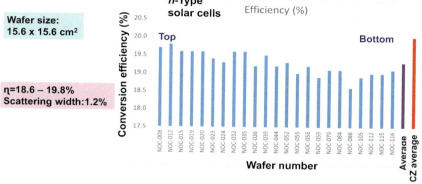

Fig. 7.26 Performance of the *n*-type solar cells prepared using wafers with a size of 15.6 × 15.6 cm² by the PERT structure in INES. Using the same solar cell structure and process to obtain the conversion efficiency of 20% for *n*-type CZ wafers, the highest conversion efficiency is 19.8% and the average conversion efficiency is 19.3% for the NOC wafers. The scattering width is only 1.2%. (Referred from Fig. 19 in K. Nakajima, in: D. Yang (Ed.), Chapter 12, Section Two, Hand Book of "Photovoltaic Silicon Material", Springer, Berlin Heidelberg, pp.1–32, 2017, On line.)

As the O concentration in ingots grown by the NOC method is lower than that in CZ ingots [11], this method is effective for the fabrication of B-doped *p*-type solar cells because there are few B—O complexes and oxide precipitates in the ingots [61–63]. Using *p*-type ingots prepared by the NOC method to determine the distribution of conversion efficiency of solar cells, the Fukushima Renewable Energy Institute (FREA) in National Institute of Advanced Industrial Science and Technology (AIST) standard solar cells are fabricated on the *p*-type wafers with a size of 15.6 × 15.6 cm². The structure of the solar cells is based on the Al-BSF one [41]. The cell thickness is 175 μm. All processing steps used for such FREA standard cells are industrial scale. The first processing step is to form the texturing surface using potassium hydroxide (KOH) solution. The phosphorous emitter is performed with thermal diffusion step using an industrial tube furnace. The sheet resistivity is 90 $\Omega/\square$. High passivation quality is performed by SiN_x deposited using an industrial plasma PECVD. The front electrode, rear electrode and BSF are made during a firing step after screen printing of Ag and Al paste, respectively. The solar cell structure is shown in Fig. 7.27 together with the distribution of conversion efficiency of *p*-type

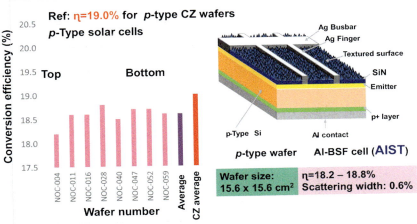

Fig. 7.27 Al-BSF structure and the distribution of conversion efficiency of p-type solar cells using p-type NOC wafers with a size of 15.6 × 15.6 cm² prepared by FREA AIST. Using the same solar cell structure and process to obtain the conversion efficiency of 19.0% for a p-type CZ wafer, all solar cells grown by the NOC method have higher conversion efficiency than 18.2%. The scattering width (0.6%) is very narrow because of high uniformity of the ingot. *(Referred from Fig. 20 in K. Nakajima, in: D. Yang (Ed.), Chapter 12, Section Two, Hand Book of "Photovoltaic Silicon Material", Springer, Berlin Heidelberg, pp.1–32, 2017, On line.)*

solar cells using an ingot grown before the furnace cleaning by the NOC method [60]. Using the same solar cell structure and process to obtain the conversion efficiency of 19.0% for p-type CZ wafers, all solar cells have higher conversion efficiency than 18.2%. The scattering width of 0.6% is very narrow because of high uniformity of the ingot. To improve the distribution of conversion efficiency, a p-type single ingot grown after the furnace cleaning shown in the upper right in Fig. 7.28 is used, which has a convex growing interface in the growth direction. The convex ingot is intentionally made to reduce dislocations using the low pulling and cooling rates at the end of the growth. The maximum diameter and height of the ingot is 35.0 and 8.0 cm, respectively. Fig. 7.28 shows the solar cell performance of wafers grown after the furnace cleaning [10]. Using the same solar cell structure and process to obtain the conversion efficiency of 19.1% for p-type CZ wafers, the highest conversion efficiency of 19.14% and the average conversion efficiency of 19.0% are obtained for the NOC wafers. The conversion efficiency is the same as that for CZ solar cells.

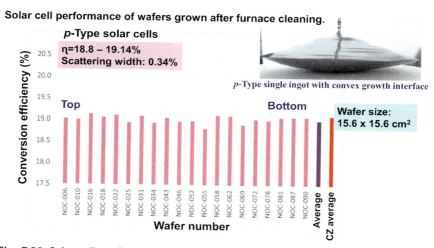

Fig. 7.28 Solar cell performance of wafers cut from the p-type ingot shown in the upper right. The highest conversion efficiency was 19.14% and the average conversion efficiency was 19.0%. Its scattering width was only 0.34%. Using the cast furnace, the NOC method enabled the preparation of Si ingots for solar cells with conversion efficiency and yield as high as those of CZ solar cells. (Referred from Fig. 3 in K. Nakajima J. Cryst. Growth 468 (2017) 705.)

The conversion efficiency becomes much higher and quite uniform comparing with the results before the furnace cleaning shown in Fig. 7.27. The scattering width of the conversion efficiency is only 0.34% which is quite narrow because of high uniformity of the ingot. It is demonstrated, using the cast furnace, that solar cells with the conversion efficiency and yield as high as those of the CZ solar cells can be realized by the NOC method. Fig. 7.29 shows the frequency of the conversion efficiency of p-type solar cells prepared by the HP cast method [64,65] and the NOC method [59]. Multicrystalline ingots are used to prepare the solar cells for the HP cast method and single ingots are used to prepare the solar cells for the NOC method. The distribution width of the conversion efficiency of the NOC solar cells is much narrower than that of the HP cast method. The yield of the regular solar cells with the conversion efficiency higher than 18.8% is 100% for the NOC method when the cutting loss is not considered. The yield is much larger than that of the HP cast method. The peak of the conversion efficiency is 1% higher than that of the HP cast method because of the efficient texture structure and high-quality wafers for the NOC method.

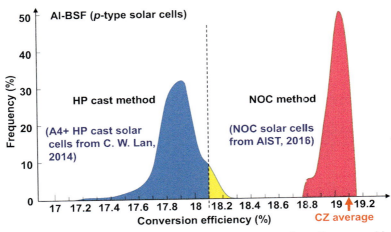

Fig. 7.29 Frequency of the conversion efficiency of p-type solar cells prepared by the HP cast method [64,65] and the NOC method. For the NOC method, the distribution was quite sharp and the yield was quite large comparing with that for the HP cast method. Multicrystals were used to prepare the solar cells for the HP method and single crystals were used to prepare the solar cells for the NOC method. *(Referred from Fig. 4 in K. Nakajima J. Cryst. Growth 468 (2017) 705.)*

The conversion efficiency strongly depends on the structure of solar cell. The simulated conversion efficiency of PERC solar cells is 1% higher than that of Al-BSF solar cells [57]. Fig. 7.30 shows the frequency of the conversion efficiency of p-type solar cells which have a structure of PERC prepared by FREA AIST. The wafers used for the PERC solar cells are cut from the same ingot used for the Al-BSF solar cells. The frequency of the PERC solar cells is shown together with those of the p-type solar cells prepared by the HP cast method and the NOC method which are shown in Fig. 7.30. For the PERC solar cells, the highest conversion efficiency is 19.8%, which is 0.7% higher than that of the Al-BSF solar cells. The average conversion efficiency is 19.64%. The scattering width of the conversion efficiency is only 0.3%, and the distribution width of the conversion efficiency of the PERC solar cells is quite narrow.

Even though the O concentration in ingots grown by the NOC method is slightly higher than that in cast ingots, the yield of the p-type regular solar cells with the conversion efficiency higher than 18.8% is larger than that of the HP cast method as shown Fig. 7.29. The yield of the n-type solar cells without O complexes will be much larger than that of the HP cast method. Using the cast furnace with high productivity, the NOC

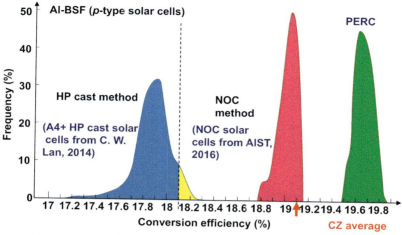

Fig. 7.30 Frequency of the conversion efficiency of the p-type PERC solar cells together with those of the Al-BSF p-type solar cells prepared by the HP cast method and the NOC method, which are shown in Fig. 7.29. The highest conversion efficiency is 19.8%, which is 0.7% higher than that of the Al-BSF solar cells. The average conversion efficiency is 19.64%. The scattering width of the conversion efficiency is only 0.3%, and the yield of the PERC solar cells is quite high.

method has a possibility to prepare Si single ingots for solar cells with conversion efficiency and yield as high as those of CZ solar cells. The present method is promising as an advanced cast method for obtaining Si single ingots with a large volume and high-efficiency solar cells with a high yield.

7.6 Key points for improvement of the NOC method

There are two kinds of application for the NOC method to obtain a Si single-crystalline ingot. For the first application, the NOC method can be used as a pulling method while an ingot is grown inside a Si melt during pulling. The NOC furnace has two zone heaters on its side and bottom as shown in Figs. 7.3 and 7.4, and it can be basically used like a CZ furnace which has two zone heaters on its side and bottom. To obtain an ingot with same quality as that of a CZ ingot, the ingot should be dislocation-free for this application as explained in Section 8.2. For the second application, the NOC method can be used as an advanced cast method without contact with the crucible wall. To obtain an ingot with higher quality and uniformity than those of an ingot grown by the HP cast method or the mono-like cast method, the ingot should have low stress and low dislocation-density for this application. The ingot should also have same large size for the diameter and volume and same low O concentration as those of the HP cast ingot or the mono-like cast ingot.

The NOC method has several different points from the CZ method [35]. In the CZ method, an ingot is grown above the surface of a Si melt and the growing interface is generally concave in the growth direction [37]. The growth condition is not under near equilibrium at the growing interface. The growth rate is determined by the pulling rate of the ingot above the surface of the Si melt [35–37]. The diameter ratio of a growing ingot is generally set to be 0.3. In contrast to the CZ method, the NOC method can control the size and shape of a growing ingot mainly by controlling those of the low-temperature region in a Si melt. The shape of the growing interface is generally convex in the growth direction, which depends on the pulling rate, cooling rate, melt depth and temperature gradient in the melt. The growth rate is determined by the expansion rate of the low-temperature region in the Si melt [38], and the expansion rate of the low-temperature region is mainly determined by the cooling rate of the Si melt and the temperature gradient in the Si melt [38]. Therefore, a high growth rate can be attained by a high cooling rate and a low temperature gradient in the deep melt because the ingot grows inside the Si melt and toward the crucible wall or bottom with higher temperature than that of the low-temperature region [31–34]. The O concentration of ingots is relatively lower than that of ingots grown by the CZ method because convection is possibly suppressed to create a low-temperature region in the Si melt and the small convection prevents the reaction between the crucible wall and the melt [11]. The distribution of defects in NOC wafers is more homogeneous than that for CZ wafers [51]. The NOC method has several merits owing to the presence of a large low-temperature region inside a Si melt and the growth inside the low-temperature region [15].

The NOC method has several different points from the HP cast and mono-like cast methods. In the HP cast and mono-like methods, an ingot is grown during contacting a crucible wall, and large stress exerted in the ingot is always caused from the crucible wall because Si expands by 11% during solidification. The large stress becomes the occurrence factor to generate many dislocations. Random nucleation occurs on the crucible wall and many grains generate from it. These grains cause the multi-crystallization and degrade the uniformity of the ingot. The growth rate is determined by the cooling rate and temperature gradient in the Si melt.

The diameter ratio of a growing ingot is 1.0 which means the same size as the crucible diameter, but large parts of the ingot near the crucible wall must be cut off because of large amount of impurities and defects in these parts.

In contrast to the HP cast and mono-like methods, the NOC method can grow an ingot without contacting with the crucible wall and largely reduce stress in the ingot. No nucleation occurs from the crucible wall because of the existence of the low-temperature region inside a Si melt and the higher melt temperature near the crucible wall. Only nucleation site is a seed crystal on the surface of the Si melt. A uniform single ingot with the low dislocation and defect densities can be easily grown by this method. The dislocation density is expected to be on the order of 10^2– 10^3 cm^{-2} in the center of the ingot without using a necking technique. The NOC method is expected to be used as an advanced cast method because it can manufacture a high-quality single ingot with a larger diameter ratio than 0.9, a large diameter similar in the crucible size and a large volume.

In the case to use the NOC method as a pulling method like the CZ method, dislocation-free ingots should be grown by this method. The necking technique is one of the key technologies for this purpose [14]. The effect of buoyancy from a Si melt should be minimized to effectively use the necking technique for such growth inside the Si melt as shown in Fig. 8.3. Another technology is to keep each temperature of the top of a seed crystal and the melt surface directly below the seed crystal almost same just when the seed crystal contacts with the melt surface as shown in Fig. 8.4.

In the case to use the NOC method as an advanced cast method, a Si single ingot with a large volume and a large diameter should be prepared using a cast furnace. The quality of the ingot should be enough high to obtain high-efficiency solar cells in high productivity. The dislocation density in the ingot can be reduced to be enough low to obtain the high-efficiency solar cells with high yield as explained in Section 7.4.4. As the diameter ratio more than 0.9 can be realized, the NOC method has a possibility to obtain an industrial-scale single ingot using a crucible with an industrial size such as a HP cast crucible. The large ingot can be grown without contacting to the crucible wall by establishing a large low-temperature region in the Si melt (see Section 7.4.2). A Si single ingot with a square shape is better to reduce the production cost of the ingot. To obtain a square-shaped ingot, the temperature gradient should be controlled to be small in front of the growing interface by establishing a wider low-temperature region and using a larger crucible (see Section 8.3).

Ingots should contain low O and C concentrations to obtain high-performance solar cells. The O concentration in an ingot should be as low as that in an ingot grown by the cast method. Several technologies to reduce the O concentration in the ingot are tried to be used such as the Ba-doped silica crucible and the liquid-inert crucible [47–50,66]. Ba doped silica crucible with inner surface coated by cristobalite is used to reduce the O solubility in a Si melt [66] (see Section 3.1.2). For the liquid-inert crucible, the inner surface of the crucible is treated to have a melt-phobic effect resulting in liquid inert surface [50,59]. This crucible is effective to suppress the reaction between a Si melt and a quartz crucible and reduce the O concentration in the melt and ingot. As the result, the evaporation of SiO_2 from the melt is suppressed and the CO concentration also decrease because CO is synthesized by the SiO_2 gas and graphite materials, resulting in reduction of the C concentration in the grown ingot [50].

The metal impurities in ingots are still a basic problem for the NOC method. As-grown lifetime is limited by a combination of getterable impurities (Cu, Ni, Fe and Cr) [40]. Their main origins are the environment around crucible and the quartz which contains metals of several ten of ppm. These fast-diffusing metallic impurities can be mitigated by P gettering and a minority carrier lifetime can be improved to several ms [40]. Thus, Si ingots grown by the NOC method have the potential to achieve as-grown ms lifetimes by more stringent impurity controls. An advanced technology should be developed to largely reduce the metal impurities by designing the clean furnace.

Si single-crystalline ingots grown by the NOC method has the potential to achieve high bulk minority carrier lifetime and high conversion efficiencies at low cost due to its low structural defect density. The NOC method has a high potential to realize solar cells with the performance and yield as high as those of CZ solar cells while keeping high productivity using a cast furnace. The NOC method will be more effective to prepare n-type solar cells when dislocation-free ingots will be realized. The potential of the NOC method is demonstrated by another researchers [57]. This method is very much interested by a semiconductor company on the view point of manufacturing a larger ingot using their existing limited-size furnaces and confirming the difference of the defect formation mechanism.

312 Crystal Growth of Si Ingots for Solar Cells Using Cast Furnaces

To effectively keep the large low-temperature region in the Si melt, the balance between the supply of heat flow from the side heater and the cutting-off of heat flow from the bottom heater is very important. The size of the insulating plate is the very important factor to intentionally control the size of the low-temperature region and the amount of ΔT. To systematically determine the heat flow balance and the insulator effect, the theoretical simulation is very effective as shown in Fig. 7.11, as an example. The systematic determination of the temperature distribution in the Si melt is an important subject to precisely understand the effects of the insulating plate set on the crucible bottom. Such research should be performed to achieve a steady-state crystal growth. The theoretical simulation is quite important as an urgent target to develop this method as a next novel growth technology in industry.

In future, a Si ingot grown by the NOC method will be much more effective to obtain a large-scale dislocation-free Si single ingot using a cast furnace by effectively controlling the soft touching process of a seed with the Si melt surface to completely remove the thermal shock and dislocations. The temperature gradient in the NOC ingot is much smaller comparing with the CZ ingot because the large part of the ingot still inside the melt. This point will largely influence the distribution of crystal defects in the ingot. The largest merit of the NOC method is the possibility of fewer defects in the large-scale dislocation-free Si single ingot using a low-cost cast furnace. The confirmation of this possibility should be the most urgent subject for widely expanding the future NOC ingot as an advanced cast and pulling method.

References

[1] M.P. Bellmann, E.A. Meese, M. Syvertsen, A. Solheim, H. Sorheim, L. Arnberg, J. Cryst. Growth 318 (2011) 265.
[2] K. Nakajima, W. Pan, Y. Nose, Japanese Patent No. 4292300, 2009.
[3] Y. Nose, I. Takahashi, W. Pan, N. Usami, K. Fujiwara, K. Nakajima, J. Cryst. Growth 311 (2009) 228.
[4] N. Usami, I. Takahashi, K. Kutsukake, K. Fujiwara, K. Nakajima, J. Appl. Phys. 109 (2011) 083527.
[5] D.F. Bliss, Evolution and application of the Kyropoulos crystal growth method, in: R.S. Feigelson (Ed.), 50 Years Progress in Crystal Growth: A Reprint Collection, Elsevier Amsterdam, 2004, pp. 29–33.
[6] K. Nakajima, R. Murai, K. Morishita, K. Kutsukake, N. Usami, J. Cryst. Growth 344 (2012) 6.
[7] S.E. Demina, E.N. Bystrova, M.A. Lukanina, V.M. Mamedov, V.S. Yuferev, E.V. Eskov, M.V. Nikolenko, V.S. Postolov, V.V. Kalaev, Opt. Mater. 30 (2007) 62.

[8] P.S. Ravishankar, Sol. Energy Mater. 12 (1985) 361.
[9] A. Nouri, Y. Delannoy, G. Chichignoud, L. Lhomond, B. Helifa, I. Lefkeir, K. Zaidat, J. Cryst. Growth 460 (2017) 48.
[10] K. Nakajima, S. Ono, Y. Kaneko, R. Mura, K. Shirasawa, T. Fukuda, H. Takato, S. Castellanos, M.A. Jensen, A. Youssef, T. Buonassisi, F. Jay, Y. Veschetti, A. Jouini, J. Cryst. Growth 468 (2017) 705.
[11] K. Nakajima, S. Ono, R. Murai, Y. Kaneko, J. Electron. Mater. 45 (2016) 2837.
[12] K. Nakajima, R. Murai, K. Morishita, K. Kutsukake, J. Cryst. Growth 372 (2013) 121.
[13] K. Nakajima, R. Murai, K. Morishita, Jpn. J. Appl. Phys. 53 (2014) 025501.
[14] W.C. Dash, J. Appl. Phys. 30 (1959) 459.
[15] K. Nakajima, R. Murai, K. Morishita, K. Kutsukake, Proceedings of the 39th IEEE Photovoltaic Specialists Conference (39th PVSC), Tampa, Florida, US, June 16–21, 2013, pp. 174-176.
[16] K. Nakajima, R. Murai, S. Ono, K. Morishita, M. Kivambe, D.M. Powell, T. Buonassisi, Jpn. J. Appl. Phys. 54 (2015) 015504.
[17] K. Nakajima, S. Ono, H. Itoh, J. Cryst. Growth 99 (2018) 55–61.
[18] V.V. Kalaev, I.Y. Evstratov, Y.N. Makarov, J. Cryst. Growth 249 (2003) 87.
[19] K. Fujiwara, K. Maeda, N. Usami, G. Sazaki, Y. Nose, A. Nomura, T. Shishido, K. Nakajima, Acta Mater. 56 (2008) 2663.
[20] P. Rudolph, M. Czupalla, B. Lux, F. Kirscht, C. Frank-Rotsch, W. Miller, M. Albrecht, J. Cryst. Growth 318 (2011) 249.
[21] S. Pizzini, A. Sandrinelli, M. Beghi, D. Narducci, F. Allegretti, S. Torchio, G. Fabbri, G.P. Ottaviani, F. Demartin, A. Fusi, J. Electrochem. Soc. 135 (1988) 155.
[22] T.F. Ciszek, J. Cryst. Growth 66 (1984) 655.
[23] C. Reimann, M. Trempa, T. Jung, J. Friedrich, G. Müller, J. Cryst. Growth 312 (2010) 878.
[24] K. Nakajima, K. Morishita, R. Murai, N. Usami, J. Cryst. Growth 389 (2014) 112.
[25] S. Togawa, K. Izunome, S. Kawanishi, S. Chung, K. Terashima, S. Kimura, J. Cryst. Growth 165 (1996) 362.
[26] M.P. Bellmann, E.A. Meese, L. Arnberg, J. Cryst. Growth 312 (2010) 3091.
[27] S. Togawa, X. Huang, K. Izunome, K. Terashima, S. Kimura, J. Cryst. Growth 148 (2010) 70.
[28] V.V. Voronkov, R. Falster, J. Cryst. Growth 273 (2005) 412.
[29] K. Nakajima, K. Morishita, R. Murai, K. Kutsukake, J. Cryst. Growth 355 (2012) 38.
[30] B.L. Sopori, J. Electrochem. Soc. 131 (1984) 667.
[31] K. Nakajima, K. Kutsukake, K. Fujiwara, K. Morishita, S. Ono, J. Cryst. Growth 319 (2011) 13.
[32] K. Fujiwara, W. Pan, N. Usami, K. Sawada, M. Tokairin, Y. Nose, A. Nomura, T. Shishido, K. Nakajima, Acta Mater. 54 (2006) 3191.
[33] T.F. Li, H.C. Huang, H.W. Tsai, A. Lan, C. Chuck, C.W. Lan, J. Cryst. Growth 340 (2012) 202.
[34] K. Kutsukake, H. Ise, Y. Tokumoto, Y. Ohno, K. Nakajima, I. Yonenaga, J. Cryst. Growth 352 (2012) 173.
[35] J. Czochralski, Z. Phys. Chem. 92 (1917) 219.
[36] S.N. Rea, J. Cryst. Growth 54 (1981) 267.
[37] W. Zulehner, J. Cryst. Growth 65 (1983) 189.
[38] K. Nakajima, K. Morishita, R. Murai, J. Cryst. Growth 405 (2014) 44.
[39] K. Izunome, X. Huang, S. Togawa, K. Terashima, S. Kimura, J. Cryst. Growth 151 (1995) 291.
[40] M. Kivambe, D.M. Powell, S. Castellanos, M.A. Jensen, A.E. Morishige, K. Nakajima, K. Morishita, R. Murai, T. Buonassisi, J. Cryst. Growth 407 (2014) 31.

[41] K. Nakajima, S. Ono, Y. Kaneko, R. Murai, K. Shirasawa, T. Fukuda, and H. Takato, Proceedings of the 43th IEEE Photovoltaic Specialists Conference (43th PVSC), Portland, Oregon, US, June 5—10, 2016, pp. 68-72.

[42] S. Castellanos, M. Kivambe, M.A. Jensen, D.M. Powell, K. Nakajima, K. Morishita, R. Murai, T. Buonassisi, ScienceDirect, Energy Procedia 92 (2016) 779.

[43] C. Creskovich, S. Prochazka, J. Am. Ceram. Soc. 60 (1977) 471.

[44] N. Machida, Y. Suzuki, K. Abe, N. Ono, M. Kida, Y. Shimizu, J. Cryst. Growth 186 (1998) 362.

[45] E. Kuroda, H. Kozuka, J. Cryst. Growth 63 (1983) 276.

[46] T. Saitoh, X. Wang, H. Hashigami, T. Abe, T. Igarashi, S. Glunz, S. Rein, W. Wettling, I. Yamasaki, H. Sawai, H. Ohtuka, T. Warabisako, Sol. Energy Mater. Sol. Cells 65 (2001) 277.

[47] S. Sakuragi, T. Shimasaki, G. Sakuragi, H. Nanba, Proceedings of 19th European Photovoltaic Solar Energy Conference and Exhibition, Palais des Congrès, Paris, France, June 7—11, 2004, pp.1197—1200.

[48] S. Sakuragi, Proceedings of 19th European Photovoltaic Solar Energy Conference and Exhibition, Palais des Congrès, Paris, France, June 7—11, 2004, pp.1201—1204.

[49] Y. Horioka, S. Sakuragi, Japanese Patent No. 4854814.

[50] T. Fukuda, Y. Horioka, N. Suzuki, M. Moriya, K. Tanahashi, S. Simayi, K. Shirasawa, H. Takato, J. Cryst. Growth 438 (2016) 76.

[51] J. Schön, A. Youssef, S. Park, L.E. Mundt, T. Niewelt, S. Mack, K. Nakajima, K. Morishita, R. Murai, M.A. Jensen, T. Buonassisi, M.C. Schubert, J. Appl. Phys. 120 (2016) 105703.

[52] M.A. Jensen, V. LaSalvia, A.E. Morishige, K. Nakajima, Y. Veschetti, F. Jay, A. Jouini, A. Youssef, P. Stradins, T. Buonassisi, Energy Procedia 92 (2016) 815.

[53] A. Youssef, J. Schön, S. Mack, T. Niewelt, K. Nakajima, K. Morishita, R. Murai, T. Buonassisi, and M. C. Schubert, Proceedings of the 43th IEEE Photovoltaic Specialists Conference (43th PVSC), Portland, Oregon, US, June 5—10, 2016, pp. 1303-1307.

[54] H. Takato, I. Sakata, R. Shimokawa, Jpn. J. Appl. Phys. Pt. 2, Lett. 40 (2001) L1003.

[55] Y.C. Wu, A. Lan, C.F. Yang, C.W. Hsu, C.M. Lu, A. Yang, C.W. Lan, Cryst. Growth Des. 16 (2016) 6641.

[56] F. Jay, D. Muñoz, T. Desrues, E. Pihan, V.A. Oliveira, N. Enjalbert, A. Jouini, Sol. Energy Mater. Sol. Cells 130 (2014) 690.

[57] J. Hofstetter, C. del Cañizo, H. Wagner, S. Castellanos, T. Buonassisi, Prog. Photovoltaics Res. Appl. 24 (2016) 122.

[58] M. Rinio, A. Yodyunyong, S. Keipert-Colberg, Y.P.B. Mouafi, D. Borchert, A. Montesdeoca-Santana, Prog. Photovoltaics 19 (2011) 165.

[59] J. Nelson, J. Nelson (Eds.), Physics of Solar Cells, Imperial College Press, London, 2003. Chap. 7.

[60] K. Nakajima, Growth of crystalline Silicon for solar cells: noncontact crucible method, in: D. Yang (Ed.), Chapter 12, Section Two, Crystalline Silicon Growth, Hand Book of "Photovoltaic Silicon Material", Springer, Berlin Heidelberg, 2017, pp. 1—32 (On line).

[61] G. Coletti, P. Manshanden, S. Bernardini, P.C.P. Bronsveld, A. Gutjahr, Z. Hu, G. Li, Sol. Energy Mater. Sci. Cells 130 (2014) 647.

[62] D. Macdonald, A. Cuevas, Sol. Energy Mater. Sol. Cells 65 (2001) 509.

[63] A. Herguth, G. Schubert, M. Kaes, G. Hahn, Proceedings of the 21st European Photovoltaic Solar Energy Conference and Exhibition (EUPVSEC 530), Dresden, Germany, September 4—8, 2006.

[64] C.W. Lan, in: The 6th World Conference on Photovoltaic Energy Conversion (WCPEC-6), Kyoto International Conference Center, Kyoto, Japan, November 23-27, 2014.

[65] C.W. Lan, C. Hsu, K. Nakajima, Multicrystalline Silicon crystal growth for photovoltaic applications, in: T. Nishinaga, P. Rudolph (Eds.), Handbook of Crystal Growth, Bulk Crystal Growth: Basic Techniques Vol. II, Part A, Elsevier, Amsterdam, 2015, pp. 373−411.

[66] X. Huang, T. Hoshikawa, S. Uda, J. Cryst. Growth 306 (2007) 422.

CHAPTER 8

Future technologies of Si ingots for solar cells

8.1 Proper grain size and stress in Si multicrystalline ingots

Random grain boundaries can reduce stress in a multicrystalline ingot, but they become origins of dislocations as shown in Figs. 4.13 and 4.14. Coherent or twin grain boundaries cannot reduce stress, but they don't become origins of dislocations as shown in Figs. 4.14 and 4.15. To obtain a multicrystalline ingot with a low dislocation density, it is very important to know which is better, a high density of random grain boundaries or a high density of twin grain boundaries in the ingot. This depends on how much stress remains in the ingot. In an ingot with large stress, the stress should be reduced by introduction of random grain boundaries, but a large amount of dislocations remains in the ingot such as an ingot grown by the HP cast method, even though the random grain boundaries can prevent movement of dislocations. In an ingot with small stress, it is better to introduce the twin grain boundaries which don't strongly affect the minority carrier lifetime or the conversation efficiency of solar cells. The dislocation density near a random grain boundary becomes lower as the coherency of the grain boundary increases like twin boundaries as shown in Fig. 4.15. To largely reduce the stress in an ingot, the ingot should be grown without contacting the crucible wall like the NOC method.

The grain size is an important factor to determine the proper density of random grain boundaries. As shown in Fig. 4.24, an ingot with small grains has small stress and many random grain boundaries and an ingot with large grains has large stress and few random grain boundaries. These concepts are used for the HP cast method and for the mono-like cast method, respectively. There is a strong relation between the density of random grain boundaries and the dislocation density in an ingot grown by the HP cast method. Random grain boundaries can prevent movement of dislocations

Crystal Growth of Si Ingots for Solar Cells Using Cast Furnaces
ISBN 978-0-12-819748-6
https://doi.org/10.1016/B978-0-12-819748-6.00008-6

Copyright © 2020 Elsevier Inc.
All rights reserved.

317

318   Crystal Growth of Si Ingots for Solar Cells Using Cast Furnaces

and become almost harmless by hydrogenation. Such effects are very lucky for the HP cast method. But the ingot still has many random grain boundaries. There is a trade-off between reduction of stress and reduction of random grain boundaries. The grain size and the distribution of small grains strongly concern in this trade-off. A large grain or a single crystal is effective for texture structure to increase 1% efficiency. It will be a next target for the mono-like cast method. However, many random grain boundaries or many small grains should be intentionally introduced in an ingot near the crucible wall to reduce the stress and dislocations in a large mono-like grain in the center of the ingot. The proper grain size or the proper grain distribution will be required to reduce both dislocations and random grain boundaries (Session 4.8.1).

8.2 Novel technologies for dislocation-free single ingots with large diameter and volume

8.2.1 Soft touching technology for dislocation-free single ingots

To realize a dislocation-free Si ingot with a large diameter and a large volume without using the CZ method, several technologies are developing. The one of the most important points for these technologies is how to realize a dislocation-free ingot without using the Dash technique [1]. The soft touching of a seed crystal on a Si melt surface is one of the most promising technologies to realize such a dislocation-free ingot. The soft touching should be performed at the almost same temperature closed to the meting point of Si to prevent generation of dislocations due to the thermal shock as shown in Fig. 8.4. Usually the temperature difference between the seed and the Mp of Si is more than 100 °C. A process to gradually decrease the descent speed of a seed is proposed to reduce the thermal shock [2]. It is better to slightly melt the bottom surface of the seed before the soft touching. Such melting method is proposed using a resistance heater to melt the lower tip of the seed crystal before contacting it with the melt surface [3]. The molten cap of the bottom of the seed is effective to prevent the generation of dislocations due to the thermal shock.

The merit of this technology is that a thicker seed crystal can be used to keep a single ingot with a large diameter and volume. A heavily B-doped Si seed may another candidate for the soft touching because this seed can prevent the generation of edge dislocations propagating along the growth direction in an undoped Si crystal, but misfit dislocations are formed near

Future technologies of Si ingots for solar cells　319

the growing interface because of large lattice misfit between the seed and crystal [4]. The grown crystal should be also heavily doped by B to reduce the lattice misfit. These technologies for the soft touching are effective for the growth of heavier dislocation-free Si ingots using a thick seed.

To realize the soft touching to reduce thermal stress, a Si single crystal with a low dislocation density (10^4-10^5 cm^{-2}) is grown by a modified FZ method on a large and thick seed with 10 cm diameter and 2 cm thickness without using the Dash technique [1]. In this method, an additional coil is added below the large radiofrequency (RF) inductor which is set between the seed and the feed rod and the direct contact between a melt pool on the upper seed surface and a liquid drop hanging at the feed rod is realized [1]. The dislocation density strongly depends on the temperature homogeneity and the shape of the growing interface. The flat interface is better than the concave one. To prevent generation of dislocation networks in the unmelted part of the seed, the seed is heated near 900 °C [1]. The improvement of the instability of the growing interface is one of the most important problems by obtaining the better temperature homogeneity near the growing interface (see Section 8.4).

8.2.2　Novel technologies for dislocation-free single ingots using the NOC method

The NOC method has several merits for the following two applications. The first one is the advanced cast method without contact to the crucible wall. An ingot with a large diameter and a large volume can be realized inside a Si melt because of the high diameter ratio of 0.9 [5]. Dislocation-free ingots are not required in this case. The dislocation density in the center of ingots grown by the NOC method is usually on the order between 10^3 and 10^4 cm^{-2} without necking. The millisecond lifetime is always obtained after gettering owing to high purity without Si_3N_4 coating. The yield of the regular solar cells with high conversion efficiency is much larger than that for the HP cast method [6]. The NOC method has the potential to be used as an advanced cast method to obtain Si single ingots for solar cells with high conversion efficiency and high yield. The second application is the pulling method of an ingot grown inside a Si melt. The low O concentration in the ingot is profitable for p-type solar cells. Dislocation-free ingots are required in this case. The NOC method has a potential to realize dislocation-free ingots by combination of the CZ method as shown in Fig. 8.3.

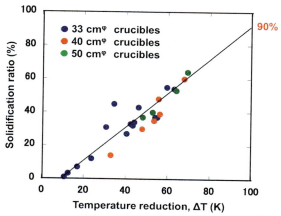

Fig. 8.1 Solidification ratio as a function of ΔT. The solidification ratio almost linearly increased with ΔT regardless of the crucible diameter. The solidification ratio is expected to be 90% when ΔT will be 100 K by widely expanding the low-temperature region in the Si melt. *(From Kazuo Nakajima et al., Growth of Si single ingots with large diameter ($\sim$45 cmϕ) and high yield of high conversion efficiency ($\geq$ 18 %) cells using a small crucible ($\sim$50 cmϕ) by noncontact crucible method, 2016 IEEE 43rd Photovoltaic Specialists Conference (PVSC), © 2016 IEEE.)*

For practical use, a large solidification ratio is very important to inexpensively realize a flat plate Si single ingot with a large diameter and a large volume using the cast furnace. Fig. 8.1 shows the solidification ratio as a function of ΔT. The solidification ratio almost linearly increased with ΔT regardless of the crucible diameter [7]. This tendency is very effective to obtain an ingot with a large volume without contact with the crucible wall. The solidification ratio does not saturate even when ΔT becomes larger because the low-temperature region deeply expands inside the Si melt as ΔT increases. The solidification ratio will be 90% when ΔT will be 100 K by widely expanding the low-temperature region in the Si melt. Fig. 8.2 shows a schematic illustration of an example of a cast ingot which can be grown by the NOC method using a cast furnace [6]. The ingot has a flat-plate surface and a flat-plate bottom as shown by the grown ingot (top size:23 cm × 23 cm and thickness:7 cm) in the upper left corner of Fig. 8.2. The ingot has a large diameter and a large volume. As the horizontal growth rate is higher than 1 mm min^{-1} and the vertical growth rate is between 0.3 and 0.6 mm min^{-1} [8], such a flat plate single ingot can be easily grown by the NOC method. The high buoyancy from the Si melt helps the large ingot to remain in the Si melt. To obtain a longer ingot with a large diameter, the ingot should be slowly pulled upward from the Si melt while the ingot continues to grow inside the low-temperature region.

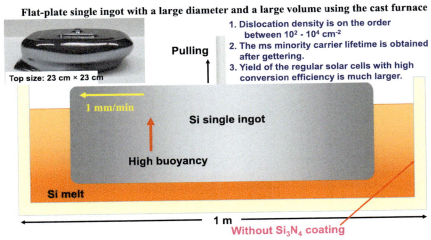

Fig. 8.2 Schematic illustration of a cast ingot which can be grown by the NOC method using a cast furnace. The ingot has the same flat-plate surface and bottom as shown the grown ingot in the upper left side. The horizontal growth rate is higher than 1 mm min^{-1} and the vertical growth rate is between 0.3 and 0.6 mm min^{-1} [1], so such a flat plate single ingot can be easily grown by the NOC method. *(Referred from Fig. 5 in K. Nakajima J. Cryst. Growth 468 (2017) 705.)*

A deeper and wider crucible is better to grow such the longer ingot using a much larger amount of melt. For such an ingot grown by the NOC method, the low dislocation density, the millisecond lifetime and the large yield of regular solar cells will be obtained. For industrial use using a modified conventional casting furnace with a pulling mechanism, the flat plate ingot with a diameter of 1 m and a thickness of 20 or 30 cm will be required. Moreover, the O and C concentrations in the ingot can be much reduced to use the Liquinert quartz crucible [9–13]. This crucible uses a barium oxide (BaO) coating layer which is made by barium hydroxide spread and fired all over the inner surface of the crucible instead of Si$_3$N$_4$ coating [13]. More precisely, this technology is based on a series of non-wettable processes which consist of coating of non-bubble quartz layer (<4 mm) on the inner surface of crucible wall, coating of alkali-earth hydroxide (e. g. BaO) on the quartz layer and heating the coating materials at higher temperature to generate devitrification and keep pin holes smaller [11].

The most important point to expand the NOC method as an advanced cast method is to realize a dislocation-free ingot with large volume. There are several growth technologies about this subject. One of them is the

An important point to expand this method as an advanced cast method
To realize a dislocation-free ingot with a large volume by developing a necking technique.

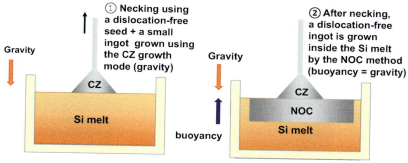

The necking technique should be performed under a balance between buoyancy and gravity which are formed by the volume and weight of the grown ingot.

Fig. 8.3 Necking technique under a balance between buoyancy and gravity which are formed by the volume and weight of the grown ingot. At first, necking is performed using a dislocation-free seed, then a small ingot is grown using a CZ growth mode to generate gravity. After necking, a dislocation-free ingot is grown inside the Si melt by the NOC method during keeping the balance between buoyancy and gravity. The balance can be performed by controlling the pulling rate and cooling rate.

necking technique under a balance between buoyancy and gravity which are formed by the volume and weight of the grown ingot as shown in Fig. 8.3. At first, the Dash necking is performed using a dislocation-free seed [14]. Then, a small ingot is grown from the seed during pulling the ingot up using a CZ growth mode while keeping the ingot growth inside the melt. The small ingot is used to generate gravity to prevent the effect of buoyancy. When the pulling direction is an [100], dislocations with Burgers vectors of b/2 [110] propagate in an {111} plane, grow out and terminate at the seed surface. After necking, a dislocation-free ingot with a large diameter is simultaneously grown from the small ingot inside the Si melt by the NOC method during keeping the balance between buoyancy and gravity. The balance can be performed by controlling the pulling rate and cooling rate. Another technology is the precise and soft touching technology as shown in Fig. 8.4. (see Section 8.2.1) The development of these technologies is the key point to industrially use the NOC method as an advanced CZ method.

Fig. 8.5 shows each yield of materials for the HP cast and NOC methods when a NOC ingot with a 68 cm diameter will be grown. 10 wafers can be obtained from a cross section of the ingot. The final yield of

Another important point to expand this method as an advanced cast method
To realize a dislocation-free ingot by developing the precise touching tech.

Fig. 8.4 Precise and soft touching technology to realize a dislocation-free large ingot. The touching of the seed to the Si melt surface should be performed at the almost same temperature closed to the meting point of Si to prevent generation of dislocations due to the thermal shock. Development of such technology is the key point to expand the NOC method as an advanced cast method.

Each yield of p-type materials for the HP cast and NOC methods

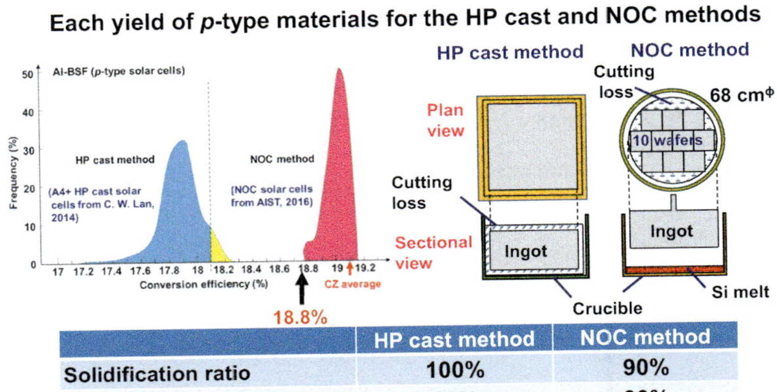

	HP cast method	NOC method
Solidification ratio	100%	90%
Cutting loss	20%	30%
Yield of high quality wafers/Si source ($\eta \geq 18.8\%$)	<10%	63-81%

Fig. 8.5 Each yield of materials for the HP cast and NOC methods when a NOC ingot with a 68 cm diameter will be grown. 10 wafers can be obtained from a cross section of the ingot. The final yield of the regular solar cells with the conversion efficiency higher than 18.8% is expected to be between 63% and 81% for the NOC method when that for the HP cast method is less than 10%. *(Referred from Fig. 24 in K. Nakajima et al. in: D. Yang (Ed.), Chapter 12, Section Two Crystalline silicon growth, Hand Book of "Photovoltaic Silicon Material", Springer, Berlin Heidelberg, 2017, On line.)*

the regular solar cells with the conversion efficiency higher than 18.8% is expected to be between 63% and 81% for the NOC method when that for the HP cast method will be less than 10%. The final yield depends on the cutting loss, the shape of ingot and the amount of re-useable materials.

The defect formation mechanism of the NOC method may be different from that of the CZ method because of the near equilibrium growth inside a Si melt and the long diffusion of point defects in a hot ingot inside the Si melt. The NOC method can be compared to the CZ method in Fig. 8.6. In the CZ method, the diameter ratio is usually 0.3. The growing interface is concave in the growth direction [15]. This method is effective for the growth of a longer ingot using the continuous-feeding CZ (CCZ) technology to keep the melt height constant [16]. In the NOC method, the diameter ratio is near 0.9. The growing interface is convex in the growth direction. This method is effective for the growth of a wider ingot as shown in Fig. 8.2. In the NOC method, the growing interface is inside the melt and in the state of near equilibrium condition. Vacancies and interstitial atoms formed at the growing interface have the equilibrium concentrations. The formation energies of the neutral vacancy and interstitial atom are 3.85 ± 0.15 and $4.8 + 0.6/-0.4$ eV, respectively,

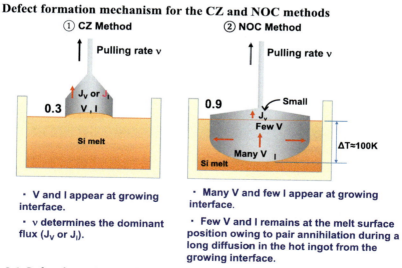

Fig. 8.6 Defect formation mechanism of the NOC method compared to that of the CZ method. It may be very different from that of the CZ method because of near equilibrium growth in a Si melt. In the CZ method, the diameter ratio is usually 0.3. In the NOC method, the diameter ratio is near 0.9.

and the vacancy formation energy in P-doped crystal is much smaller [17]. Therefore, the equilibrium concentration of vacancy is slightly higher than that of interstitial atom at the growing interface (the ratio of vacancy/interstitial atom:1−1.4) [18]. The migration energies of the neutral vacancy and interstitial atom are 0.45 ± 0.04 and 0.49 ± 0.05 eV, respectively [17]. The activation energy of self-diffusion is given by the sum of the formation energy and the migration energy [17]. The activation energies of self-diffusion of the neutral vacancy and interstitial atom are estimated as 4.11−4.49 eV and 4.84−5.94 eV, respectively. Pair annihilation between vacancies and interstitial atoms occurs in a hot ingot inside the melt through a long diffusion from the growing interface to the melt surface position during a large temperature reduction near 100 K. The reaction constant of the pair annihilation process depends on the energy barrier of the process and the temperature [19]. The concentrations of vacancies and interstitial atoms may decrease from the growing interface toward the melt surface position owing to the pair annihilation, which also depend on the diffusion mechanism of each point defect. The defect formation mechanism of the NOC method may be different from that of the CZ method. These hypotheses should be experimentally confirmed using dislocation-free NOC ingots. Schön et al. reported that the distribution of crystal defects of NOC wafers was more homogeneous than that of CZ wafers because the morphology or impurity decoration of oxide precipitates in the NOC wafers seemed to be different from that in the CZ wafers [20].

In the NOC method, a conventional silica crucible can be used in a modified conventional cast furnace with a pulling mechanism. Therefore, by development of above mentioned technologies, the NOC method can be effectively used to prepare an ingot with a large diameter, a large volume, quite few dislocations and a low O concentration at a low cost. Thus, the NOC method has high potential for actual use in industrial production. Fig. 8.7 shows each merit of the growth methods for p-type solar cells in the mega solar market. When the flat plate single ingot with a large diameter and a large volume shown in Fig. 8.2 will be grown using the cast furnace to keep the high productivity, the NOC method is expected to be an advanced cast method to realize uniform large Si single ingots with enough quality to obtain solar cells using a cast furnace. This method is concerned about preparation of a Si larger ingot using an improved existing furnace and confirmation of the defect formation mechanism.

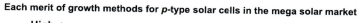

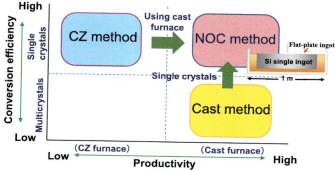

Fig. 8.7 Each merit of the growth methods for *p*-type solar cells in the mega solar market. When the flat plate single ingot with a large diameter and a large volume is grown using the cast furnace to keep the high productivity, the NOC method is expected to be an advanced cast method to realize uniform large Si single ingots with sufficient quality to obtain solar cells with high conversion efficiency and high yield. *(Referred from Fig. 26 in K. Nakajima et al. in: D. Yang (Ed.), Chapter 12, Section Two Crystalline silicon growth, Hand Book of "Photovoltaic Silicon Material", Springer, Berlin Heidelberg, 2017, On line.)*

8.3 Growth of square-shaped ingots using the NOC method

8.3.1 Basic points of growth of square-shaped ingots

The cast ingot is manufactured to have a square shape using a square-shaped crucible. The CZ ingot has a circular cross section, but the periphery area of circular cross section is usually cut off to make square-shaped wafers for solar cells. The material loss of about 36% is occurred for the circular ingot. In such situation, square-shaped single ingots are required even for the NOC method using a cast furnace. Using the NOC method, Si single ingots with a diameter as large as 90% of the crucible diameter can be grown [5,21]. The best use of this merit can be effectively made when preparing square Si single ingots and designing square wafers for solar cells.

The growth of square or rectangular Si single ingots by the CZ method and the FZ method [22] has been reported by several researchers [23–27]. Rudolph et al. prepared rectangular Si single bulk crystals using a TMF to stabilize the radial temperature gradient in the Si melt [24,26]. To obtain a large square single ingot for solar cell applications, the side-face width of the four-cornered pattern appearing on the top surface of the ingot should be much larger [28]. The side-face width is defined as the side length of the largest square in the four-cornered pattern on the top surface of the ingot. The size of the square ingot is basically determined by the side-face width. Rudolph et al. named the side-facet width for the side-face width [24,29]. For the CZ growth method, Kuroda et al. reported that the growth of square crystals could be explained by supercooling occurring in the radial direction of the ingots and the radial temperature gradient near the growing interface [26]. Miller et al. used numerical modeling to report that a small radial temperature gradient near the melt surface and a large vertical temperature gradient below the crystals were both important for the growth of square crystals by forming {110} faces [27]. Rudolph et al. used the TMF to obtain a small radial temperature gradient for the growth of Si bulk crystals with a quadratic cross section, and they obtained an ingot with a maximum area of 9.1×9.1 cm^2 and a side-face width of 5.6 cm [24,29]. This side-face width was not large enough to obtain significantly larger ingots with a square shape. Factors that affect the side-face width such as radial temperature gradient, degree of supercooling and quality of crystals should be made clear.

The NOC method can be used to prepare square Si single ingots with a four-cornered pattern without the use of the TMF [28]. The ingots are grown inside a low-temperature region in a Si melt using crucibles without Si_3N_4 coating. The size of the square part of an ingot can be clearly determined from the side-face width of the four-cornered pattern. For obtaining a large square Si single ingot with a large side-face width, it is very important to establish a large low-temperature region in a Si melt while maintaining a small initial temperature reduction, ΔT_{in} (K). ΔT_{in} is used to spread the crystal from the seed along the surface of the Si melt and required to obtain the ingot with a large flat top surface. Square Si single ingots with sizes of 9.4×9.7 and 10.9×11.0 cm^2 is obtained, which has no fan-shaped {110} faces and has diagonal lengths of up to 91% of the crucible diameter (see Section 8.3.2).

8.3.2 Growth of Si single ingots with four-cornered pattern

The size and shape of an ingot are affected by those of the low-temperature region. To obtain a large low-temperature region, a large temperature reduction, ΔT of 70 K is established using a crucible with a 50 cm diameter. ΔT is required for the complete growth of the ingot. In this case, the diameter of the low-temperature region is about 45 cm, which is inferred from the size of the largest ingot grown [5]. To form a nucleation site, a seed crystal is set in contact with the surface of a Si melt above the melting point of Si. Silica crucibles without a Si_3N_4 coating are used to obtain *p*- and *n*-type single ingots [28]. The depth of Si melt is 3.5—4.0 cm. To control the initial size of an ingot and obtain a flat top surface in the low-temperature region, a Si melt is cooled by ΔT_{in} without pulling the ingot. For growth of a circular single ingot, ΔT_{in} required is between 20 and 30 K, and ΔT required is between 30 and 50.2 K.

Fig. 8.8A and B show a circular *n*-type ingot grown using a crucible with a diameter of 33 cm and the four-cornered pattern on its top surface, respectively [28]. ΔT_{in} for the growth is 30 K and ΔT is 45 K. The mass, diameter and height are 5320 g, 19.5 cm and 8.2 cm, respectively. Fig. 8.8C shows a crystallographic image of a four-cornered pattern grown from a Si (100) seed crystal, along with the top surface of a circular Si bulk crystal. The top surface of the ingot clearly shows a four-cornered pattern with four {111} faces with macro-steps (ms {111}) and four {110} side faces [30,31]. The ms {111} faces make an angle of 54.7° with the (100) seed plane. As shown in Fig. 8.8B, the growth in the <100> direction stops

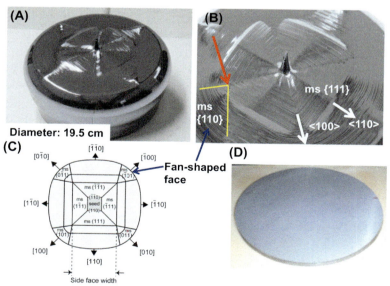

Fig. 8.8 (A) *n*-Type ingot grown using a crucible with a diameter of 33 cm. The diameter of the ingot is 19.5 cm. (B) Four-cornered pattern on the top surface of the ingot. (C) Crystallographic image of a four-cornered pattern grown from a (100) seed crystal. (D) As-cut cross section cut 1.4 cm below the top surface. *(Referred from Fig. 1 in K. Nakajima et al. Jpn. J. Appl. Phys. 53 (2014) 025501.)*

at the point shown by the red arrow, but the {110} side faces of the four-cornered pattern continue to grow through this point. From this point, fan-shaped {110} faces with macro-steps (ms {110}) appear on the four corners of the four-cornered pattern, which borders on two {111} faces with macro-steps, as shown by the yellow lines. The {110} side-face width of the four-cornered pattern is determined at this point as shown in Fig. 8.8C. The side-face width of the ingot is 6.2–6.5 cm. Fig. 8.8D shows an as-cut cross section cut 1.4 cm below the top surface. No grain boundaries are observed in the cross section, and the surface orientation of the entire cross section is found to be {100} using the EBSD method. The EPD of the ingot is $2.0-3.2 \times 10^4$ cm^{-2}.

The NOC method can be employed to prepare square n-type Si single ingots with side faces of 9.4×9.7 and 10.9×11.0 cm^2 [28]. Fig. 8.9A shows an n-type single ingot with a square shape grown using a crucible with a diameter of 17 cm. ΔT_{in} for the growth is 10.9 K. The mass, size and

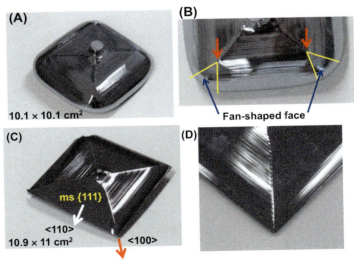

Fig. 8.9 (A) n-Type square-shaped single bulk crystal grown using a crucible with a diameter of 17 cm. It has a size of 10.1×10.1 cm^2 (with a diagonal length of 14.3 cm) and a height of 2.5 cm. The (110) side-face width is 6.2 cm. (B) <100> corners of the ingot with fan-shaped faces. (C) n-Type single bulk crystal with a perfect square shape grown using a crucible with a diameter of 17 cm. The dimension of the square is 10.9×11.0 cm^2 (with a (110) side-face width of 11 cm and a diagonal length of 15.5 cm). (D) <100> corner of the ingot that has a perfect square shape without any fan-shaped faces. *(Referred from Fig. 3 in K. Nakajima et al. Jpn. J. Appl. Phys. 53 (2014) 025501.)*

height of the ingot are 520 g, 10.1 × 10.1 cm^2 (with a diagonal length of 14.3 cm) and 2.5 cm, respectively. The maximum diagonal length of the square ingots is as large as 91% of the crucible diameter. The side–face width is 6.2 cm. The EPD of the ingot is 6–8 × 10^3 cm^{-2}. Fig. 8.9B shows <100> corners of the ingot with fan-shaped faces. The growth in the <100> direction stops at the points indicated by the red arrows, and fan-shaped {110} faces with macro-steps (ms {110}) appear from these points as shown by the yellow lines. Fig. 8.9C shows an n-type single ingot with a perfect square shape grown using a crucible with a diameter of 17 cm. The perfect square shape means that the side length of the top surface of the ingot is equal to the side-face width. The single ingot with a perfect square shape has a four-cornered pattern without any fan-shaped {110} faces. ΔT_{in} for the growth is 10.5 K. The mass, size and height of the ingot are 260 g, 10.9 × 11.0 cm^2 (with a diagonal length of 15.5 cm) and 1.0 cm, respectively. The side-face width is 10.9–11.0 cm. The EPD of the ingot is 1.2 × 10^3–3.8 × 10^4 cm^{-2}. Fig. 8.9D shows a <100> corner of the ingot that has a perfect square shape without any fan-shaped {110} faces. For the growth of a large square single ingot with a large side-face width, it can be inferred by the comparison between Fig. 8.9B and D that it is very important to obtain a perfect <100> corner. The top surface of each ingot shows a four-cornered pattern with four {111} faces with macro-steps (ms {111}) and four {110} side faces.

The initial stage of growth from a seed crystal can be directly observed through the window of the furnace chamber. Fig. 8.10A and B show the growing interface of the ingot 12 and 20 min after the start of growth from the seed crystal, respectively [28]. The observed ingot is the same one with a perfect square shape as shown in Fig. 8.9C and D. The initial growing interface of the {110} side faces is curved as shown in Fig. 8.10A, but it gradually becomes straight, as shown in Fig. 8.10B. Fig. 8.10C and D show crystallographic images corresponding to Fig. 8.10A and B, respectively. As shown in Fig. 8.10C, the crystal first begins to grow rapidly in the <100> direction because there are many kinks on the {100} face. Then, the {110} side faces grow more slowly and their growing interface is curved because the {110} face has many steps and but fewer kinks. As shown in Fig. 8.10D, the crystal grows while clearly maintaining a square shape. The growth in the <100> direction stops very close to the crucible wall. Then, the crystal grows in the <110> direction and finally the growth of the {110} side faces also ends. Thus, an ingot is obtained with a clear four-cornered pattern without any fan-shaped {110} faces, and the side-face width of

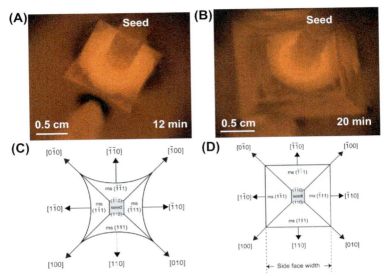

Fig. 8.10 (A) Growing interface of the square ingot shown in Fig. 8.9C 12 min after the start of growth from the seed crystal. (B) Growing interface of the same ingot 20 min after the start of growth from the seed crystal. (C) Crystallographic image of the four-cornered pattern shown in (A). (D) Crystallographic image of the four-cornered pattern shown in (B). *(Referred from Fig. 5 in K. Nakajima et al. Jpn. J. Appl. Phys. 53 (2014) 025501.)*

the four-cornered pattern is finally determined. The stepwise-lateral growth in the <110> direction on the closed packed {111} faces is preferable for the Si growth.

8.3.3 Factors affecting growth of square Si single ingots

The important factors in obtaining Si single ingots with a square top face are known to be the supercooling of a Si melt in the radial direction of an ingot and a low radial temperature gradient [22,24,27,29]. The size of the square face is determined by the side-face width of the four-cornered pattern on the top surface of the ingot. Fig. 8.11 shows the side-face widths (cm) of several ingots grown by the NOC method as a function of $1/\Delta T_{in}$ (K^{-1}). The side-face widths of the square and circular single ingots with a four-cornered pattern on their top surface are measured to determine the relationship. ΔT_{in} is necessary to grow an ingot along the surface of the Si melt without pulling growth. The side-face width increases almost linearly with $1/\Delta T_{in}$. Thus, the side-face width is inversely proportional to ΔT_{in} from the viewpoint of decreasing the temperature

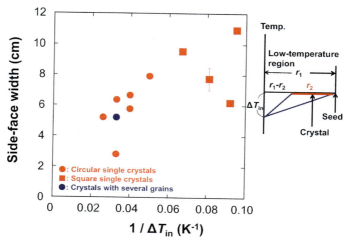

Fig. 8.11 Side-face width (cm) of several bulk crystals grown by the NOC method as a function of $1/\Delta T_{in}$ (K^{-1}), where ΔT_{in} (K) is the initial temperature reduction. The side-face width increased with $1/\Delta T_{in}$. The relationship between the parameters ΔT_{in}, r_1, and r_2 is also illustrated. (Referred from Fig. 6 in K. Nakajima et al. Jpn. J. Appl. Phys. 53 (2014) 025501.)

gradient near the growing interface as discussed below in detail [32]. A large value of $1/\Delta T_{in}$ is required to obtain large square single crystals.

Kuroda et al. [26] reported that the side-face width, d (cm) of a square crystal was proportional to the supercooling, ΔT_s (K) that occurred in the radial direction of an ingot and inversely proportional to the radial temperature gradient in the melt near the growth interface, G_T^r (K cm^{-1}) i.e.,

$$d \propto \Delta T_s / G_{T^r} \tag{8.1}$$

In the NOC method, ΔT_{in} is the temperature reduction that is used to induce crystal growth along the surface of the Si melt from the seed crystal. The low-temperature region in the upper central part of the Si melt becomes larger across the surface of the Si melt as ΔT_{in} increases [30]. Therefore, the temperature difference is proportional to ΔT_{in}, and the radial temperature gradient is almost proportional to ΔT_{in}. Therefore, the side-face width is also proportional to $1/\Delta T_{in}$, as shown in Fig. 8.11.

The radial temperature gradient across the crucible near the growing interface for the NOC growth can be expressed as

$$G_{T^r} = \Delta T_{in}/(r_1 - r_2) \quad r_1 > r_2 \tag{8.2}$$

Here, r_1 (cm) is the radius of the low-temperature region across the crucible at ΔT_{in} and r_2 (cm) is the radius or half-diagonal length of the ingot. The relationship between these parameters is also illustrated in Fig. 8.11 r_1 is larger than r_2 because the growing interface is always behind the front of the low-temperature region across the crucible. As r_2 approaches r_1 during growth, the growth of the side face becomes more difficult because G_T^r increases. The radial temperature gradient in the low-temperature region can be roughly estimated using r_2 which is measured from a grown crystal instead of r_1-r_2, because at the final stage of growth, r_2 can be considered to be close to r_1. For the square ingot with the largest side-face width of 11 cm shown in Fig. 8.9C, the radial temperature gradient is estimated to be 1.3 K cm^{-1}. This value is comparable to previously reported values [24,26,27,29]. For circular ingots with a smaller side-face width, the radial temperature gradient is estimated to be 3–4 K cm^{-1}.

By comparing the ingots shown in Fig. 8.8A, 8.9A and 8.9C, the relationship between the side-face width and shape of the ingots can be deduced. The side-face width of the ingot shown in Fig. 8.8A is the smallest among all the ingots, while the side-face width of the ingot shown in Fig. 8.9C is the largest. When the side-face width is small, the growth along the <100> direction cannot continue for a long time, and fan-shaped {110} faces with macro-steps (ms {110}) appear on the four corners of the four-cornered pattern. The four corners become round as the crystal grows, and finally a circular ingot is obtained as shown in Fig. 8.8A. When the side-face width is large, the growth along the <100> direction can continue for a longer time and fan-shaped {110} faces with macro-steps (ms {110}) does not appear on the four corners, until the corners of the ingot approach near the crucible wall. Therefore, a single bulk crystal with a perfect square shape can be obtained when the side length of the top surface of the ingot becomes equal to the side-face width as shown in Fig. 8.9C. The ingot shown in Fig. 8.9A does not have a perfect square shape because the side-face width is not large enough to impede the growth of the fan-shaped {110} faces.

Factors that affect the side-face width may include the quality of the crystal such as the dislocation density. The relationship between the side face width and the dislocation density is experimentally obtained [28]. The side-face width tends to increase as the EPD decreases. This means that the side-face width is affected by the dislocation density because the face growth is affected by the quality of the ingot.

8.3.4 Key points for improvement of the growth of square-shaped ingots

As indicated by Eqs. (8.1) and (8.2), the most important point to obtain large square ingots with a large side-face width is to maintain a small radial temperature gradient near the growth interface even when the supercooling in the melt increases. However, the radial temperature gradient near the growth interface generally increases as the supercooling in the melt increases. The development of a technology is required to maintain such a small radial temperature gradient even when the supercooling is large. This concept is consistent with that of Miller et al. [27]. In the NOC method, the radial temperature gradient near the growing interface is proportional to ΔT_{in}. Therefore, ΔT_{in} should be kept small even when the low-temperature region becomes larger. To keep ΔT_{in} small, a wide and uniform temperature profile in the Si melt should be realized using a crucible with a larger diameter or designing a furnace with a suitable temperature profile. A small ΔT_{in}, a large low-temperature region, and a low EPD are essential for obtaining large square Si single ingots using the NOC method [28].

As the growth rate of the NOC method increases with the cooling rate, the limit of the cooling rate depends on the heat capacity of the furnace. For the cast furnace, the high cooling rate sometimes causes overcooling below the setting temperature. For example, during the surface growth of the ingot shown in Fig. 7.15A, the cooling is occasionally stopped at $\Delta T_{in} = 10, 15, 20, 25$ and 30 K to prevent the overcooling of the furnace. The control of the overcooling of the furnace is also effective for keeping the initial diameter constant during crystal growth as shown in Fig. 7.15C. The diameter of the ingot remains constant during pulling even using a high cooling rate of 0.4 K min^{-1}. The furnace should be designed to have a heat capacity suitable to increase the cooling rate to above 0.4 K min^{-1}, at which the horizontal growth rate is higher than 2 mm min^{-1}.

8.4 Ga-doped Si multicrystalline ingots

Ga-doped p-type Si single and multicrystalline wafers have higher minority carrier lifetime comparing with that of B-doped Si wafers, and they show no light-induced degradation of carrier lifetime because the generation of C_i-O_i complexes is strongly suppressed in Ga-doped Si crystal [33–36] (see Section 2.5.2). For Ga-doped Si wafers, the lifetime degradation is suppressed even with high O content [37]. Moreover, as-grown Ga-doped Si multicrystalline wafers slightly degraded, but no degradation is observed in impurity-gettered wafers by P diffusion and hydrogen passivation because the P diffusion results in reduction of trap density [34,35,38]. Therefore, Ga–doped Si solar cells show no light-induced degradation even though the O concentration is high, on the other hand, B-doped Si solar cells show degradation except for the quite low O concentration [39]. A high-temperature oxidation near $800\,^\circ C$ is effective to obtain no light degradation of Si solar cells using Ga-doped wafers [37]. Al doped Si solar cells also show no light induced efficiency degradation [40]. The degradation does not occur for the bulk lifetime of In-doped p-type Si wafers, but the degradation occurs for Al-doped Si wafers with Al-related defects [40,41]. Thus, the Ga or In doped Si is suitable material for high-efficiency solar cells because B is more active to make a complex with impurities as compared with Ga. However, its resistivity varies widely over the length of the crystal because Ga has the small segregation coefficient of $0.007-0.01$ [42−44].

Al, Ga and In can be used as p-type dopant like B. However, the segregation coefficients of Al (0.0029 [40]), Ga and In (0.0004−0.0005 [45−48]) are much lower than B (0.8 [49]) in Si, which cause a high variation of resistivity along the growth direction (see Section 2.5.1). The homogeneous distribution of dopants is more difficult to be obtained for the Al, Ga and In elements. A few Ga atoms evapcrated from Si melt surface during growth [36]. The evaporation rates of B and Ga from a Si melt is 8×10^{-6} cm s^{-1} and 2×10^{-3} cm s^{-1}, respectively [36]. In is a volatile dopant and the evaporation rate of In from a Si melt is estimated to be 1.6×10^{-4} cm s^{-1} [47]. The resistivity of Ga-doped Si is larger than that of B-doped Si because Ga is not fully ionized at Ga concentration higher than 10^{16} atoms cm^{-3} at 300 K [50]. The electrical activity of In in Si also decreases from 100% to 20% as the doping level increases from 10^{14} cm^{-3} to 10^{17} cm^{-3} because of formation of micro-precipitates [48].

As the segregation coefficient of Ga and B are changed by B codoping, the distribution of resistivity in a Ga and B-codoped Si crystal is smaller than in a simply Ga-doped Si crystal. There is strong interaction between Ga and B in a Si crystal, which increases the segregation coefficient of Ga and decreases the segregation coefficient of B in the codoped Si crystal [42]. In this case, the bond and strain energies related to uptaking Ga into the Si crystal decreases by the interaction between Ga and B.

Ga-doped Si multicrystalline ingots with a mass of $3.5-75$ kg are prepared using different types of cast furnace [34,35,44]. The variation of resistivity becomes smaller as the ingot size becomes larger, and the small and large ingots have the higher diffusion length and lifetime in the middle parts of both ingots [35]. For the CZ growth, keeping the growth rate constant is important to control the Ga concentration in a growing ingot because the variation of the Ga concentration is caused by fluctuation of the growth rate in conjunction with very small segregation coefficient of Ga [36]. Moreover, a rapid operation to add Ga directly to a Si melt is effective to suppress Ga evaporation and control the Ga concentration in the ingot [36,50]. To control the variation of dopants in a Si single ingot during growth, the CCZ process with a double-crucible method is already developed and a CCZ feeder is provided by an equipment company for Si ingot growth using low segregation coefficient dopants such as Ga and P [51]. In the CCZ method, more latent heat due to a high growth rate is effectively released at the growing interface using a water-cooled jacket [16].

In the Ga-doped Si crystal, FeGa complexes are usually generated. Nærland et al. precisely studied about FeGa, which are more energetically stable than FeB [52]. No defect states can be detected by DLTS in the Ga-doped crystal. The orthorhombic defect configuration is the most stable configuration in Fe-containing Ga-doped Si at RT. The capture cross section ratio of FeGa is 220, and the estimated capture cross sections are $\sigma_n^{FeGa} = 1.9 \times 10^{-14}$ cm^2 and $\sigma_p^{FeGa} = 7.9 \times 10^{-17}$ cm^2. The minority carrier lifetime in as-grown wafers is dominated by low levels of FeGa related defect complexes with an energy level of $E_v + 0.09$ eV. The Fe$_i$ defect exhibits the higher recombination activity under low-level injection and the FeGa defect exhibits the higher recombination activity under high-level injection. From other reports for FeGa, the energy level of $E_v + 0.24$ eV is determined by DLTS, the equilibrium binding energy is 0.47 eV and the hole capture cross section at $E_v + 0.24$ eV is $\sigma_p^{FeGa} = 2 \times 10^{-14}$ cm^2 or $\sigma_p^{FeGa} = 3 \times 10^{-15}$ cm^2 (at 120 K) [53].

A broad range of doping concentration from 1×10^{16} to 2×10^{17} cm^{-3} is suited for efficiencies above 20% if a random pyramid passivated emitter and rear cell (RP-PERC) structure is used for CZ wafers [54]. This point seems to be promising for the growth of Ga doped multicrystalline ingots. Thus, the cast growth of Ga-doped Si multicrystalline ingots will be one of advanced technologies even though the variation of Ga still exists in a Si ingot during growth. To keep the Ga doping concentration or the resistivity more constant, NeoGrowth crystallization method is proposed, which involves providing a Si melt feed into a free-standing puddle of melt maintained on top of a large area seed crystal [55] (see Section 8.2.1). The melt feed with Ga dopant during growth has a possibility to keep the p-type resistivity more constant within a range of 1.5−2.0 Ωcm [55]. The dislocation density is in the 10^3–10^5 cm^{-2} range, and the O$_i$ and C$_s$ concentrations are 5×10^{17} cm^{-3} and 7.8×10^{16} cm^{-3} in the middle center of the ingot [55].

8.5 Si-Ge multicrystalline ingots

8.5.1 Novel growth technology for SiGe multicrystalline ingots and SiGe solar cells

The Si-Ge binary system has the phase diagram which shows complete miscibility in solid state. The composition of a SiGe ingot can be freely selected in the range from Si to Ge. A SiGe multicrystalline ingot with any compositional distributions from Si to Ge can be easily prepared by the cast method. On the other hand, it requires a special growth technology to obtain a SiGe single ingot with uniform composition [56]. The dopant segregation coefficient varies corresponding the composition of SiGe [46]. The EDX mapping of Si and Ge elements on a cross section of a SiGe multicrystalline ingot shows that the structure is mixture of Si-rich blade-like and Ge-rich matrix crystals [57,58]. A SiGe multicrystalline ingot with microscopic compositional distribution will be one of the future materials for solar cells. The fundamental idea is to nonuniformly disperse a small amount of Ge in a multicrystalline Si ingot for controlling macroscopic physical properties. Such a SiGe multicrystalline ingot can be easily prepared for the Si-Ge binary system. Physical properties of the SiGe multicrystalline ingot can be widely tuned by controlling the microscopic compositional distribution while fixing average composition [59].

The concept and design of SiGe solar cells prepared using a multicrystalline ingot is shown in Fig. 8.12 [60,61]. The absorption coefficients are shown for Si, Ge, and SiGe with microscopic compositional distribution of the average Ge composition of 0.3%, which is estimated by the following relation,

$$\beta = \alpha_{Si}(1 - x) + \alpha_{Ge}x \qquad (8.3)$$

where β, α_{Si}, and α_{Ge} are the absorption coefficients of Si$_{1-x}$Ge$_x$, Si and Ge, respectively, x is the Ge composition in SiGe. The absorption range of SiGe increases in the long wavelength region, and it also largely increases near the wavelength of 1.2 μm by adding a small amount of Ge. The effective increase of the absorption coefficient in the wide spectral region by the small addition of Ge can be experimentally confirmed by spectroscopic

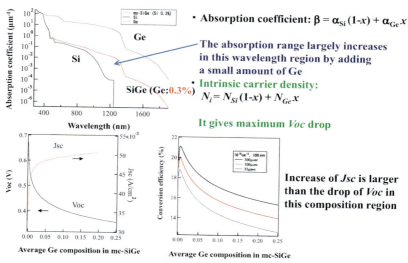

Fig. 8.12 Concept and design of SiGe solar cells prepared using a multicrystalline ingot with microscopic compositional distribution. The absorption coefficient, open circuit voltage and conversion efficiency of the SiGe solar cells are shown. (*Referred from Figs. 2, 4 and 5 in N. Usami et al. Jpn. J. Appl. Phys. 44 (2005) 857.*)

ellipsometry [62]. Typically, the absorption coefficient of SiGe multicrystals at the wavelength of 0.6–0.8 μm is 1.2 times larger than that of Si multicrystals. This increase is significantly effective on the internal quantum efficiency when the diffusion length is shorter [62]. The intrinsic carrier density of SiGe is estimated by the following relation,

$$N_i = N_{i,Si}(1-x) + N_{i,Ge}\,x, \qquad (8.4)$$

where N_i, $N_{i,Si}$ and $N_{i,Ge}$ are the intrinsic carrier densities of $Si_{1-x}Ge_x$, Si and Ge, respectively. This estimation gives the maximum drop of V_{oc}. Since the absorption coefficient as well as the intrinsic carrier concentration of nonuniform SiGe are not linearly dependent on the Ge composition, the results calculated using Eqs. (8.3) and (8.4) are totally different from those of uniform SiGe. The absorption coefficient of nonuniform SiGe is much larger than that of uniform SiGe with a same Ge composition particularly for the longer wavelengths [60].

For the SiGe multicrystalline solar cell with microscopic compositional distribution, J_{sc} rapidly increases by adding a small amount of Ge, but V_{oc} decreases by adding Ge. V_{oc} is controlled by the presence of Ge with a much smaller band gap than Si even when the amount of Ge is very small. However, there is a region where the conversion efficiency of the SiGe solar cells has a maximum point for the average Ge composition because the increase of J_{sc} is larger than the drop of V_{oc} in this compositional region as shown by the lower right figure in Fig. 8.12. Fig. 8.13 shows a schematic diagram of the SiGe multicrystalline solar cell with microscopic compositional distribution. The SiGe multicrystal has columnar structure. It has many Ge rich-regions in Si-rich matrix to enhance the absorption performance in the longer wavelength region by adding a small amount of Ge. The band structure of the Si/SiGe heterointerface is type-II as shown by the upper right figure in Fig. 8.13, so electrons can easily move from Ge-rich regions to the Si-rich matrix. The Ge composition of the Ge-rich regions should be relatively lower not to form deep wells in the band structure. The band gap shifts to narrower by strain in the Ge-rich regions as shown by the lower right figure in Fig. 8.13, so adding a large amount of Ge is not required to enhance the absorption performance in the longer wavelength region.

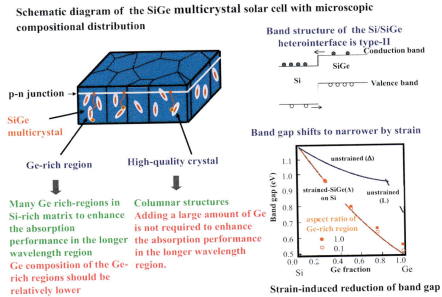

Fig. 8.13 Schematic diagram of the SiGe multicrystalline solar cell with microscopic compositional distribution. The band structure of the Si/SiGe heterointerface is type-II, and the band gap shifts to narrower by strain in the Ge-rich regions.

To prepare a SiGe multicrystalline solar cell with microscopic compositional distribution, a SiGe ingot with columnar structure was grown in a Si_3N_4 coated silica crucible with 7.0 cm diameter using a Bridgman-type vertical furnace [63]. Even though the {111} face is dominant for the Si multicrystalline ingot as shown in Fig. 1.6A, the {110} face is dominant for the SiGe multicrystalline ingot, even for the average Ge composition of only 1% as shown in Fig. 1.6B (see Sections 1.2.2 and 1.3.2). The area of grains with the {110} face as preferential orientation increases as the average Ge composition increases [63]. The Σ3 grain boundary is dominant for both Si and SiGe multicrystalline ingots, and more than 50% of total grain boundaries is the Σ3 grain boundary. Fig. 8.14 shows the compositional fluctuation in a SiGe multicrystalline ingot together with that of a SiGe ingot reported by Geiger et al. [64]. The black line is the Ge composition in their SiGe ingot. The red circles are the Ge compositions in the SiGe multicrystalline ingot with microscopic compositional distribution. The compositional fluctuation in the SiGe multicrystalline ingot is larger than ±20% around the average composition. The compositional fluctuation is larger than that reported by Geiger et al. and this gives the larger absorption coefficient even though the average Ge composition is smaller.

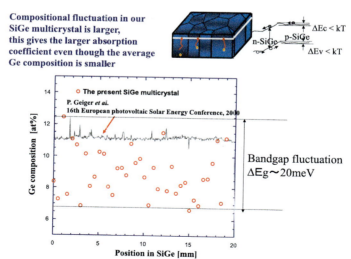

Fig. 8.14 Compositional fluctuation in the SiGe multicrystalline ingot together with that of a SiGe ingot reported by Geiger et al. [35]. The black line is the Ge composition in their SiGe ingot. The red circles are the Ge compositions in the SiGe multicrystalline ingot with microscopic compositional distribution.

However, the Ge composition in Ge-rich regions is not so large. The band structure has small fluctuation and the depth of wells is shallower than k T as shown by the upper figure in Fig. 8.14. Therefore, the trapping of minority carriers is not a serious problem in the SiGe multicrystal solar cell.

Fig. 8.15 shows the wavelength dependence of the internal quantum efficiency of Si and SiGe multicrystalline solar cells [62], which can remove the effect of reflection from the solar cell surface. The average Ge composition is only 1%. The internal quantum efficiency of the SiGe multicrystalline solar cell is much larger than that of the Si multicrystalline solar cell over all the wavelength regions. Especially, its remarkable increase in the longer wavelength can be obtained. Much larger short circuit photocurrent can be expected for the SiGe multicrystalline solar cell. Fig. 8.16 shows (A) the short circuit photocurrent, (B) open circuit voltage, (C) filling factor and (D) conversion efficiency of a SiGe multicrystalline solar cell with microscopic compositional distribution as a function of the average Ge composition [62]. The short circuit photocurrent drastically increases with increasing the average Ge composition from 0.5% to 5%, and abruptly decreases at the average Ge composition of 10%. However, the open circuit voltage and the filling factor do not largely decrease within the compositional region below 5%. When the average Ge composition is 10%, the current density and the open circuit voltage suddenly decrease. The filling factor is more than 0.7 except for the average Ge composition of

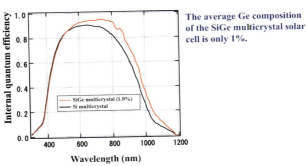

Fig. 8.15 Wavelength dependence of the internal quantum efficiency of Si and SiGe multicrystalline solar cells. The average Ge composition is only 1%. The internal quantum efficiency of the SiGe multicrystalline solar cell is much larger than that of the Si multicrystalline solar cell. *(Referred from Fig. 3 in W. Pan et al. J. Appl. Phys. 96 (2004) 1238.)*

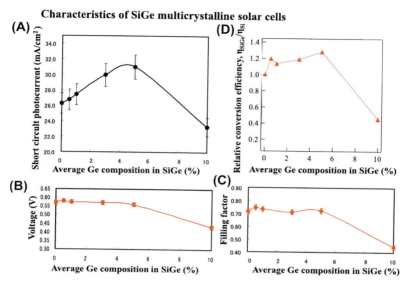

Fig. 8.16 (A) Short circuit photocurrent, (B) open circuit voltage, (C) filling factor and (D) conversion efficiency of a SiGe multicrystalline solar cell with microscopic compositional distribution as a function of the average Ge composition. The conversion efficiency has a maximum point at the average Ge composition around 5%. *(Referred from Fig. 2 in W. Pan et al. J. Appl. Phys. 96 (2004) 1238.)*

10%. Therefore, the conversion efficiency has a maximum point at the average Ge composition around 5%. The conversion efficiency of the SiGe multicrystalline solar cell is 12.3%, and it is 1.2–1.3 times higher than that of the Si multicrystalline solar cell prepared by the same furnace, owing to the drastic increase of the short circuit photocurrent by introduction of a small amount of Ge to multicrystalline Si [62]. On the other hand, microcrystalline SiGe thin-film solar cells are studied because it does not exhibit light-induced degradation like amorphous Si, but the highest conversion efficiency is only 5.6% when the Ge concentration is 20% [65]. Thus, the short circuit photocurrent, open circuit voltage, filling factor and conversion efficiency of SiGe solar cells reported by Geiger et al. and Isomura et al. [64,65] are much lower comparing to those of the SiGe multicrystalline solar cell with microscopic compositional distribution because the Ge composition of their SiGe solar cells is larger than 10% and it is obviously too large to realize a crystal with high quality. The density of misfit dislocations is low in SiGe multicrystalline ingots with the average Ge composition lower than 5%. Misfit dislocations largely increase in SiGe multicrystalline ingots with the average Ge composition of 10%, which can be known by TEM images of these crystals [63]. The dislocations cause the degradation of the characteristics of the SiGe multicrystalline solar cells when the average Ge composition is larger than 10%.

The SiGe multicrystalline solar cells with the average Ge composition of 0.5% have the response in the longer wavelength region up to 1.35 µm, and the SiGe multicrystalline solar cells with the average Ge composition of 1.0% have the response up to 1.40 µm, while the Si multicrystalline solar cells have the response shorter than 1.28 µm [62]. Using the same solar cell process to obtain the conversion efficiency of 16% for p-type Si multicrystalline wafers, the highest conversion efficiency of 17% is obtained for SiGe multicrystalline wafers [62]. In this case, the average Ge composition should be kept below 5%. These results support that the SiGe multicrystals with microscopic compositional distribution can increase the absorption coefficient in the longer wavelength region. Thus, the SiGe multicrystalline solar cells with microscopic compositional distribution have a potential for further improvement of the conversion efficiency by effectively utilizing the longer wavelength region of the solar spectrum without affecting the open circuit voltage. Such SiGe multicrystalline solar cells with microscopic compositional distribution will be one of the advanced technologies for Si high-performance solar cells. Moreover, the dominant {110} surface of the SiGe multicrystalline wafers is more effective to reduce the surface reflection comparing with the dominant {111} surface of Si multicrystalline wafers [63]. The only problem is the cost of Ge which is higher than that of Si. The effective use of small amount of Ge is expected for the SiGe multicrystalline solar cells with microscopic compositional distribution.

The performance of SiGe solar cells with microscopic compositional distribution can be confirmed by theoretical investigation [61]. Generally, the introduction of Ge to Si negatively affects the performance of solar cells owing to the reduction band gap. However, the SiGe solar cells have power conversion superior to Si solar cells in a limited compositional window on the Si-rich side [60,61]. In the window, the increase of J_{sc} overcompensates the decrease of V_{oc}, and controls the overall conversion efficiency. For a large compositional variation or amplitude of Ge at a given average Ge composition in SiGe, the window appears only when the diffusion length is not so large because the increase of J_{sc} is larger than the drop of V_{oc} only in a crystal with the relatively low diffusion length as shown in Fig. 8 12 [61]. The width of the window is greatly affected by the compositional distribution in SiGe.

8.5.2 Ge doped Si multicrystalline ingots

Ge doping in Si multicrystalline ingots is one of novel technologies to reduce crystal defects and improve the performance of solar cells. The fracture toughness of Si multicrystalline wafers is enhanced by Ge doping because Ge atoms accumulate at grain boundaries and hinder the crack propagation over the grain boundaries [66]. The formation of the O precipitates (Ge-related complexes) with small sizes and high density contribute to the enhanced mechanical strength because of their dislocation pinning-up effect, and the Ge content of 10^{18} cm^{-3} is the lower limit for observing such effect [67]. The Ge doped ingots show a lower dislocation density and a homogeneous distribution of dislocations by the induced strain field and the lower mobility of dislocations due to the pinning effect [68]. The Ge doped crystals can be used as seed plates for the mono–like cast method because of the solid solute hardening effect [69]. The penetration depth of dislocations which are generated by the surface pressure caused by feedstock particles on the seed plates can be largely reduced using these crystals [69] (see Section 6.1). The Ge doping retards the reaction of B–O defects as the Ge concentration increases because Ge atoms capture O_i atoms and generate a large energy barrier for diffusion of O_i to form O_{2i} [70] (see Section 2.5.3). Ge in the vicinity of B (Ge-B complexes) increases the energy barrier of the formation and dissociation of Fe–B pairs because the Ge doping increases the diffusion barrier of Fe_i [70]. Without the prior thermal process, the Ge (larger size) co-doping with B (smaller size) can introduce the compressive stress to compensate the tensile stress by B-doping, and it can suppress O precipitation and remove misfit dislocations [71]. The higher Ge doped CZ Si wafers show a significant lower thermal donor generation after thermal anneal at 450 °C [72].

Su et al. [73] studied the effects of Ge doping in detail. The segregation coefficient of Ge in a Si melt into a Si crystal is about 0.56. For B-doped p-type Si ingots, the concentration of FeB complexes in a Ge-doped Si multicrystalline ingot is one-order lower than that in an undoped one because of a higher diffusion barrier for F_i in the Ge doped Si ingot, resulting in the lower FeB concentration. The minority carrier lifetime in the Ge-doped Si ingot is higher than that in the undoped one because of a lower FeB complexes. The Ge doping can decrease the length of the red–zone with low minority carrier lifetime in the bottom of the Si multicrystalline ingot because of the higher diffusion barrier for F_i. Ge doping also prevents movement of dislocations and reduces the dislocation density in the Si ingot.

For every future technologies of Si ingot growth using cast furnaces, the common demands are the growth of a high-quality ingot with low crystal defects especially dislocations, low impurities, low stress and uniform crystal structure and the growth of an industrial large-scale ingot at a low cost. The dendritic cast and HP cast methods have a possibility to grow such multicrystalline ingots and the mono-like cast and NOC methods have a possibility to grow such single ingots by developing much more advanced technologies. Especially, as the solar cell performance of single ingots is better than that of multicrystalline ingots, the improvement of quality is quite important for multicrystalline ingots and the improvement of scale-up is an urgent subject for single ingots. I think the future dominant technology will be basically determined on this point.

References

[1] H.-J. Rost, R. Menzel, D. Siche, U. Juda, S. Kayser, F.M. Kießling, L. Sylla, T. Richter, J. Cryst. Growth 500 (2018) 5.

[2] T. Izumi, Method for Pulling a Single Crystal, November 30, 1999. United States Patent Number: 5,993, 539.

[3] K.-M. Kim, S. Chandrasekhar, Non-Dash Neck Method for Single Crystal Silicon Growth, Mar. 23, 1999. United States Patent Number: 5,885,344.

[4] T. Taishi, X. Huang, I. Yonenaga, K. Hoshikawa, J. Cryst. Growth 275 (2005) e2147.

[5] K. Nakajima, S. Ono, R. Murai, Y. Kaneko, J. Electron. Mater. 45 (2016) 2837.

[6] K. Nakajima, S. Ono, Y. Kaneko, R. Mura, K. Shirasawa, T. Fukuda, H. Takato, S. Castellanos, M.A. Jensen, A. Youssef, T. Buonassisi, F. Jay, Y. Veschetti, A. Jouini, J. Cryst. Growth 468 (2017) 705.

[7] K. Nakajima, S. Ono, Y. Kaneko, R. Murai, K. Shirasawa, T. Fukuda, H. Takato, Proceedings of the 43th IEEE Photovoltaic Specialists Conference (43th PVSC), Portland, Oregon, US, June 5-10, 2016, pp. 68-72.

[8] K. Nakajima, K. Morishita, R. Murai, J. Cryst. Growth 405 (2014) 44.

[9] S. Sakuragi, T. Shimasaki, G. Sakuragi, H. Nanba, in: Proc. Of the 19th European Photovoltaic Solar Energy Conference, 2004, pp. 1197−1200.

[10] S. Sakuragi, in: Proc. Of the 19th European Photovoltaic Solar Energy Conference, 2004, pp. 1201−1204.

[11] Y. Horioka, S. Sakuragi, Coating Method of Quartz Crucibles for Si Crystal Growth, November 11, 2011. Japanese Patent Number: JP-4,854,314.

[12] T. Fukuda, Y. Horioka, N. Suzuki, M. Moriya, K. Tanahashi, S. Simayi, K. Shirasawa, H. Takato, J. Cryst. Growth 438 (2016) 76.

[13] K. Fujiwara, Y. Horioka, S. Sakuragi, Energy Sci. Eng. 3 (2015) 419.

[14] W.C. Dash, J. Appl. Phys. 30 (1959) 459.

[15] V.V. Kalaev, I.Y. Evstratov, Y.N. Makarov, J. Cryst. Growth 249 (2003) 87.

[16] W. Zhao, L. Liu, J. Cryst. Growth 458 (2017) 31.

[17] M. Suezawa, N. Fukata, Y. Iijima, I. Yonenaga, Jpn. J. Appl. Phys. 53 (2014) 091302.

[18] V.V. Voronkov, J. Cryst. Growth 59 (1982) 625.

[19] R. Habu, T. Iwasaki, H. Harada, A. Tomiura, Jpn. J. Appl. Phys. 33 (1994) 1234.

[20] J. Schön, A. Youssef, S. Park, L.E. Mundt, T. Niewelt, S. Mack, K. Nakajima, K. Morishita, R. Murai, M.A. Jensen, T. Buonassisi, M.C. Schubert, J. Appl. Phys. 120 (2016) 105703.

[21] K. Nakajima, R. Murai, K. Morishita, K. Kutsukake, J. Cryst. Growth 372 (2013) 121.

[22] G. Ratnieks, A. Muiznieks, A. Mühlbauer, J. Cryst. Growth 255 (2003) 227.

[23] K. Nakajima, R. Murai, K. Morishita, K. Kutsukake, in 39th IEEE Photovoltaic Specialists Conference (39th PVSC), Tampa, Florida, US, June 16−21, 2013.

[24] P. Rudolph, J. Jpn. Assoc. Cryst. Growth 39 (2012) 8.

[25] A. Muiznieks, A. Rudevics, K. Lacis, H. Riemann, A. Lüdge, F.W. Schulze, B. Nacke, Proc. Int. Scientific Colloq. Modelling for Material (2006) 89.

[26] E. Kuroda, S. Matsubara, T. Saitoh, Jpn. J. Appl. Phys. 19 (1980) L361.

[27] W. Miller, C. Frank-Rotsch, M. Czupalla, P. Rudolph, Cryst. Res. Technol. 47 (2012) 285.

[28] K. Nakajima, R. Murai, K. Morishita, Jpn. J. Appl. Phys. 53 (2014) 025501.

[29] P. Rudolph, M. Czupalla, B. Lux, F. Kirscht, C. Frank-Rotsch, W. Miller, M. Albrecht, J. Cryst. Growth 318 (2011) 249.

[30] W. Zulehner, J. Cryst. Growth 65 (1983) 189.

[31] W.N. Borle, S. Tata, S.K. Varma, J. Cryst. Growth 8 (1971) 223.

[32] J.C. Brice, J. Cryst. Growth 6 (1970) 205.

[33] A. Khan, M. Yamaguchi, Y. Ohshita, N. Dharmarasu, K. Araki, J. Appl. Phys. 90 (2001) 1170.

[34] M. Dhamrin, H. Hashigami, T. Saitoh, Prog. Photovoltaics Res. Appl. 11 (2003) 231.

[35] M. Dhamrin, K. Kamisako, T. Saitoh, T. Eguchi, T. Hirasawa, I. Yamaga, Prog. Photovoltaics Res. Appl. 13 (2005) 597.

[36] T. Hoshikawa, T. Taishi, S. Oishi, K. Hoshikawa, J. Cryst. Growth 275 (2005) e2141.

[37] T. Saitoh, X. Wang, H. Hashigami, T. Abe, T. Igarashi, S. Glunz, S. Rein, W. Wettling, I. Yamasaki, H. Sawai, H. Ohtuka, T. Warabisako, Solar energy mater, Sol. Cell. 65 (2001) 277.

[38] M. Dhamrin, C. Schmiga, K. Kamisako, T. Saitoh, Solar energy mater, Sol. Cell. 90 (2006) 3179.

[39] S.W. Glunz, S. Rein, J. Knobloch, W. Wettling, T. Abe, Prog. Photovoltaics Res. Appl. 7 (1999) 463.

[40] S. Yuan, X. Yu, X. Gu, Y. Feng, J. Lu, D. Yang, Superlattice. Microst. 99 (2016) 158.

[41] J. Schmidt, K. Bothe, Phys. Rev. B 69 (2004) 024107.

[42] S. Uda, X. Hung, M. Arivanandhan, R. Gotoh, The 5th Inter. Symp. On Advanced Sci. and Tech. of Si Mater. Kona, Hawaii, USA, November 10−14, 2008.

[43] F.A. Trumbore, Bull Syst. Tech. J. 39 (1960) 212.

[44] R. Søndenå, H. Haug, A. Song, C.-C. Hsueh, J.O. Odden, P.V. Silicon, in: The 8th Intern. Conf. On Crystalline Silicon Photovoltaics, AIP Conf. Proc. 1999, 2018, p. 130016.

[45] W. Dietze, W. Keller, A. Mühlbauer, in: J. Grabmaier (Ed.), Crystals−Growth, Properties and Applications, vol. 5, Springer-Verlarg, 1982.

[46] Z.A. Agamaliev, Z.M. Zakhrabekova, V.K. Kyazimova, G.K. Azhdarov, Inorg. Mater. 52 (2016) 244.

[47] S. Haringer, A. Giannattasio, H.C. Alt, R. Scala, Jpn. J. Appl. Phys. 55 (2016) 031305.

[48] X. Yu, X. Zheng, K. Hoshikawa, D. Yang, Jpn. J. Appl. Phys. 51 (2012) 105501.

[49] X. Yu, D. Yang, Growth of crystalline silicon for solar cells: czochralski Si, in: D. Yang (Ed.), Chapter 13, Section Two, Crystalline Silicon Growth, Hand Book of "Photovoltaic Silicon Material", Springer, Berlin Heidelberg, 2018, pp. 1−45 (On line).

[50] T. Hoshikawa, X. Huang, K. Hoshikawa, S. Uda, Jpn. J. Appl. Phys. 47 (2008) 8691.

[51] The Pamphlet of "GT Advanced Technologies, Crystal Growth".

[52] T.U. Nærland, S. Bernardini, H. Haug, S. Grini, L. Vines, N. Stoddard, M. Bertoni, J. Appl. Phys. 122 (2017) 085703.

[53] A.A. Istratov, H. hieslmair, E.R. Weber, Appl. Phys. A 69 (1999) 13.

[54] S.W. Glunz, S. Rein, J.Y. Lee, W. Warta, J. Appl. Phys. 90 (2001) 2397.

[55] N. Stoddard, J. Russell, E.C. Hixson, H. She, A. Krause, F. Wolny, M. Bertonl, T.U. Naerland, L. Sylla, Prog. Photovolt. Appl. 26 (2018) 324.

[56] K. Nakajima, T. Kusunoki, Y. Azuma, N. Usami, K. Fujiwara, T. Ujihara, G. Sazaki, T. Shishido, J. Cryst. Growth 240 (2002) 373.

[57] K. Nakajima, N. Usami, K. Fujiwara, Y. Murakami, T. Ujihara, G. Sazaki, T. Shishido, Solar energy mater, Sol. Cell. 72 (2002) 93.

[58] K. Nakajima, N. Usami, K. Fujiwara, Y. Murakami, T. Ujihara, G. Sazaki, T. Shishido, Solar energy mater, Sol. Cell. 73 (2002) 305.

[59] N. Usami, K. Fujiwara, T. Ujihara, G. Sazaki, Y. Murakami, K. Nakajima, H. Yaguchi, Jpn. J. Appl. Phys. 41 (2002) L37.

[60] N. Usami, K. Fujiwara, W. Pan, K. Nakajima, Jpn. J. Appl. Phys. 44 (2005) 857.

[61] N. Usami, W. Pan, K. Fujiwara, M. Tayanagi, K. Ohdaira, K. Nakajima, Solar energy mater, Sol. Cell. 91 (2007) 123.

[62] W. Pan, K. Fujiwara, N. Usami, T. Ujihara, K. Nakajima, R. Shimokawa, J. Appl. Phys. 96 (2004) 1238.

[63] K. Fujiwara, W. Pan, N. Usami, K. Sawada, A. Nomura, T. Ujihara, T. Shishido, K. Nakajima, J. Cryst. Growth 275 (2005) 467.

[64] P. Geiger, P. Raue, G. Hahn, P. Fath, E. Bucher, E. Buhrig, H. J. Moller, Proc. Of the 16th European Photovoltaic Solar Energy Conference, 2000, p.150.

[65] M. Isomura, K. Nakahata, M. Shima, S. Taira, K. Wakisaka, M. Tanaka, S. Kiyama, Solar energy mater, Sol. Cell. 74 (2002) 519.

[66] P. Wang, X. Yu, Z. Li, D. Yang, J. Cryst. Growth 318 (2011) 230.

[67] J. Chen, D. Yang, X. Ma, Z. Zeng, D. Tian, L. Li, D. Que, L. Gong, J. Appl. Phys. 103 (2008) 123521.

[68] M.P. Bellmann, T. Kaden, D. Kressner-Kiel, J. Friedl, H.J. Möller, L. Arnberg, J. Cryst. Growth 325 (2011) 1.

[69] M. Trempa, M. Beier, C. Reimann, K. Roβhirth, J. Friedrich, C. Löbel, L. Sylla, T. Richter, J. Cryst. Growth 454 (2016) 6.

[70] X. Zhu, X. Yu, D. Yang, J. Cryst. Growth 401 (2014) 141.

[71] J. Zhao, P. Dong, J. Zhao, X. Ma, D. Yang, Superlattice. Microst. 99 (2016) 35.

[72] J.M. Rafi, J. Vanhellemont, E. Simoen, J. Chen, M. Zabala, D. Yang, ECS Trans. 50 (2012) 177.

[73] J. Su, G. Zhong, Z. Zhang, X. Zhou, X. Huang, J. Cryst. Growth 416 (2015) 57.

Index

Note: 'Page numbers followed by "f" indicate figures and "t" indicate tables'.

A

Advanced cast method, 308–309.
    See also Dendritic cast method
Advanced Industrial Science and
    Technology (AIST), 304–306
Al-BSF solar cell. *See* Aluminum
    back-surface-field solar cell
    (Al-BSF solar cell)
ALD. *See* Atomic layer deposition
    (ALD)
Aluminum (Al), 43–44
Aluminum back-surface-field solar cell
    (Al-BSF solar cell), 90, 307
Aluminum oxide (Al_2O_3), 65
Amorphous Si heterojunction solar cell
    technology (α-Si:H/c-Si),
    252–253
Annealing period, 146–147
Antimony (Sb), 43–44
Area fractions of grain orientations,
    210–211
Argon (Ar), 1
    gas flow, 101–102, 108–109
Atomic layer deposition (ALD), 65

B

B_O complexes, 91–94
BaO coating. *See* Barium oxide
    coating (BaO coating)
Barium (Ba)-doped silica
    crucible, 311
Barium oxide coating (BaO coating),
    320–321
Bismuth (Bi)-crystal model, 37–38
BN. *See* Boron nitride (BN)
Boron nitride (BN), 141
{100}/{100} boundary planes, 239
Brandon criterion, 22
Bridgman method, 225–226
Bunching effect, 111–113

C

Calcium (Ca), 43–44
Carbon (C), 40
    atomic impurities, 39–42
    concentration, 241
    contamination, 130–131
Carbon monoxide (CO), 40
Cast furnaces, growth of silicon
    ingots using
    electrical properties and solar cells of Si
    single ingots, 299–308
    growth of Si ingots using Si_3N_4 coated
    crucibles, 278–280
    growth of Si single ingots using NOC
    method, 281–299
    key points for improvement of NOC
    method, 308–312
    low-temperature region in Si melt,
    266–278
    NOC method, 259–266
CCZ technology. *See* Continuous-
    feeding CZ technology (CCZ
    technology)
Cerium (Ce), 43–44
Chromium (Cr), 43–44, 302
Clean grain boundaries, 71
Cobalt (Co), 43–44
Coefficient of variation of grain size
    (CV_{GS}), 199–200
Coherent grain boundaries, 32–33,
    120–121
Coincidence model, 24
Coincidence site lattice (CSL),
    18, 20f, 70
Concave growing interface, 291–292
Constitutional supercooling, 7–9
Continuous-feeding CZ technology
    (CCZ technology), 324–325
Convection of melt, 297
Conventional silica crucible, 325

349

Index

Conventional wafer, 198
Conversion efficiency, 252–253, 304–307
Convex bottom of ingot, 291–293, 295–297
Copper (Cu), 43–44
  Cu-Kα radiation, 138–139
Copper silicide (Cu_3Si), 137
Cristobalite, 107
Crucible diameter, 274–277
Crystal
  defects, 65–66
    in Si multicrystalline ingots, 118–133
  theoretical estimation for characterization, 94–95
CSL. *See* Coincidence site lattice (CSL)
Cylindrical carbon heater, 264–265
CZ
  method, 1, 226, 262–263, 309
  Si crystal, 87–88
  wafers, 254

D

Dash necking, 321–322
Deep-level transient spectroscopy (DLTS), 48–49, 66, 242–243
Defect formation mechanism of NOC method, 262–263, 324–325
Dendrite crystals, 157f, 274
  arrangement, 169–175
  growth mechanisms of, 161–164, 162f–163f
  growth of, 156–161
  ingot growth controlled by, 165–169
  preferential or rapid growth direction, 159f
  of silicon, 158–160, 158f
Dendritic cast method, 12–14, 196.
    See also Mono-like cast method
  conversion efficiency of solar cells, 182f–183f
  generation of dislocations, 175–179
  key points
    for impact on development of seed-assisted cast methods, 189–190
    for improvement, 187–189

motivation to development, 155–156
pilot furnace for manufacturing industrial scale ingots, 165–169, 184f
quality and solar-cell performance of Si ingots, 179–184
Denuded zones, 215–216
Depleted regions, 215–216
Di-iron trioxide (Fe_2O_3), 137
Diameter ratio, 274–277, 276f, 287–290
Diffusion
  coefficient, 45–46
  length, 74–77, 300
    estimation of minority carriers, 94–95
Disk-shaped carbon heater, 264–265
Disk-shaped graphite plate, 268–269, 269f
Dislocation(s), 236–237
  clusters
    in HP cast method, 206–207
    in Si multicrystalline ingots, 121–126
  density, 195–196, 236–237, 249–251, 317–319, 333, 344
  dislocation-free single ingots using NOC method, 319–325
    novel technologies for, 318–325
    soft touching technology for, 318–319
  generation of dislocations during growth, 33–36
  in HP cast method, 204–206
  recombination activity of, 72–73
  shear stress effects on dislocations formation, 36–38
  in Si multicrystalline ingots, 121–126
  in Si single ingots, 292–295
DLTS. *See* Deep-level transient spectroscopy (DLTS)
Dopants, 88–89
  distribution in Si multicrystalline ingots, 12–14

E

EBIC method. *See* Electron beam induced current method (EBIC method)

EBSD. *See* Electron backscattering diffraction (EBSD)

Edge-defined film-fed growth (EFG), 52

EDX. *See* Energy-dispersive X-ray spectroscopy (EDX)

Effective lifetime (T_{eff}), 63—64

Effective recombination lifetime, 63—64

EFG. *See* Edge-defined film-fed growth (EFG)

EL imaging technique. *See* Electroluminescence imaging technique (EL imaging technique)

Electrical quality of Si ingots, 64

Electroluminescence imaging technique (EL imaging technique), 66

Electron backscattering diffraction (EBSD), 15—17
    orientation maps, 165, 165f—166f

Electron beam induced current method (EBIC method), 25—26

Energy-dispersive X-ray spectroscopy (EDX), 135

EPD
    of dislocations, 292—293, 293f
    as function of distance, 294f
    in ingots, 204

Extended process (EXT process), 87

External gettering, 79

F

Facet dendrite, 158—160
    growth, 156—158

Facet-facet grooves, 29—30

[111] facets, 158—161

Fast diffusing metals, 45

Fast solidification process, 52—53

Fast-diffusing metal impurities, 80—81

Feedstock particles, 198

Fixed diamond abrasive, 225—226

Float-zone method (FZ method), 76

Floating cast method, 259—260

Force balance, 12

Four-cornered pattern, square-shaped ingots, 327—331, 328f

Fourier transform-infrared spectrometry (FTIR), 39—40, 137—138

Fukushima Renewable Energy Institute (FREA), 304—306

Functional grain boundaries, 232—233

Furnace cleaning, 301

Furnace design
    to growing Si ingot inside Si melt, 263—266
    with hot zone, 101—104

Fused quarts particles, 202—203

FZ method. *See* Float-zone method (FZ method)

G

Gallium (Ga)-doped Si multicrystalline ingots, 334—337

GBs. *See* Grain boundaries (GBs)

Germanium (Ge), 10

Gettering
    and hydrogenation, 78—82
    and low-temperature annealing for Fe impurities, 82—86

Gibbs free energy, 50

Grain
    formation in Si multicrystalline ingots, 116
    in Si crystals, 15—17
    size, 195, 199—200, 225—226
    structures, 207—211

Grain boundaries (GBs), 120—121
    character of Si multicrystalline ingots, 118—121
    energy, 230—232
    formation in Si multicrystalline ingots, 116—118
    generation and annihilation, 211—215
    $\Sigma 3$ grain boundaries, 158—161, 158f—159f, 178—179, 209, 232—233
    {111} $\Sigma 3$ grain boundaries, 234
    {510} $\Sigma 13$ grain boundaries, 235—236
    $\Sigma 13$ grain boundary, 235—236
    {310} $\Sigma 5$ grain boundary, 232—233

352   Index

Grain boundaries (GBs) (*Continued*)
  grooves, 11–12
  recombination activity, 68–72
  in Si crystals
    coherent grain boundaries, 32–33
    misorientation effects of grain
      boundaries, 22–25
    random grain boundaries, 25–26
    small and large angle grain
      boundaries, 17–22
    twin boundary, 26–29
    twin formation, 29–32
  small angle, 246–248
3-Grain ingots, 226
Graphite plates, 168–169, 172–174,
  173f, 178–179
Growing interface
  degree of supercooling estimation
    in Si melt near, 4–6
  and grooves, 9–12

H

Hafnium (Hf), 137
Heat flow of Kyropoulos method,
  260–261, 261f
Heterojunction (HET), 252–253
HF. *See* Hydrogen fluoride (HF)
High performance cast method (HP cast
  method), 195–196, 308–310.
    See also Mono-like cast method
  behavior and control in ingots,
    204–207
    dislocation clusters,
      206–207
    dislocations, 204–206
  control of grain size, grain orientation
    and grain boundaries, 196–204
    non-Si particles, 202–204
    other assisted Si seeds with different
      shapes, 201–202
    Si particles or chips placed on
      bottom of crucibles,
      196–200
  electrical properties, 216–218
  key points for improvement,
    219–221
  solar cells, 218–219

structure and defects in Si ingots,
  207–216
  generation and annihilation of grain
    boundaries, 211–215
  grain structures, 207–211
  precipitations, 215–216
High-dislocation areas, 250–251
HP cast method. *See* High performance
  cast method (HP cast method)
Hydrogen fluoride (HF), 300
Hydrogen nitric acid (HNO_3), 300
Hydrogenation, 78–82, 195–196,
  251–252

I

ICP-MS. *See* Inductively coupled
  remote plasma mass spectroscopy
  (ICP-MS)
Impurities, 137
  concentration, 249–251
  iron (Fe), 48–51
  metallic impurities and precipitates,
    43–48
  oxygen (O), 39–42
  recombination activity of, 66–68
  in Si multicrystalline ingots, 118–133
  in Si single ingots, 295–299
In-situ observation system, 110–113,
  156–164
Indium (In), 8–9
Inductively coupled remote plasma mass
  spectroscopy (ICP-MS), 44–45
Industrial scale ingots, 165–169, 184f
INES. *See* National Solar Energy
  Institute (INES)
Insulating plate effect on growth of Si
  ingots, 273–278
Insulation partition block, 226–228
Insulator diameter, 274–277
Insulator plate, 267–268, 277–278
Internal gettering, 80
Internal quantum efficiency (IQE), 78
Interstitial carbon (C_i), 40
Interstitial copper (Cu_i), 45–46
Interstitial nickel (Ni_i), 45–46
Interstitial oxygen atoms (Oi atoms),
  39–40

Index **353**

Interstitial transition metal
impurities, 67
IQE. *See* Internal quantum efficiency (IQE)
Iron (Fe), 43—44
impurities and precipitates, 48—51

J

Junction gap, 228—229

K

Kyropoulos method
heat flow, 260—261, 261f
NOC method comparison with, 260—261

L

Laue scanner, 15—17, 229
Lifetime degradation, 93
Light-element impurities, 42
Light-induced degradation, 334—335
Liquid-inert crucible, 42, 299, 311
Liquinert crucible, 42
Liquinert quartz crucible, 42, 148, 320—321
Low-temperature region in Si melt, 266—278
effect of ΔT_d on ingot size and, 271—272
growth parameters and crystal sizes for grown ingots, 275t
heat flow, 268f
effect of insulating plate on growth of Si ingots, 273—278
method to establishing, 268—269
minority carrier lifetime of wafers, 278t
n-type ingot, 273f
relationship between ΔT and, 269—271, 270f
simulated temperature distributions in Si melt, 277f
temperatures in melt without and with insulator plate, 301t

M

Magnesium (Mg), 137
Magnetic-field-applied Czochralski method (MCZ method), 297—299
Manganese (Mn), 43—44
Maple wafers from JA Solar, 226
Massachusetts Institute of Technology (MIT), 301
MCZ method. *See* Magnetic-field-applied Czochralski method (MCZ method)
Melt convection, 131
Metal contamination, 225—226
Metallic impurities, 43—48, 68—69, 90—91
in ingots, 311
in Si multicrystalline ingots, 126—129
slowly-diffusing, 80—81
Metallic precipitates, 215—216
Metallurgical-grade Si, 105—106
Methane (CH_4), 116
Micro-twins, 249—251
Microwave photoconductive decay method (μ-PCD method), 64
Minority carrier lifetime
loading carrier lifetime, 74—77
in mono-like part of ingot, 252
of Si single ingots, 299—303
Misorientation, 228—232, 234
effects of grain boundaries, 22—25
MIT. *See* Massachusetts Institute of Technology (MIT)
Molybdenum (Mo), 43—44
MONO method, 226
Mono-cast method, 226
Mono-like cast method, 225—226, 308—310. *See also* Dendritic cast method; High performance cast method (HP cast method)
behavior of dislocations and precipitates in mono-like ingots, 236—249
artificially controlled defect technique, 247f
concept to control of dislocations using grain boundaries, 244—246

354 Index

Mono-like cast method (*Continued*)
 control of dislocations using large-angle boundaries, 246–249
 formation of precipitates and propagation of dislocations, 241–243
 generation and propagation of dislocations from Si seed plates, 236–241
 non-perfect highly symmetrical grain boundaries, 246–249
 propagation and multiplication of dislocations in upper part of ingots, 243–244
 controlling large single grain, 226–229
 growth of mono-like ingots, 226–228
 multiple Si single-crystal seed plates, 227f
 seed plates and gaps between seeds, 228–229
 growth and control of small grains appearing from crucible wall, 230–236
 key points for improvement, 254–255
 quality of Si ingots using, 249–254
Mono-like method, 226
Mono-like wafers, 226
Multi-crystals, 1–2
Multicrystalline
 ingots, 304–306
 Si ingots, 17

N

n-type ingots, 301
n-type Si solar cell performance, 89–91
n-type solar cells, 303, 304f
Nanoscale metallic precipitates, 215–216
National Solar Energy Institute (INES), 252–253
Necking technique, 310, 321–322, 322f
Neodymium (Nd), 43–44
NeoGrowth crystallization method, 336–337
Nickel (Ni), 43–44

Nickel silicide (NiSi$_2$), 45–46
Nitrogen (N), 40–41
 atomic impurities, 39–42
 concentration, 130–131, 241
 gas flow, 108–109
NOC method, 12–14, 259–266
 comparison with Kyropoulos method, 260–261
 furnace design to growing Si ingot inside Si melt, 263–266
 key points for improvement, 308–312
 novel technologies for dislocation-free single ingots using, 319–325
 principle, 262–263
 Si single ingot growth using, 281–299
 solar cell performance preparing by, 303–308
 square-shaped ingot growth using, 326–334
 trigger for, 259–260
Non-destructive X-ray synchrotron-based technique, 128
Non-perfect highly symmetrical grain boundaries, 244–245
 control of dislocations using large-angle and, 246–249
Non-Si particles, 202–204
Normalized recombination strength, 65–66
Nucleation, 259–260
 energy, 202
 for Si multicrystalline ingots, 115–116
 site
  of dendrite crystals, 155, 166–169, 172–173, 187–188
  of seed crystal, 310

O

Optical measurement method, 77–78
Optical properties of Si crystals, 77–78
Oxide precipitates, 86–87
Oxygen (O)
 atomic impurities, 39–42
 concentration, 241, 251
  in ingot, 295, 296f, 297, 298f, 304–306, 309, 311
 precipitates, 42–43

P

p-type CZ wafers, 304–306
p-type ingots, 135, 304–306
  solar cell performance of wafers cut
    from, 306f
p-type solar cells, 319
  Al-BSF structure and distribution of
    conversion efficiency, 305f
  frequency of conversion efficiency, 307f
  Si solar cell performance, 89–91
Pair annihilation, 324–325
Parallel twins, 158–164, 158f–159f, 189
Passivated emitter, rear totally diffused
    structure (PERT structure),
    252–253
Passivated emitter and rear contact
    solar cell (PERC solar cell),
    90, 307, 308f
Passivated emitter rear locally-diffused
    solar cell (PERL solar cell),
    90–91
μ-PCD method, 300
PDG. See Phosphorous diffusion
    gettering (PDG)
PECVD. See Plasma-enhanced chemical
    vapor deposition (PECVD)
PERC solar cell. See Passivated emitter
    and rear contact solar cell
    (PERC solar cell)
PERL solar cell. See Passivated emitter
    rear locally-diffused solar cell
    (PERL solar cell)
PERT structure. See Passivated emitter,
    rear totally diffused structure
    (PERT structure)
Phase-field model, 22
Phosphorous (P), 12–14
Phosphorous diffusion gettering (PDG),
    65, 302–303
Phosphoryl chloride ($POCl_3$), 44–45
Photoluminescence method (PL
    method), 40–41, 77
Pilot furnace for manufacturing
    industrial scale ingots,
    165–169, 184f
PL method. See Photoluminescence
    method (PL method)

Plasma-enhanced chemical vapor
    deposition (PECVD), 81
Platinum (Pt), 80–81
Poly-Si slabs, 201
Polycrystallization, 8–9
Potassium (K), 137
Potassium hydroxide (KOH), 304–306
Precipitates, 43–48
  iron (Fe), 48–51
  oxygen (O), 42–43
  in Si multicrystalline ingots, 126–129
Precipitations, 215–216
Pulling method, 308–310, 319

Q

QSSPC method. See Quasi steady state
    photoconductivity method
    (QSSPC method)
Quartz crucible, 101–102
  Liquinert, 42, 148, 320–321
  unidirectional growth using, 105–108
Quasi steady state photoconductivity
    method (QSSPC method), 64
Quasi-single cast method, 226

R

Radial heat flux, 226–228
Radial temperature gradient, 332
Radiofrequency inductor (RF inductor),
    319
Random grain boundaries, 25–26, 32,
    70, 72, 169–170, 178–179,
    239–241, 317
Random pyramid passivated emitter and
    rear cell (RP-PERC), 336–337
Rapid thermal processing (RTP), 128
Recombination centers, 66–73
  dislocations, recombination activity of,
    72–73
  grain boundaries, recombination activity
    of, 68–72
  impurities, recombination activity of,
    66–68
Red zone, 197, 225–226, 250–251
Resistivity of Si single ingots, 299–303
RF inductor. See Radiofrequency
    inductor (RF inductor)

356 Index

Rock curve imaging, 237
Room temperature (RT), 48—49
Rough-facet grooves, 30—31
RP-PERC. *See* Random pyramid
    passivated emitter and rear cell
    (RP-PERC)
RT. *See* Room temperature (RT)
RTP. *See* Rapid thermal processing
    (RTP)

S

S2 wafers from GCL, 226
Sapphire, 260
Sawing technique, 106
Scandium (Sc), 43—44
Scanning electron microscopy (SEM),
    68—69, 135
Scanning infrared polar-iscope
    (SIRP), 52
Scattering width, 303
Scheil's equation, 12—14
Seed assisted growth, 197
Seed manipulation for artificially
    controlled defect technique
    (SMART), 246—251
Seed plates and gaps between seeds,
    228—229
Seed-assisted cast
    furnace, 103—104
    methods, 189—190
Seeded cast method, 226
Self-crucible formation, 134
SEM. *See* Scanning electron
    microscopy (SEM)
Shear stress effects on dislocations
    formation, 36—38
SiCN precipitates, 241
Side-face width
    of square and circular single ingots,
    331—332
    of square-shaped crystal, 286—287
Silica. *See* Silicon dioxide (SiO_2)
Silica rod, 226—228
Silicon (Si) crystals
    crystallographic structure and defects of
    crystal defects, 38—39
    determination, 14—15
    dislocations, 33—38

    grain boundaries in Si crystals, 17—33
    grains in Si crystals, 15—17
  electrical properties
    diffusion length and minority loading
    carrier lifetime, 74—77
    recombination centers, 66—73
    of Si crystals with defects, 65—66
  equilibrium segregation coefficients of
    impurities and dopants, 53—54
  growth
    degree of supercooling, 2—9
    dopant distribution in Si multicrystal-
    line ingots, 12—14
    growing interface and grooves, 9—12
  impurities and activities in defects
    Fe impurities and precipitates, 48—51
    metallic impurities and precipitates,
    43—48
    O, C and N atomic impurities,
    39—42
    O precipitates, 42—43
  methods to measuring electrical
    properties, 63—65
  optical measurement method and
    optical properties, 77—78
  processes to control electrical and
    optical properties
    gettering and hydrogenation, 78—82
    gettering and low-temperature
    annealing for Fe impurities, 82—86
    thermal donors, 87—88
    TR process, 86—87
  Si single and multi-crystals, 1—2
  solar cells
    B_O complexes, 91—94
    dopants, 88—89
    performance of *p*-and *n*-type Si solar
    cells, 89—91
  solubility limits of light-element
    impurities in, 54
  strain and stress, 52—53
  theoretical estimation for crystals
    characterization, 94—95
Silicon (Si) ingot, 43—44
  quality using mono-like cast method,
    249—254
    dislocation density, micro-twins and
    impurity concentration, 249—251

electric properties, 251–252
  solar cells, 252–254
structure and defects in, 207–216
Silicon (Si) multicrystalline ingots
  crystal defects and impurities in,
      118–133
    dislocations and dislocation clusters,
      121–126
    metallic impurities and precipitates,
      126–129
    SiO_2, Si_3N_4 and SiC precipitates,
      129–133
  dopant distribution in, 12–14
  electrical properties, 143–144
    and solar cells, 143–145
  grain boundaries
    character, 118–121
    formation, 116–118
  grain size and stress in, 317–318
  grains formation, 116
  growth of, 110–118, 147–148
    in-situ observation system
      development, 110–113
    shape of growing interface during
      growth, 113–115
    unidirectional growth, 101–109
  large-scale ingots growth in industry,
    145–147
  nucleation for, 115–116
  Si_3N_4 coating materials, 133–143
Silicon (Si) single ingots
  electrical properties, 299–308
    resistivity and minority carrier
      lifetime of Si single ingots,
      299–303
    solar cells, 303–308
  growth using NOC method,
    281–299
    dislocations in, 292–295
    impurities in, 295–299
    with large diameter and diameter
      ratio, 287–290
    rate, 281–287
    shape of growing interface,
      290–292
    surface or horizontal growth
      length, 283f

Silicon carbide (SiC), 40–41
  particles, 241
  precipitate, 129–133
Silicon dioxide (SiO_2), 2–4, 133–134
  powders, 203–204
  precipitate, 129–133
Silicon nitride (Si_3N_4), 1, 40–41
  coating, 137–138
  coating materials, 133–143
    identification Si_3N_4 particles,
      135–139
    process to form Si_3N_4 transformers or
      precipitates, 139–141
    Si multicrystals growth from Si melt,
      133–135
    wetting properties of Si_3N_4 coating,
      141–143
  growth of Si ingots using Si_3N_4 coated
    crucibles, 278–280
  particles, 241
  precipitate, 129–133
Silicon oxycarbide ($Si_xC_yO_z$), 241
Silicon oxynitride ($Si_xO_yN_z$), 239–241
Silicon–germanium (Si–Ge) multicrys-
  talline ingots, 337–345
  compositional fluctuation, 340f
  Ge doped Si multicrystalline ingots,
    344–345
  novel growth technology for SiGe solar
    cells and, 337–343, 338f
  wavelength dependence of internal
    quantum efficiency, 341f
Silver (Ag), 43–44
Simulated temperature distributions,
  277–278
{310} single crystal, 232–233
Single-layer Si beads, 201–202
Sinoite (Si_2N_2O), 142
Sinton method, 65
SIRP. See Scanning infrared polar-iscope
  (SIRP)
Small grain
  appearing from crucible wall, 230–236
  regions, 179–182, 187
SMART. See Seed manipulation for
  artificially controlled defect
  technique (SMART)

358 Index

Sodium (Na), 137

Soft touching technology for dislocation-free single ingots, 318–319

Solar cells, 64, 218–219, 252–254
Al-BSF, 90, 307
in HP cast method, 218–219
n-type, 89–91, 303, 304f
p-type, 89–91
PERC, 90, 307, 308f
performance preparing by NOC method, 303–308
PERL, 90–91
of Si multicrystalline ingots, 144–145

Solid-liquid interface, 291–292

Solid-solid-liquid tri-junction (SSL-TJ), 11–12

Solidification, 105–106
ratio, 320–321

SPV method. See Surface photovoltage method (SPV method)

Square-shaped ingot growth using NOC method, 326–334
basic points, 326–327
factors affecting, 331–333
with four-cornered pattern, 327–331, 328f
key points for improvement of, 334

SSL-TJ. See Solid-solid-liquid tri-junction (SSL-TJ)

Stacking faults, 39

Strain of Si crystals, 52–53

Stress, 106–107
of Si crystals, 52–53

Structure control, 155, 166–167, 169–170, 182–184, 188–190, 195

Structure-related dislocations, 34

Sub-band-gap
luminescence, 66
PL spectroscopy, 78

Submicron surface textures, 144

Substitutional carbon (C_s), 40

Supercooling, 2–4, 155–161, 166–167, 172–174, 187–188, 265–266, 271–272, 287
degree of supercooling
constitutional supercooling, 7–9
estimation in Si melt near growing interface, 4–6

experiments of Si melt, 2–4
experiments to determining, 6–7

Surface photovoltage method (SPV method), 74–75, 300

Swirl patterns, 299–300

Synchrotron radiation X-ray white beam topography, 237

Synchrotron-based nanoprobe mapping, 73

T

Tabula Rasa process (TR process), 86–87, 302–303

Tantal (Ta), 43–44

TEM. See Transmission electron microscopy (TEM)

Temperature gradient, 102–103, 226–228

Temperature reduction of Si melt, 265–266

Thermal conductivity, 168–170, 172–174, 202

Thermal donors, 87–88

Thermal shock, 318

Thorium (Th), 43–44

3-D visualization of reconstructed ingot image, 212

Tilt grain boundaries, 244, 246–248

TJ. See Tri-junction (TJ)

TMFs. See Traveling magnetic fields (TMFs)

TR process. See Tabula Rasa process (TR process)

Transition metal impurities, 43–44, 47–48, 71, 127–128

Transmission electron microscopy (TEM), 25

Traps, 66

Traveling magnetic fields (TMFs), 107

Tri-junction (TJ), 29

Twin boundaries, 26–29, 317

Twin formation, 29–32

Twin nucleation, 31, 120

Twin-related dendrite. See Facet dendrite

Two-dimensional nucleation (2-D nucleation), 2, 4–5

U

U-grade wafers from SAS, 226
Undercooling. *See* Supercooling
Unidirectional growth of Si
multicrystalline ingots
Ar and N gas flow, 108–109
furnace design with hot zone,
101–104
unidirectional growth using quartz
crucibles, 105–108

V

Vanadium (V), 80–81
Vertical Bridgmen method, 226
Virtus wafers from Renasolar, 226
Volatile components, 44–45
Voronoi diagram, 17

W

Wavy perturbation, 9
Wetting properties of Si_3N_4 coating,
141–143

X

X-ray beam-induced current technique
(XBIC technique), 128
X-ray diffraction (XRD), 138–139
X-ray fluorescence (XRF), 73
μ-XRF microscopy, 128
X-ray rocking curve (XRC), 24–25

Z

Zirconium (Zr), 43–44
Zone heaters, 264
Zyarock, 2–4

Printed in the United States
By Bookmasters